T0300987

LIFE IN ANTARCTIC DESERTS AND OTHER COLD DRY ENVIRONMENTS

Astrobiological Analogs

The McMurdo Dry Valleys form the largest relatively ice free area on the antarctic continent. The perennially ice covered lakes, ephemeral streams, and extensive areas of exposed soil are subject to low temperatures, limited precipitation, and salt accumulation. The dry valleys thus represent a region where life approaches its environmental limits. This unique ecosystem has been studied for several decades as an analog to environments on other planets, particularly Mars. For the first time, the detailed terrestrial research of the dry valleys is brought together here, presented from an astrobiological perspective. Chapters include a discussion on the history of research in the valleys, a geological background of the valleys, setting them up as analogs for Mars, followed by chapters on the various subenvironments in the valleys such as lakes, glaciers, and soils. Includes concluding chapters on biodiversity and other analog environments on Earth.

PETER T. DORAN is a Professor of Earth and Environmental Sciences and active research scientist at the University of Illinois at Chicago. He is a veteran of numerous expeditions to the Arctic and Antarctic in the pursuit of studying climate and ecosystem change and astrobiology. Doran is currently the lead investigator of NASA's Environmentally Non-Disturbing Under-ice Robotic ANtarctiC Explorer (ENDURANCE). He is on various NASA review panels and working groups focused on future exploration of Mars and other locations of astrobiological interest and is a sitting member of the NASA Subcommittee on Planetary Protection. He has published more than 50 peer-reviewed scholarly articles. He is a member of American Geophysical Union, the Union of Concerned Scientists, Committee on Space Research, the Geological Society of America, and the American Quaternary Association.

W. BERRY LYONS received his Ph.D. in chemical oceanography from the University of Connecticut in 1979. He has since carried out many research studies in the Antarctic and recently stepped down as the lead investigator of the McMurdo Dry Valleys Long Term Ecological Research (MCM LTER) project, one of the 26 LTER sites supported by the U.S. National Science

Foundation. He is a U.S. representative on the Geosciences Scientific Group of the Scientific Committee for Antarctic Research (SCAR) and also serves as Treasurer for the International Association of GeoChemistry. He has coauthored over 200 scientific publications, many of them focused on antarctic and polar themes. He is currently a Professor in the School of Earth Sciences and former Director of the Byrd Polar Research Center at the Ohio State University. His primary research interests include aqueous geochemistry, environmental geochemistry, and the impact of climate change and anthropogenic activities on biogeochemical processes.

DIANE M. MCKNIGHT received her Ph.D. in Environmental Engineering from the Massachusetts Institute of Technology in 1979. She was a research hydrologist with the U.S. Geological Survey in Denver, Colorado, from 1979 to 1996, where she studied the biogeochemistry of aquatic ecosystems. She has been researching the Antarctic and Arctic since 1987, where she has conducted research on stream ecosystems as part of the McMurdo Dry Valleys Long Term Ecological Research (MCM LTER) project and alpine lakes as part of the Niwot Ridge Long Term Ecological Research (NWT LTER) project. Since 1996, she has been a Professor in the Department of Civil, Environmental and Architectural Engineering and a Fellow of the Institute of Arctic and Alpine Research at the University of Colorado, Boulder. She has coauthored over 180 scientific publications, many of them focused on polar and alpine themes. Her primary research interests include aquatic ecology of algae in lakes and streams, and biogeochemical studies of natural organic material and trace metals in freshwater systems.

Cambridge Astrobiology

Series Editors

Bruce Jakosky, Alan Boss, Frances Westall, Daniel Prieur and Charles Cockell

Books in the series

1. Planet Formation: Theory, Observations, and Experiments
   Edited by Hubert Klahr and Wolfgang Brandner
   ISBN 978-0-521-86015-4

2. Fitness of the Cosmos for Life: Biochemistry and Fine-Tuning
   Edited by John D. Barrow, Simon Conway Morris,
   Stephen J. Freeland and Charles L. Harper, Jr.
   ISBN 978-0-521-87102-0

3. Planetary Systems and the Origin of Life
   Edited by Ralph Pudritz, Paul Higgs and Jonathan Stone
   ISBN 978-0-521-87548-6

4. Exploring the Origin, Extent, and Future of Life: Philosophical, Ethical and Theological Perspectives
   Edited by Constance M. Bertka
   ISBN 978-0-521-86363-6

LIFE IN ANTARCTIC DESERTS AND OTHER COLD DRY ENVIRONMENTS

Astrobiological Analogs

Edited by

PETER T. DORAN
University of Illinois at Chicago

W. BERRY LYONS
Ohio State University

DIANE M. McKNIGHT
University of Colorado at Boulder

CAMBRIDGE
UNIVERSITY PRESS

CAMBRIDGE
UNIVERSITY PRESS

University Printing House, Cambridge CB2 8BS, United Kingdom

One Liberty Plaza, 20th Floor, New York, NY 10006, USA

477 Williamstown Road, Port Melbourne, VIC 3207, Australia

314-321, 3rd Floor, Plot 3, Splendor Forum, Jasola District Centre, New Delhi - 110025, India

103 Penang Road, #05-06/07, Visioncrest Commercial, Singapore 238467

Cambridge University Press is part of the University of Cambridge.

It furthers the University's mission by disseminating knowledge in the pursuit of education, learning and research at the highest international levels of excellence.

www.cambridge.org
Information on this title: www.cambridge.org/9780521889193

First published 2010

A catalogue record for this publication is available from the British Library

Library of Congress Cataloging in Publication data
Life in Antarctic deserts and other cold dry environments : astrobiological analogues / edited by Peter T. Doran, W. Berry Lyons, Diane M. McKnight.
p. cm. – (Cambridge astrobiology)
ISBN 978-0-521-88919-3 (Hardback)
1. Biotic communities–Antarctica. 2. Ecology–Antarctica. 3. Antarctica–Environmental conditions. 4. Exobiology. I. Doran, Peter T. II. Lyons, W. Berry. III. McKnight, Diane M. IV. Title. V. Series.
QH84.2.L54 2010
577.0911′6–dc22

2009051397

ISBN 978-0-521-88919-3 Hardback

Additional resources for this publication at
www.cambridge.org/uk/catalogue/catalogue.asp?isbn=9780521889193

Contents

Contributors

Dale T. Andersen
SETI Carl Sagan Center, 515 N. Whisman Road, Mountain View,
CA 94043, USA

Jenny Baeseman
International Arctic Research Center, University of Alaska, Fairbanks,
AK 99775–7340, USA

Liz Bagshaw
Bristol Glaciology Centre, Geographical Sciences, University of Bristol,
Bristol, BS8 1SS, UK

J. E. Barrett
Department of Biological Sciences, Virginia Polytechnic Institute and
State University, Blacksburg, VA 24061, USA

Nathalie A. Cabrol
NASA Ames Research Center, Space Science and Astrobiology Division,
MS 245–3, Moffett Field, CA 94035–1000, USA

Michael H. Carr
U.S. Geological Survey, Menlo Park, CA 94025, USA

Peter T. Doran
Earth and Environmental Sciences, University of Illinois at Chicago,
845 W. Taylor St., Chicago, IL 60430, USA

Christine Foreman
Center for Biofilm Engineering and Department of Land Resources and
Environmental Sciences, Montana State University, Bozeman,
MT 59717, USA

Andrew G. Fountain
Department of Geology, Portland State University, Portland, OR 97207–0751, USA

Michael N. Gooseff
Department of Civil and Environmental Engineering, Pennsylvania State University, University Park, PA 16802, USA

Ian Hawes
World Fish Center, PO Box 77, Gizo, Solomon Islands

James W. Head, III
Department of Geological Sciences, Brown University, Providence, RI 02912, USA

Brian D. Lanoil
Department of Environmental Sciences, University of California, Riverside, CA 92521, USA

Pascal Lee
Mars Institute, P.O. Box 6, NASA Research Park, Moffett Field, CA 94035–0006, USA

W. Berry Lyons
Byrd Polar Research Center, Ohio State University, 1090 Carmack Rd, Scott Hall, Columbus, OH 43210, USA

David R. Marchant
Department of Earth Sciences, Boston University, Boston, MA 02215, USA

Christopher P. McKay
NASA Ames Research Center, MS 245–3, Moffett Field, CA 94035, USA

Diane M. McKnight
Institute of Arctic and Alpine Research, 1560 30th Street, Campus Box 450, Boulder, CO 80309, USA

Jill Mikucki
Department of Earth Sciences, Dartmouth College, Hanover, NH 03750, USA

James A. Nienow
Department of Biology, Valdosta State University, Valdosta, GA 31698, USA

Michael A. Poage
Geoscience Department, Indiana University of Pennsylvania, Indiana, PA 15705, USA

John C. Priscu
Department of Land Resources and Environmental Sciences, Montana State University, Bozeman, MT 59717, USA

Carol R. Stoker
NASA Ames Research Center, Space Science and Astrobiology Division, MS 245–3, Moffett Field, CA 94035–1000, USA

Henry J. Sun
Division of Earth and Ecosystem Sciences, Desert Research Institute, Las Vegas, NV 89119, USA

Cristina Takacs-Vesbach
Department of Biology, University of New Mexico, Albuquerque, NM 87131, USA

Martyn Tranter
Bristol Glaciology Centre, Geographical Sciences, University of Bristol, Bristol, BS8 1SS, UK

David S. Wettergreen
Carnegie Mellon University/Robotics Institute, 5000 Forbes Avenue, Pittsburgh, PA 15213, USA

Lydia Zeglin
Department of Biology, University of New Mexico, Albuquerque, NM 87104, USA

1

Introduction

PETER T. DORAN, W. BERRY LYONS,
AND DIANE M. MCKNIGHT

The dry valleys of East Antarctica are at first glance a barren landscape. This was certainly Robert Falcon Scott's impression when he was the first to visit the dry valleys in 1903. As his expedition marched down what is now called Taylor Valley, he commented in his journal "we have seen no living thing, not even a moss or lichen" and "It is certainly the valley of the dead; even the great glaciers which once pushed through it have withered away" (Scott, 1905). A party from Scott's second expedition, led by senior geologist Griffith Taylor, also visited the valleys in 1911 (Taylor, 1922). Another 45 years elapsed before other visitors came to the valleys when Operation High Jump established logistics bases at nearby McMurdo Station and Scott Base in 1956. These bases provided relatively easy access to the valleys by tracked vehicles and helicopters across the McMurdo Sound to these previously hard-to-get-to areas. Afterwards, the New Zealand national program carried out all kinds of natural science research in the valleys, largely based out of the busy Lake Vanda station which supported three manned over-winter investigations (Harrowfield, 1999). Early biological work in the dry valleys was also carried out by the U.S. program in the 1960s by now well-known ecologists Gene Likens, Charles Goldman, and John Hobbie who founded long-term monitoring programs at Hubbard Brook, Lake Tahoe, and Toolik Lake in Alaska (respectively). The National Science Foundation established a Long Term Ecological Research (LTER) site in the dry valleys in 1993 which has become one of the main sources of biological data and ecological understanding from the dry valleys. The McMurdo LTER has enhanced the connections to extreme environment research and also astrobiology, the development of

Life in Antarctic Deserts and Other Cold Dry Environments: Astrobiological Analogs, ed. Peter T. Doran, W. Berry Lyons and Diane M. McKnight. Published by Cambridge University Press.

which is outlined below. Today, hundreds of people a year visit the dry valleys, mostly for science, but some for tourism as well.

An early chronology of the dry valleys as a Mars analog

Some of the earliest field research performed in the dry valleys was a direct result of the Mariner 4 space probe which orbited Mars in July 1965. Images returned from this mission for the first time showed Mars to be a cratered, cold, and dry planet. The revelations from Mariner 4 drove Norm Horowitz from the Jet Propulsion Laboratory to be perhaps the first to consider the dry valley soils as suitable models for what the surface of Mars may be like. Horowitz, with colleagues Roy Cameron and Jerry Hubbard, initiated a study of soil microbiology in the dry valleys. They collected about 500 bags of soil from around the valleys which are still held today in cold storage at the National Aeronautics and Space Administration (NASA) Ames Research Center by Chris McKay. They wrote a series of high-profile papers (Horowitz et al., 1969, 1972) which started the myth of sterile soils in the dry valleys. Of the samples they analyzed, about 14% were reported to be lifeless. One sample (726, which is now depleted) did not show organic matter but did show signs of life. Gil Levin (an instrument principal investigator on the Viking Biology mission to Mars) latter asked John Cronin at Arizona State University to do a state of the art organic analysis on this sample and still found no organic matter.

The first grantee of NASA's exobiology program was Wolf Visniac of Yale Medical School. Visniac developed a device named the "Wolf trap" that could detect microorganisms living in the soil of another planet. Vishniac used his device on some of the "sterile" samples of Horowitz et al., and showed that at least some of them contained viable microbes. Vishniac also traveled to the dry valleys to test concepts for the Wolf trap. Although his device was initially selected to be included on the Viking mission to Mars, budget constraints forced it to be cut from the project before the landers were finished. Vishniac continued his interest in the dry valley soils and fell to his death while on a sample-collecting expedition on December 10, 1973, in Tyrol Valley, Upper Wright Valley (Dick and Strick, 2004).

Prior to Vishniac's death, he met Imre Friedmann, a microbiologist from Florida State University. Friedmann had been unsuccessful in acquiring research money to go to the dry valleys to look for endolithic bacteria similar to those he was investigating in hot deserts. Reviewers were skeptical that he would find any microbes at all. In frustration, Friedmann asked Wolf Vishniac to collect sandstone samples for him and send them back for analysis. When Vishniac was found dead, there was a bag in his pack labeled "samples for Imre Friedmann." Vishniac's wife Helen was mailed all of the personal effects and she forwarded the samples to Friedmann, who discovered endoliths in some samples and wrote a paper which was published in *Science*. Friedmann became a prominent figure in exobiological research in the dry valleys and elsewhere until his death on June 11, 2007.

Chris McKay, now of NASA Ames Research Center (an author on Chapters 4 and 9) was a scientist working with Imre Friedmann from 1980 to 1986. McKay had seen a talk by Mike Carr (author on Chapter 5) at Ames about the Viking mission imagery. Carr had speculated that some of the deposits on Mars were remnants of ancient lakes. McKay asked if the lakes had ice covers and Carr's answer was yes, they must have been frozen solid. During a trip to Antarctica with Friedmann, McKay met Robert Wharton from Virginia Tech. Wharton was conducting his Ph.D. research on the perennially ice-covered lakes in the dry valleys. McKay became interested in how the lakes could maintain liquid water in such a cold environment. This meeting started a collaboration which led to the publication of the first paper (McKay et al., 1985) making the connection between the dry valley lakes and purported lakes on Mars in the past. This connection continues to be made (e.g., Wharton et al., 1989, 1995; Doran et al., 1998), and the fact that these lakes harbor a viable ecosystem year round in this harsh climate makes them a frequently cited example of an extreme environment on Earth.

As mentioned above, the early 1990s saw the establishment of the McMurdo LTER, for which Robert Wharton was the first principal investigator. It is interesting to note that a number of studies begun in the dry valleys in relation to the exobiology research have been incorporated into the LTER. For instance, an important study (Squyres et al., 1991) on the ice-covered lake sediment dynamics was led by Steven Squyres, who later became the Science Lead for NASA's Mars Exploration Rovers. The Mars connection with the dry valleys intensified following the establishment of the Planetary Analog joint program between NASA and the U.S. National Science Foundation (NSF). Under this program, Wharton was awarded a grant to develop and test a remotely operated vehicle (ROV) which helped with operational aspects and algorithm development for subsequent Pathfinder/Opportunity missions to Mars. This project also involved Dale Andersen and Carol Stoker who are authors on Chapter 9. All of the meteorological records at the lakes in Taylor Valley, including the longest continuous meteorological record at Lake Hoare, were initiated as part of exobiological research to study the formation of the lake ice covers. In fact, the common concern for keeping these measurements running was part of the motivation for the scientists studying different aspects of the dry valleys to join forces and form the LTER.

As scientists with diverse backgrounds studying ecosystems in temperate regions have become involved in research in the dry valleys, their interest has been captured by the potential implications of their research results for understanding Mars. The chapters in this book represent a synthesis of our current understanding of the dry valleys from a martian analog perspective that also informs our overall understanding of life in extreme environments on Earth.

Summary of chapters in this book

In this book, Chapters 2 through 7 each discuss different environmental components of the dry valleys in relation to Mars. The last two chapters look at microbial diversity in general and other analog sites on Earth.

Chapter 2 by David Marchant and Jim Head, looks for geomorphic analogies between Mars and the McMurdo Dry Valleys. By using the dry valleys to "calibrate" the climatic significance of certain geomorphic features, they have been able to make conclusions about the climate that formed similar features on Mars, and also to speculate about past climate change on Mars. In the dry valleys, three microclimate zones (coastal thaw, inland mixed, and stable upland) are defined on the basis of atmospheric temperature, soil moisture, and relative humidity. These zones are sensitive to changing climate, which can impact distribution and morphology of features at the macroscale (e.g., slopes and gullies); mesoscale (e.g., polygons and debris-covered glaciers); and microscale (e.g., salt weathering and surface pitting). Marchant and Head conclude by stating that through examining the relationships between the climatic zones, geomorphic features, and soil organisms in the dry valleys, some inference may be gained on the habitat of potential martian biota of the past.

Chapter 3 by Barrett et al. describes the soils in the dry valleys and the importance of water, both past and present, on their biogeochemistry and ecology. Although the authors readily acknowledge that colder temperatures and much lower atmospheric pressures make the martian environment quite different, they point out that past conditions on Mars, when liquid water could have been present, make the dry valleys interesting paleoanalogs. Liquid water is the dominant driver of both biological and geochemical processes in the dry valley soils; these processes respond rapidly to temporal climatic events that produce liquid water which can be transported within dry valley landscapes. Soils in both locations have formed under extremely cold and arid conditions and have high salt concentrations, yet the dry valley soils lack the low pH weathering products recently observed on Mars (e.g., Squyres et al., 2006). Aeolian features and cryogenic features such as patterned ground are observed in both the dry valleys and Mars, suggesting that similar processes affected soil development and evolution.

Barrett et al. provide an excellent discussion of the contemporary processes along hydrological margins affecting soil processes in the dry valleys and then use this information to interpret geochemical features that relate to past geologic times when the hydrological conditions were much different from those of today. This "present is the key to the past" approach provides a more complex view of the past history of the dry valleys. The authors suggest that this approach may be useful in the interpretation of landscape and hydrological history where liquid water once existed on the surface of Mars.

Sun et al. in Chapter 4 focus on life in the near surface of the rocks of the dry valleys. The cryptoendolithic microbial ecosystem consists of cyanobacterial or algal primary producers, fungal consumers, and bacterial decomposers that utilize sun and water within the top few millimeters of dry valley sandstones. This chapter provides a thorough review of the cryptoendolithic community, including a description of the various species present, the physicochemical environment, adaptations, turnover time and productivity rates and amino acid racemization and pseudoracemization. This chapter concludes with a discussion of the significance of cryptoendoliths for the possibility of life on Mars. The authors argue that the cryptoendolithic microbial ecosystem on Earth has shown that life is more robust than previously realized, and that rocks may have been the final habitat for life on the surface of Mars.

Gooseff et al. in Chapter 5 review our knowledge of dry valley stream ecosystems and the potential similarity of the fluvial features observed on the martian landscape. The chapter describes these features and their size and scale in both locations. A major conclusion is that due to their similarities, the martian stream systems in the past may also have been greatly dependent on shallow, subsurface processes, as the dry valley streams currently are. The authors also summarize the ecology of the dry valley streams, many of which contain cyanobacterial mats that grow during the several weeks when streams flow and then are "freeze dried" during the winter. The authors point out the importance that the hyporheic zone of these streams plays in the overall biogeochemical processes, such as weathering of minerals and nutrient uptake and transport. These authors conclude that given the commonalities of the two systems, any probable fluvial ecosystem on Mars was undoubtedly similar to those observed today in Taylor and Wright Valleys.

Mikucki et al. in Chapter 6 discuss the dry valley lakes and ponds as analogs for past water bodies on Mars. They provide a review of evidence for past standing water on Mars before discussing the character, ecology, and history of all major lakes (Lakes Fryxell, Hoare, Bonney, Vanda, and Vida) in the McMurdo Dry Valleys. The chapter also covers the character of numerous shallow ponds in the dry valley region. Microbial mats, a ubiquitous feature of all dry valley lake environments, represent a large portion of the living biomass in the dry valleys and so are given significant attention in this chapter. The extensive growth of these lake mats creates structures that may be preserved as fossils, similar to stromatolites found on Earth. Another common feature of the dry valley lakes is that their perennial ice covers contain viable microbial communities, including photosynthetic mats.

Two other unique saline features of the dry valleys are reviewed by Mikucki et al., including Blood Falls and Don Juan Pond. Blood Falls is an iron-rich subglacial brine which sporadically discharges from the face of Taylor Glacier. Don Juan Pond is a calcium chloride brine pool in Wright Valley which is believed to remain liquid year round – despite winter temperatures below −40 °C. The eutectic point of the brine is −51.8 °C (Marion, 1997). Don Juan Pond is often cited as a good example of liquid water existing in an extremely cold region, but the water is also extremely salty, precluding much of anything from living there (whether Don Juan Pond supports life is a matter of debate according to Mikucki et al.).

In Chapter 7 Tranter et al. discuss the ecosystems observed on the surface, within and potentially beneath the glaciers in Taylor Valley. Although the "cryo-ecosystems" have been known for many years, only recently have their hydrology and ecology been described. This chapter summarizes these recent findings on these unusual ecosystems in Taylor Valley. The photos provided are especially helpful in grasping the nature of these systems. Small cryoconite holes cover about 4.5% of the ablation zones of glaciers in Taylor Valley and "cryolakes" are prominent features there. Aeolian-deposited dust initiates the formation of these features. Warmer summers can lead to the flushing of these features, transferring solutes, organic matter, and organisms from the glaciers into the streams and lakes. Thus, soil material transported to the glaciers by wind is eventually transported into the aquatic systems in Taylor Valley. As documented in this chapter, these holes and lakes on the glacier surface are "bioreactors" where inorganic materials are fixed biologically over time. The authors describe what is known about ice behavior on Mars and conclude that the present-day ice cap surfaces of Mars are incapable of producing liquid water, thus the terms in the energy balance at the surface of the glaciers in the two environments differ greatly and the martian systems are unlikely to contain similar ecosystem analogs. Life beneath the ice caps also would depend upon the presence of liquid water perhaps in part due to volcano–ice interactions. These types of interactions may have happened in the past on Mars, as suggested recently by Niles and Michalski (2009).

In Chapter 8 Takacs-Vesbach et al. describe the microbial diversity patterns observed in the McMurdo Dry Valleys and the ecological processes and conditions that regulate diversity. Microbes are found in ephemeral environments, such as streams, and highly stable environments, such as lake bottoms. The existence of these diverse water-bearing habitats primarily depends on energy inputs from solar radiation and wind. One line of evidence for the diversity of microbial life comes from the range of processes observed; for example, both nitrifying and denitrifying bacteria regulate the nitrogen cycle in the dry valleys. Another important tool is the application of modern molecular methods, which has revealed that a broad range of microbes are present in the dry valleys. The low abundance of *Archea* is a common finding across the diverse microbial habitats. Thoughts on how the diversity patterns may be relevant to Mars are imbedded in this discussion. Takacs-Vesbach et al. hypothesize that diversity in the dry valleys is caused by cumulative mutations that persist in the environment. This accumulation may occur partially because disturbance is rarely catastrophic for these slow-growing populations. Furthermore, low bacterial growth rates result in a community where competitive displacement is infrequent. Presumably, life on Mars would have been (or is presently) subjected to similar environmental pressures that would limit biotic interactions and produce similar patterns of microbial diversity. Takacs-Vesbach et al. conclude that understanding dry valleys diversity will provide major insights into fundamental ecological processes on Earth, and potentially other planets like Mars with similar ecological conditions.

Finally, Chapter 9 by Cabrol et al. presents an extensive review of other Mars analog sites on Earth. High-altitude lakes, subsurface aqueous habitats, and arctic and desert regions are the major categories discussed. The high-latitude lakes discussion focuses on evaporative lakes in the high mountainous region of the Andes. The authors argue that these lakes are good analogs for lakes that may have existed towards the end of the first 500 million years of martian history due to the low air temperature, high daily and yearly temperature fluctuations, aridity, strong evaporation, thin atmosphere, high ultraviolet radiation, ice, reduced precipitation, and volcanic and hydrothermal activity. The hydrogeologic system of the Río Tinto in Spain, because of its acidity and high iron and sulfate content, is viewed as a good analog for the system that may have been responsible for forming the deposits of Meridiani on Mars. In the Arctic, despite mean annual temperatures well below the freezing point of water and pervasive permafrost, saline springs flow year round. These springs increase the viable microbial habitat in extreme cold environments and could be analogs for saline springs on Mars even in recent history. Other arctic regions, such as Haughton Crater, are now well established as terrestrial outposts, which can be used as an analog for Mars in many regards. Finally, hot deserts on Earth are discussed in relation to their low humidity and implications for life on Earth in on hyperarid Mars.

We anticipate that the dry valleys will continue to serve as a useful and provocative analog for understanding many aspects of Mars, especially the potential for past or current life. While the focus of astrobiological comparisons has been on the possibility for microbial life on Mars, we should keep in mind that multicellular life is found in the dry valleys, in the form of nematodes, tardigrades, rotifers, and even springtails, with new species still being discovered.

References

Dick, S. J. and Strick, J. E. (2004). *The Living Universe: NASA and the Development of Astrobiology*. Piscataway, NJ: Rutgers University Press, 308 pp.

Doran, P. T., Wharton, R. A., Des Marais, D. J., and McKay, C. P. (1998). Antarctic paleolake sediments and the search for extinct life on Mars. *Journal of Geophysical Research*, **103**, 28 481–28 493.

Harrowfield, D. L. (1999). *Vanda Station: History of an Antarctic Outpost 1968–1995*. Christchurch, New Zealand: New Zealand Antarctic Society, 52 pp.

Horowitz, N. H., Auman, A. J., Cameron, R. E., et al. (1969). Sterile soil from Antarctica: organic analysis. *Science*, **164**, 1054–1056.

Horowitz, N. H., Cameron, R. E., and Hubbard, J. S. (1972). Microbiology of the Dry Valleys of Antarctica. *Science*, **176**, 242–245.

Marion, G. M. (1997). A theoretical evaluation of mineral stability in Don Juan Pond, Wright Valley, Victoria Land. *Antarctic Science*, **9**, 92–99.

McKay, C. P., Clow, G. D., Wharton, R. A., and Squyres, S. W. (1985). Thickness of ice on perennially frozen lakes. *Nature*, **313**, 561–562.

Niles, P. B. and Michalski, J. (2009). Meridiani Planum sediments on Mars formed through weathering in massive ice deposits. *Nature Geoscience*, **2**, 215–220.

Scott, R. F. (1905). *The Voyage of Discovery*. London: McMillan.

Squyres, S. W., Andersen, D. W., Nedell, S. S., and Wharton, J. R. A. (1991). Lake Hoare, Antarctica: sedimentation through thick perennial ice cover. *Sedimentology*, **38**, 363–380.

Squyres, S. W. and 17 colleagues (2006). Two years at Meridiani Planum: results from the Opportunity rover. *Science*, **313**, 1403–1407.

Taylor, T. G. (1922). *The Physiography of the McMurdo Sound and Granite Harbour Region*. London: Harrison and Sons, Ltd., 246 pp.

Wharton, Jr., R. A., McKay, C. P., Mancinelli, R. L., and Simmons, Jr., G. M. (1989). Early Martian environments: the Antarctic and other terrestrial analogs. *Advances in Space Research*, **9**(6), 147–153.

Wharton, R. A., Crosby, J. M., McKay, C. P., and Rice, J. W. (1995). Paleolakes on Mars. *Journal of Paleolimnology*, **13**, 267–283.

2

Geologic analogies between the surface of Mars and the McMurdo Dry Valleys: microclimate-related geomorphic features and evidence for climate change

DAVID R. MARCHANT AND JAMES W. HEAD, III

Abstract

The McMurdo Dry Valleys (MDV), classified as a hyperarid, cold-polar desert, have long been considered an important terrestrial analog for Mars because of their cold and dry climate and their suite of landforms that closely resemble those occurring on the surface of Mars at several different scales, despite significant differences in current atmospheric pressure. The MDV have been subdivided on the basis of summertime measurements of atmospheric temperature, soil moisture, and relative humidity, into three microclimate zones (Marchant and Head, 2007): a coastal thaw zone, an inland mixed zone, and a stable upland zone. Minor differences in these climate parameters lead to large differences in the distribution and morphology of features at the macroscale (e.g., slopes and gullies); mesoscale (e.g., polygons, viscous-flow features, and debris-covered glaciers); and microscale (e.g., rock-weathering processes/features, including wind erosion, salt weathering, and surface pitting). Equilibrium landforms form in balance with environmental conditions within fixed microclimate zones. For example, sublimation polygons indicate the presence of extensive near-surface ice in the MDV and identification of similar landforms on Mars appears to provide a basis for detecting the location of current and past shallow ice. The modes of occurrence of the limited and unusual biota in the MDV provide terrestrial laboratories for the study of possible environments for life on Mars. The range of microenvironments in the MDV are hypersensitive to climate variability, and their stability and change provide important indications of climate history and potential stress on the biota.

Life in Antarctic Deserts and Other Cold Dry Environments: Astrobiological Analogs, ed. Peter T. Doran, W. Berry Lyons and Diane M. McKnight. Published by Cambridge University Press.

Extreme hyperaridity on Mars and in the MDV underlines the importance of salts and brines on soil development, phase transitions from liquid water to water ice, and in turn, on process geomorphology and landscape evolution at a range of scales. Past and/or ongoing shifts in climate zonation are indicated by landforms that today appear in disequilibrium with local microclimate conditions in the MDV, providing a record of the sign and magnitude of climate change there. Similar types of landform analyses have been applied to Mars where microclimates and equilibrium landforms analogous to those in the MDV occur in a variety of local environments, in different latitudinal bands, and in units of different ages. Here we document the nature and evolution of microclimate zones and associated geomorphic processes/landforms in the MDV, an exercise that helps to provide a quantitative framework for assessing the evolution of landforms and climate change on Mars.

Introduction

The recognition of groups of climate-related landforms on Earth has led to the definition of different morphogenetic regions (e.g., Wilson, 1969; Baker, 2001), each defined in terms of mean annual temperature and precipitation (Fig. 2.1). A byproduct of this classification scheme is the recognition of specific equilibrium landforms: that is, those geomorphic features that are produced in equilibrium with prevailing climate conditions. A shift in the spatial distribution of equilibrium landforms over time, for example a latitudinal variation in glacial deposits, can be interpreted as a change in local and/or regional climate conditions. One region where detailed studies of equilibrium landforms can be used to shed light on climate change on Mars is the McMurdo Dry Valleys (MDV). The MDV are among the most Mars-like terrestrial environments on Earth (e.g., Anderson et al., 1972; Gibson et al., 1983; Mahaney et al., 2001; Wentworth et al., 2005; Marchant and Head, 2007), although there is a major difference in atmospheric pressure that influences the stability and mobility of liquid water on Mars. Detailed studies of the geomorphic processes operating there provide a basis for identifying and interpreting surficial landforms on Mars.

In this chapter we examine equilibrium landforms in the hyperarid polar desert of the MDV. We characterize landforms in three main microclimate zones, assessing the role of small variations in summertime temperature and precipitation in producing and sustaining different characteristic landforms at a variety of scales. We also discuss candidate martian analogs for each of the mapped landforms in the MDV. Finally, we explore how the

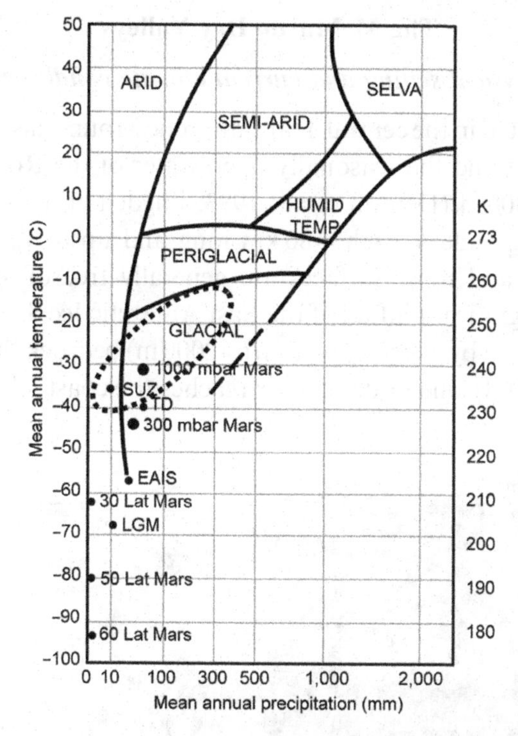

Fig. 2.1. Morphogenetic regions for landforms on Earth (adapted from Baker, 2001, and Marchant and Head, 2007). Dashed oval highlights the McMurdo Dry Valleys, which includes the stable upland zone (SUZ), the inland mixed zone (IMZ), and the coastal thaw zone (CTZ). Also plotted are present Mars conditions at 30°, 50°, and 60° latitude, as well as Mars at 300 mbar and 1000 mbar. TD, modern conditions at Taylor Dome, 35 km southwest of the MDV; EAIS, modern conditions at Vostok, interior East Antarctica (78° S); LGM, conditions during the last glacial maximum (~18 ka) in interior East Antarctica. Atmospheric pressure differs substantially between Mars and Earth and is an important factor in the presence and mobility of liquid water.

spatial distribution of landforms on both planets vary as a function of climate, and how lateral displacement of key landforms may be used to shed light on past and/or ongoing climate change. This treatment provides a basis for the environmental assessment of biological niches and habitats on both planets. Throughout this chapter, we synthesize and summarize in detail the earlier work by Marchant and Head (2007) that focused on the array of geomorphic landforms in MDV microclimate zones and the analogous landforms on Mars. Our new investigation of climate change on Mars comes from geomorphic analyses of mid-latitude craters such as Newton Crater.

The McMurdo Dry Valleys

Physical setting and current climate conditions

The MDV are located in the central Transantarctic Mountains, between the East Antarctic Ice Sheet and the seasonally open water of the Ross Sea, and show local relief of ~2800 m (Denton et al., 1993; Sugden et al., 1995a) (Fig. 2.2). East–west trending valleys, each ~80 km long and up to 15 km wide, extend across the region and most surfaces are generally free of ice. There are no vascular plants. Significant influx of the East Antarctic Ice Sheet into the MDV is currently prevented by a high-elevation (~2000 m) bedrock threshold (western part of Fig. 2.2). Only one outlet glacier reaches the coast, the Ferrar Glacier

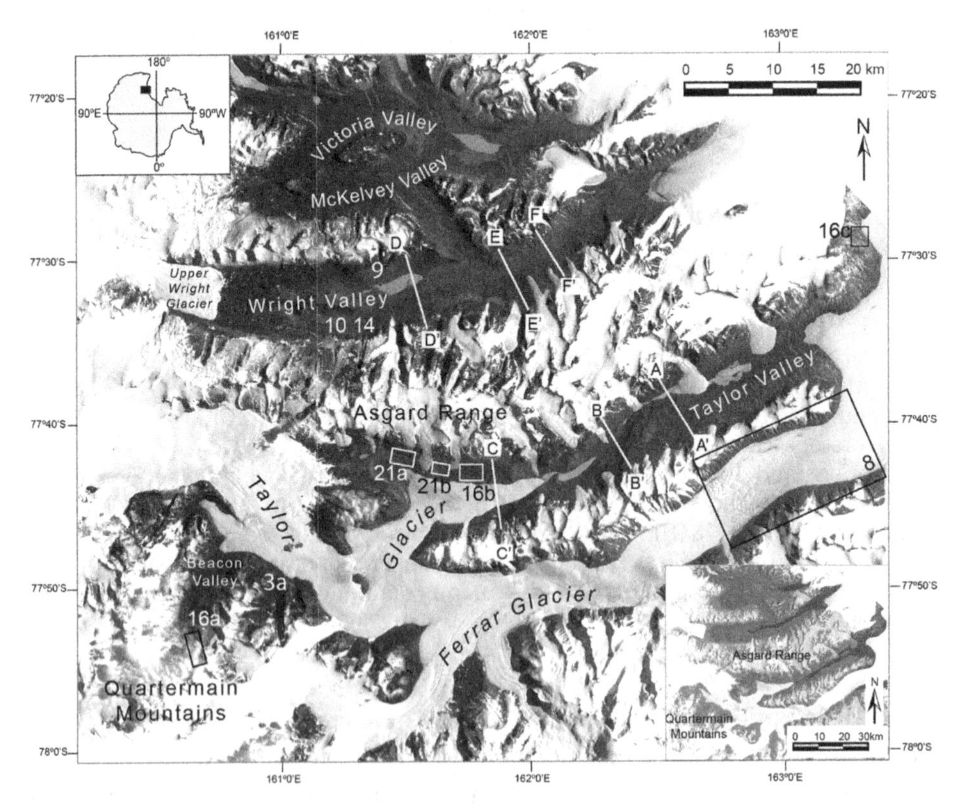

Fig. 2.2. McMurdo Dry Valleys. Location map showing major geographic features. Location of cross-valley profiles (Fig. 2.7) plotted as yellow lines; location of other figures in text shown as numbers and boxes. (Upper left inset) Black square shows location of dry valleys within Antarctica. (Lower right inset) Map showing general range for coastal thaw zone (CTZ; dark gray), inland mixed zone (IMZ; gray), and stable upland zone (SUZ; light gray) (adapted from Marchant and Head, 2007). The color version of this figure can be found at www.cambridge.org/uk/catalogue/catalogue.asp?ibsn=9780521889193.

(Fig. 2.2), and two others, Taylor Glacier and Upper Wright Glacier, terminate on land (Fig. 2.2). Local precipitation (primarily) and wind-blown snow from the polar plateau feed numerous alpine glaciers that occur on elevated benches and on plateaus between the main valleys. Below ~800 m elevation near the coast and in the central portions of the MDV, local bedrock consists of a basement complex that includes pre-Cambrian to Paleozoic age granites and gneisses. Exposed bedrock further inland consists of generally flat-lying sedimentary rocks of the Beacon Supergroup (Devonian to Triassic age sandstones, siltstones, and conglomerates) and 200- to 300-m-thick sills of Ferrar Dolerite (Jurassic age intrusives; Elliot and Fleming, 2004).

Processes that influence the geomorphology of the MDV

The three main processes that influence landforms in the MDV are katabatic winds, active-layer cryoturbation, and cold-based glaciation.

Katabatic winds

Gravity-driven katabatic winds commonly exceed $50\,\mathrm{km\,h^{-1}}$ (Schwerdtfeger, 1984; Marshall and Turner, 1997; Nylen et al., 2004). They flow off the EIAS and gather speed as they pass through the MDV. They warm adiabatically, and result in places with slightly elevated air temperature excursions that are particularly apparent during winter months (Fountain et al., 1999; Doran et al., 2002; Nylen et al., 2004). Katabatic winds transport significant snow from the polar plateau, some of which is deposited in the lee of topographic obstacles in the MDV. The fate of this snow, melting or sublimation, is a critical factor in the development of most unconsolidated landforms. The winds also entrain sand grains, scour bedrock landscapes, and produce large dunes (Selby, 1977; Malin 1984, 1987; Lancaster, 2002).

Active-layer cryoturbation

In permafrost regions, a traditional active layer is defined as the surface horizon that experiences seasonal temperature fluctuations above and below 0 °C (273 K) (Yershov, 1998; Davis, 2001). Active-layer thickness depends primarily on atmospheric temperature, and secondarily on substrate heat conduction. A "wet" active layer contains visible ice and/or liquid water, whereas a "dry" active layer contains minimal soil moisture, generally <5% gravimetric water content (GWC). In inland regions of the MDV, dry active layers are the norm, and in some places they are only a few centimeters thick (McKay et al., 1998; Kowalewski et al., 2006; McKay, 2009), while in coastal regions a wet-active layer up to ~25 cm in thickness is common (Campbell et al., 1997a, 1997b).

Cold-based glaciation

All of the alpine glaciers in the MDV are presently cold based. They flow only by internal deformation (in contrast to wet-based glaciers that flow additionally by basal sliding and regelation). Rates of bedrock erosion beneath cold-based glaciers are extremely low ($\sim 10^{-7}$ m y^{-1} beneath a typical cold-based alpine glacier in the MDV; Cuffey et al., 2000). This erosion rate is not only several orders of magnitude less than that for typical wet-based alpine glaciers (Hallet et al., 1996; Spotila et al., 2004; Koppes and Hallet, 2006; Brook et al., 2006), it is also several orders of magnitude less than that for bedrock erosion due to the impact of saltating sand grains in the MDV (Malin, 1984, 1987). Rather than eroding underlying bedrock and/or unconsolidated debris, cold-based glaciers tend to preserve underlying landscapes (e.g., Kleman and Hatterstrand, 1999; Borgstrom, 1999; Fabel et al., 2002; but see also Atkins et al., 2002). The geomorphic signature of cold-based glaciers are minute drop moraines, which form as a line of cobbles and boulders dropped at the snout of a stationary glacier front, or as a series of arcuate ridges reflecting a fluctuating ice margin (Fig. 2.3). The boulders and cobbles are entrained by rockfall onto glacier ice. Recent work has identified analogous ridges on Mars, on the flanks of the Tharsis volcanoes, and on the floors and walls of some high- and mid-latitude craters, which independently suggest the geologically recent growth and decay of cold-based glaciers there (see also below) (Head and Marchant, 2003; Head et al., 2003, 2006a, 2006b; Garvin et al., 2006; Milkovich et al., 2008; Shean et al., 2005, 2007a).

MDV microclimate zones

Early geomorphic investigations in the dry valleys region were largely centered along the coast. These relatively warm and humid sites quickly became type locales for the entire dry valleys region. The implicit assumption was that geomorphic processes observed at coastal sites could be applied throughout the dry valleys. Ultimately, this perpetuated the notion of a uniform suite of geomorphic processes operating within a single hyperarid, cold-polar desert climate for the 4000-km^2 region. Paradoxically, soil scientists early on considered the influence of microclimate on soil development (e.g., Campbell and Claridge, 1969), but geomorphologists studying the Antarctic have only recently exploited the strong connection between microclimates, geomorphic process, and landforms. An understanding of the full

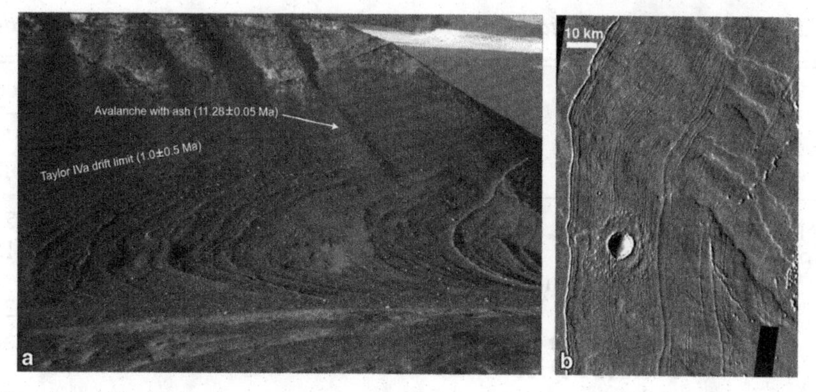

Fig. 2.3. (a) Oblique aerial photograph of lower Arena Valley, MDV, looking to the northwest (see Fig. 2.2 for location). On the valley wall and floor are 36 moraines deposited from a cold-based lobe(s) of Taylor Glacier. The moraine ridges, each as much as 1.5 m high, open to the valley mouth and reflect at least three glacial phases, termed Taylor II, III, and IVa glaciations (Denton et al., 1993; Marchant et al., 1994). Younger moraines cross-cut stratigraphically older moraines. The minimum age for the oldest moraine shown, Taylor IVa, is 1.0 ± 0.5 Ma, based on ^{10}Be cosmogenic-exposure age analyses (Brook et al., 1993). This moraine, along with several additional lower-elevation moraines, cross a well-preserved avalanche deposit (arrowed) that, in turn, rests on dolerite colluvium inclined from 28° to 32°. Sandstone bedrock, cut by thin mafic dikes, occurs near the top of the photograph. The unconsolidated avalanche deposit contains ~30% volcanic ash in the matrix (<2 mm) fraction. Volcanic crystals (sanidine) from this deposit date to 11.28 ± 0.05 Ma (^{40}Ar/^{39}Ar dating, Marchant et al., 1993a), and provide an age for the avalanche deposit. Neither the ash-avalanche deposit nor the moraines show morphologic evidence for significant post-depositional modification. Such deposits point to the long-term geomorphic stability of the stable upland zone and to the minimal role of cold-based glaciers in eroding underlying debris; field of view at valley bottom is ~3 km (from Swanger and Marchant, 2007). (b) Arcuate ridges of the distal, ridged facies on the northwest flank of Arsia Mons, Mars. The ridges are believed to represent drop moraines, analogous to those in Arena Valley, and reflect the growth, fluctuation, and decay of cold-based, tropical mountain glaciers on Arsia Mons (e.g., Head and Marchant, 2003). Emplacement of the moraine ridges has not disturbed underlying lava flows (broad ridges trending from lower right to upper left) nor impact ejecta associated with the large crater at left center (adapted from Shean et al., 2007).

range of microclimates in the MDV has only been possible with the deployment of widespread temperature-sensitive data loggers.

On the basis of measured variations in summertime environmental conditions (air and soil temperature, relative humidity, and soil moisture), the MDV have been divided into a series of microclimate zones (Marchant and

Table 2.1. *General climate data for microclimate zones in the McMurdo Dry Valleys*

		Summer[a] (annual)		
		Stable upland zone[b]	Inland mixed zone[c]	Coastal thaw zone[d,e]
Atmospheric temperature	Mean (°C)	−10 (−22)	−7 (−18)	−5 (−20)
	Mean daily maximum	−8 (−18)	−4 (−14)	−2 (−15)
	Mean daily minimum	−12 (−25)	−10 (−21)	−7 (−24)
	Positive degree days	0	3	11
Soil temperature	Mean @ 0 cm depth	−5 (−22)	–	2 (−18)
	Mean @ 10 cm depth	−6 (−21)	–	1 (−17)
RH	Mean relative humidity (%)	41 (43)	67 (55)	64 (63)

Notes:
[a] Summer months: December, January, February; annual conditions in parentheses.
[b] Beacon Valley; 1176 m a.s.l., 74 km from coast.
[c] Howard Glacier; 472 m a.s.l., 16 km from coast.
[d] Explorer's Cove; 26 m a.s.l., 4 km from coast.
[e] Soil data for coastal zone from near Lake Bonney; 60 m a.s.l., 25 km from coast. –, data not available.
Source: All data from McMurdo LTER weather stations. Available at www.mcmlter.org/queries/met/met_home.jsp

Denton, 1996; Bockheim, 1997; Doran et al., 2002; Marchant and Head, 2007). Minor changes in soil moisture and atmospheric temperature are sufficient to produce major changes in equilibrium geomorphic processes and surface topography, particularly along microclimate boundaries. Due to the crossing of geomorphic thresholds, the geomorphic impact of minor variations in soil moisture and temperature in the MDV (as in most desert regions) is disproportionately large when compared with similar shifts in humid-temperate latitudes (Langbein and Schumm, 1958; Schumm, 1965; Schumm and Lichty, 1965). We distinguish a coastal thaw zone (CTZ), an inland mixed zone (IMZ), and a stable upland zone (SUZ) (Fig. 2.2, inset) (Table 2.1), though alternative subdivisions are possible (e.g., Campbell and Claridge, 1969, 1987, 2006; Bockheim, 2002, 2003).

In the *coastal thaw zone (CTZ)*, modern summertime air temperatures show a mean season-long temperature of ~−5 °C (268 K), and a mean daily maximum of ~−2 °C (271 K). The rather high relative humidity (RH; averages ~64%) is due to the prevalence of southeasterly winds, which carry moisture from the Ross Sea (Table 2.1). Surface and subsurface ground temperatures vary considerably over relatively short horizontal distances of ~10 m due to large variations in surface albedo that arise from the presence of numerous tills capped by different lithologies in the CTZ (Hall et al., 1993; Denton and Marchant, 2000). Subsurface soil temperatures in the CTZ rise well above 0 °C (273 K) (Table 2.1 and snowfall exceeds 80 mm of water equivalent per year (Schwerdtfeger, 1984); a large fraction of the snowfall melts and infiltrates near-surface soils. The GWC of most soils in the CTZ is thus >30% (Campbell et al., 1997a, 1997b).

In the *inland mixed zone (IMZ)*, alternating winds (westerly katabatic and easterly from the Ross Sea) produce variable RH, but the mean summertime RH is close to that of the coast, ~67%. Mean summertime air temperatures are ~−7 °C (266 K), with a mean daily maximum of ~−4 °C (269 K) (Table 2.1). Snowfall is probably less than that of the coastal thaw zone but is uncertain because of the unknown quantity of wind-blown snow from the polar plateau. Near-surface soils contain <30% GWC (Campbell et al., 1997a, 1997b), except for regions alongside ephemeral streams and isolated snow patches.

In the *stable upland zone (SUZ)*, summertime RH is ~41%; this low value reflects the passage of dry, katabatic winds originating over the East Antarctic Ice Sheet. Summertime air temperatures are colder, with a mean of ~−10 °C (263 K) and a mean daily maximum of ~−8 °C (265 K) (Table 2.1). Although precipitation is limited, snow blown off the polar plateau may accumulate on small glaciers and feed perennial snowbanks. Glaciers and snowbanks lose mass by sublimation and the upper horizons of most soils contain < 5% GWC (Campbell et al., 1997a).

The importance of salts and brines

Soil salinity can locally modify geomorphic processes within these major microclimate zones by altering the abundance of liquid water in subzero conditions (e.g., Claridge and Campbell, 1968, 1977, 2005; Bockheim, 1997, 2002; Mahaney et al., 2001; Campbell and Claridge, 2006). The importance of salts in the MDV can be enhanced by several factors (Bao and Marchant, 2006; Bao et al., 2008). Because of the hyperarid environment (in which evaporation/sublimation exceeds precipitation), salts that are deposited by precipitation of snow generally become concentrated and remain near the site of deposition. The paucity of liquid water causes salts released during weathering of rocks and soils to accumulate continuously in the soils. At the ground surface, salts commonly occur as coatings and efflorescences within weathering pits (see below), and along cracks and joints. Salts are also observed below the ground surface and occur as widely distributed efflorescences and as discrete horizons sometimes up to 15 cm in thickness (e.g., Bockheim, 1997; Bao and Marchant, 2006; Campbell and Claridge, 2006).

Variations in soil moisture can be induced by areal and vertical concentrations of salts. For example, surface water has been observed at temperatures of <−4 °C (269 K) in small damp, salty hollows (e.g., Campbell and Claridge, 2006). In upper Wright Valley at Don Juan Pond, surface water remains below the solidus even at temperatures below −40 °C to −50 °C (223 K to 233 K) (e.g., Marion, 1997; Takamatsu et al., 1998; Healy et al., 2006). Water cycled laterally through hyporheic zones (regions beneath and lateral to streambeds, where mixing of shallow groundwater and surface water occur) can mobilize salts, forming saline solutions, brines, and seeps, and can locally redeposit salts in soil horizons and on the surface (e.g., McKnight et al., 1999; Lyons et al., 2005; Harris et al., 2007). Mobilization and concentration of salts can also be caused by vertical migration of water through soils (e.g., Gibson et al., 1983; Dickinson and Rosen, 2003; Wentworth et al., 2005). Significant locally elevated water contents can be caused by these variations in vertical concentrations; such variations can produce local complexities in the geochemistry and thermal states of soils, and can influence geomorphic processes and resultant landforms. Salt types include a wide range of crystalline phases of sodium, potassium and magnesium chlorides, nitrates, and sulfates. As soil age increases, salts increase in importance and abundance (e.g., Claridge and Campbell, 1977; Mahaney et al., 2001; Wentworth et al., 2005; Bao and Marchant, 2006; Bockheim and McLeod, 2006; Campbell and Claridge, 2006; Bao et al., 2008).

Mars

Current conditions and physical setting

Mars is currently a global hyperarid, cold desert similar in many ways to the MDV (Figs. 2.1, 2.4, 2.5; and see summary in Zurek, 1992, and Carr, 1996). The predominantly CO_2 atmosphere is thin, resulting in an atmospheric pressure that is less than 1/100th that of Earth (Fig. 2.4). The atmosphere is often close to saturation with water vapor, but only very tiny amounts are actually present (an average of 10 precipitable microns) due to the low temperatures, and thus the global climate is hyperarid (Fig. 2.1; e.g., Baker, 2001). Heat exchange between the surface of Mars and the thin atmosphere is much weaker than on the Earth; because of this, the elevation zonality of surface temperature is much less pronounced on Mars than in Antarctica.

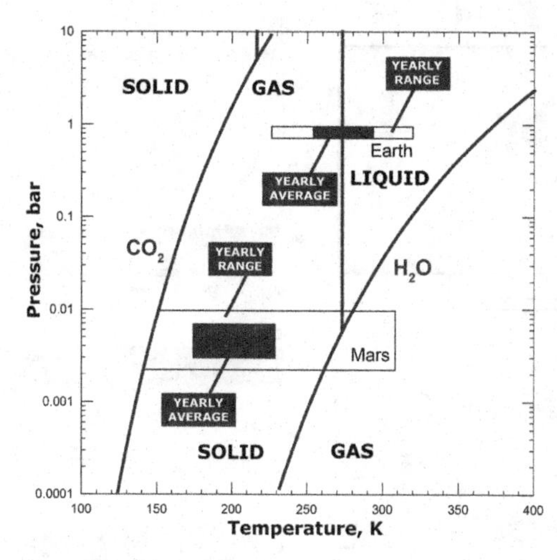

Fig. 2.4. Phase diagram in temperature-pressure coordinates for water and carbon dioxide for Earth and Mars. Larger boxes represent current overall ranges and inset boxes represent current yearly average surface temperature and near-surface atmospheric pressure for both Earth and Mars. The dimensions of the boxes represent the general latitudinal, elevational, and temporal variations. On Earth, temperatures and pressures are in the range where liquid water and ice tend to dominate geomorphic processes (see Fig. 2.1), and CO_2 comprises only about 0.035% of the atmosphere, is in the gaseous state, and is not a direct geomorphic factor. The atmosphere of Mars is composed predominantly of CO_2; during winter at high latitudes, CO_2 condenses onto the surface to form the seasonal cap, a phenomena that is reflected in the seasonal variation of atmospheric pressure at the Viking Lander sites. For H_2O on Mars, current year-average temperatures are much lower and well within the solid (ice) domain for all year-average temperatures (see Fig. 2.1). Temporal changes of temperatures cause sublimation of water ice and condensation of water vapor to produce snow and frost, such as is observed seasonally at the Viking Lander 2 site (see Figs. 2.19, 2.20) (Jones et al., 1979; Wall, 1981). Note that at the range of temperatures and pressures on Mars, conditions for metastable liquid water can occur locally for short periods of time (e.g., Hecht, 2002) but liquid water is not a major factor in regional geomorphic processes (see also Fig. 2.1). The presence of salts and brines can modify these relations. Water may be locally important in some special microenvironments (e.g., pits in surface rocks, gullies) and in the recent history of Mars (e.g., Kreslavsky et al., 2008). Changes in spin axis and orbital parameters can cause changes in the distribution of ice stability zones (e.g., Fig. 2.6). For example, longer-term changes can result in either (1) periods of much higher obliquity leading to sublimation of the polar caps and transport of water to the topics to form glacial deposits there (e.g., Forget et al., 2006), (2) periods of very low obliquity during which the atmosphere collapses and condenses onto the surface (e.g., Kreslavsky and Head, 2005), or (3) periods of time in early Mars history when Mars may have been "warm and wet" and more Earth-like (e.g., Craddock and Howard, 2002) (Fig. 2.1; points labeled 300 mbar Mars, 1000 mbar Mars) (from Marchant and Head, 2007).

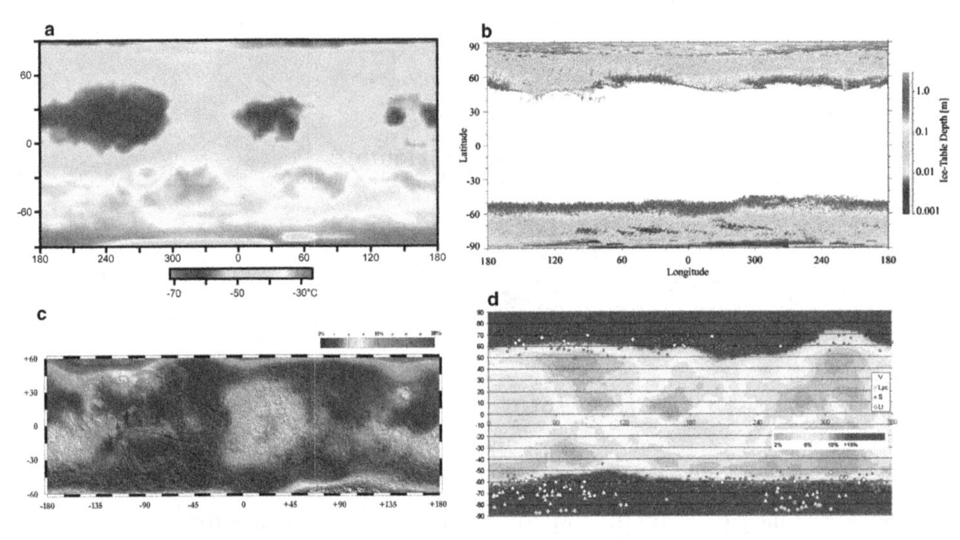

Fig. 2.5. Temperature, water stability, the presence and depth of ice on the martian surface and in the shallow subsurface, and the distribution of some ice-related landforms. (a) Year-maximum, day-average surface temperatures on Mars. Data on surface temperatures from the European Martian Climate Database (www-mars.lmd.jussieu.fr/). This data set has been generated using a general circulation model (Forget et al., 1999) but represents the observed martian climate system well (from Kreslavsky et al., 2008). (b) Predicted depth to the ice table under present climate conditions (from Mellon, 2003). (c) Observed distribution of surface and near-subsurface ice (the ice table) from Mars Odyssey GRS/NS data (from Boynton et al., 2002; Feldman et al., 2002; Mitrofanov et al., 2002). (d) Distribution of polygons on Mars superposed on map of the ground ice proportion in mass as measured by the Mars Odyssey Neutron Spectrometer (Feldman et al., 2002). V, large heterogeneous polygons; LPC, straight crack networks close to the south polar cap; S, homogeneous polygons of a size <40 m; LT, large homogeneous polygons formed by cracks associated with topography (from Mangold et al., 2004).

Surface temperatures vary diurnally and with seasons. At the equator, for example, daily mean surface temperature is ∼−58 °C (∼215 K) (Fig. 2.5), but temperatures can range from −113 °C to −93 °C (160–180 K) at night to as high as −13 °C to 7 °C (260–280 K) during the day. At the Viking Lander sites day-average air temperatures (Tillman, 1988; Zurek et al., 1992) clearly show this latitude dependence, and daily maxima and minima at the Viking Lander 2 site differ by up to ∼60 K during northern summer. Due to the obliquity of its rotational axis and the eccentricity of the orbit of Mars (e.g., Laskar et al., 2004), insolation and the exact temperature at any point on the surface of Mars are dependent on the latitude and the season; albedo, thermal inertia of the surface, and local variations in topography also influence temperatures (Hecht, 2002). Atmospheric dust can vary significantly with time; its presence increases the absorption of solar radiation in the atmosphere, decreasing radiative and conductive heat exchange with the surface, making the vertical atmospheric temperature profile more isothermal, and affecting the temperature regime of the surface. For example, the effects of global dust storms on air temperature at the VL 2 site were clearly observed. Surface wind speeds measured at the landing sites are typically a few meters per second, with daily maxima of 8–10 $m\,s^{-1}$, and gusts up to 40 $m\,s^{-1}$ (Zurek et al., 1992).

Topography, elevation, and roughness play important roles in defining the climate and their microclimate zones, just as these parameters influence climate in the MDV. On Mars, high elevations and rough topography dominate the southern heavily cratered uplands, and low elevations and smoother topography characterize the northern lowlands (e.g., Smith et al., 2001; Kreslavsky and Head, 1999, 2000, 2002a). Impact craters of all sizes, huge volcanoes, and extensive troughs and valleys (e.g., Carr, 1981, 2006) all create local microclimate environments through variations in elevation, albedo, slope, aspect, etc.

Due to the lack of surface vegetation, rocks, soils, and bedrock exposures characterize the surface of Mars in a manner similar to the MDV. Surface thermal inertia on Mars is therefore dominated by variations in thermal conductivity; blocky terrains and bedrock exposures have high thermal inertia, and granular soils and dust have low thermal inertia due to conduction of heat being largely restricted to the contact points between grains. Surface thermal inertia on Mars varies regionally and globally (e.g., Christensen et al., 2001; Jakosky and Mellon, 2001).

Liquid water is stable only where temperatures exceed 273 K (0 °C) and water partial pressure exceeds 6.1 mbar (Fig. 2.4). This is an unlikely situation in the current climate period. The water vapor content in the atmosphere is currently so low that if the atmosphere is well mixed, the partial pressure of water is 2–3 orders of magnitude below that needed to stabilize liquid water. Liquid water boils away if the total pressure is below the triple point 6.1 mbar and it evaporates if the vapor pressure of water is below the saturated vapor pressure (RH $<100\%$). Atmospheric pressure exceeds 6.1 mbar in some low areas on Mars; here liquid water would remain until rapidly consumed by evaporation and freezing (Kreslavsky and Head, 2002b). Localized conditions could also create microenvironments for liquid water. For example, if an ice-rich soil is heated at a high rate, and soil permeability was low enough so that soil pore water vapor pressure could build to 6.1 mbar, then liquid water could form if temperatures exceeded freezing. Liquid water might be formed at even lower temperatures if salts and brines were involved. Just as in the MDV, any melting conditions would be very transient because of the limited period where required temperatures are met and the rapid decrease of daytime temperatures with depth (Carr, 2006).

On Mars, the cold atmosphere is capable of holding only about 1/6500–13 000th of the water in the Earth's atmosphere, but seasonal/latitudinal temperature variations can cause interchange among several reservoirs (polar ice, the atmosphere, and the surface and soil layer). Cyclic mobilization of water (sublimation and desorption) can be caused by seasonal insolation variations, and result in redistribution latitudinally in the atmosphere, deposition on the surface, and diffusive exchange with the soil layer or regolith (e.g., Farmer, 1976; Zent et al., 1986; Zurek, 1992; Zurek et al., 1992; Haberle et al., 2001; Richardson and Mischna, 2005).

The distribution of water ice on the surface and in the near subsurface of Mars can be predicted from the current distribution of mean surface temperatures (Fig. 2.5) and the distribution of maximum surface temperatures averaged over a day is a good indication of the presence and behavior of ice. The maximum day-average surface temperatures during a year for the current environment on Mars are everywhere below freezing (Figs. 2.4, 2.5a). Thus, we can predict that at the present, Mars lies in periglacial, glacial, or hyper-arid glacial morphogenetic environments (Fig. 2.1).

Incident solar radiation, thermal inertia, and albedo have been combined to calculate mean annual temperature as a function of position on the surface of Mars and then to map this into regolith ice stability zones (e.g., Farmer and Doms, 1979; Mellon and Jakosky, 1995). These calculations suggest that ground ice is currently unstable in the near subsurface at latitudes less than about 50°, but is currently stable at higher latitudes. Recent predictive mapping of the depth to the current ice table (Fig. 2.5b) (Mellon, 2003) show that at latitudes greater than ~60°, most depths to the ice table are between 1 and 10 cm. These predictions have largely been confirmed by the Mars Odyssey Gamma Ray Spectrometer experiment package (Fig. 2.5c), which mapped the global distribution of surface and near-surface water ice (Boynton et al., 2002; Feldman et al., 2002; Mitrofanov et al., 2002).

Mean annual temperatures on Mars and their variation with microclimate zones (Fig. 2.1), as in the MDV, are important but are not the only factor. Key to the formation and evolution of landforms at all scales is the range in annual temperature, the maximum temperatures, the period of time these maximum temperatures are achieved, their relationship to the melting point of water (Fig. 2.4), the presence and influence of salts and brines, and the total surface pressure and its relationship to the triple point. While the

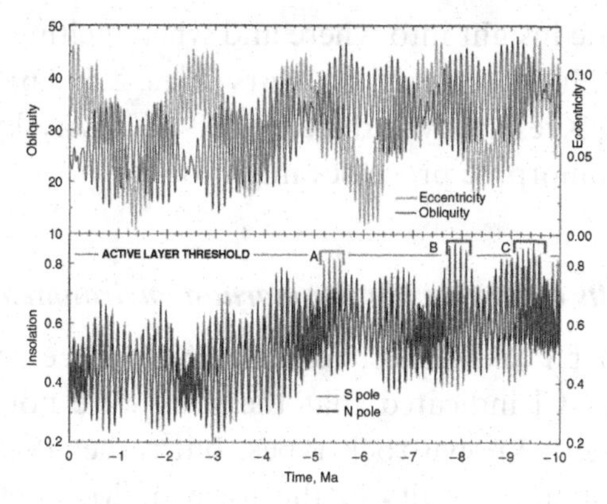

Fig. 2.6. The recent history of obliquity and eccentricity of Mars and its relationship to insolation and surface conditions (modified from Kreslavsky et al., 2008). (Top) Obliquity and eccentricity for the last 10 Ma. The data are from Laskar et al. (2004). (Bottom) The year-maximum day-average insolation at the poles, calculated from the spin axis/orbital evolution for the last 10 Ma. The thin horizontal line marks the estimated active layer threshold. Times where the insolation exceeds this threshold are predicted to result in an active layer (~5.2 Ma, ~7.9 Ma, and ~9.2 Ma).

year-average (Figs. 2.1, 2.4) and day-average (Fig. 2.5a) surface temperatures on Mars are well below the freezing point of water over the whole planet, surface temperature can exceed the freezing point in some places during a portion of the day. For example, currently the surface temperature exceeds the freezing point of water for very short periods of the day in the equatorial zone throughout the year, and in the mid latitudes during the summer, especially on equator-facing slopes (Kreslavsky et al., 2008). Viking Lander 1 and 2 site temperatures varied considerably daily and with seasons (Tillman, 1988; Zurek et al., 1992).

Although seasonal thermal cycling can be a significant factor in the surface geomorphology of Mars (e.g., Lachenbruch, 1962; Mellon, 1997; Marchant et al., 2002; Sletten et al., 2003; Mangold et al., 2004; Mangold, 2005; Levy et al., 2006, 2008a, 2008b), in the current Mars environment the summer day-average surface temperature never exceeds the melting point of water anywhere on the surface (Fig. 2.5a); thus near-surface seasonal freezing and thawing of pure water (i.e., not brines, see also below) are not predicted to occur (e.g., Paige, 2002; Costard et al., 2002; Kreslavsky and Head, 2004; Kreslavsky et al., 2008) (Fig. 2.1). However, detailed calculations of spin axis/orbital parameter behavior in the recent geological past (e.g., Laskar et al., 2004) can be used to predict climate forcing sufficient to produce microenvironments with a seasonal active layer (Fig. 2.6). Together with new altimetry data, this can provide insight into where and when melting temperatures and an active layer could have occurred on Mars (Figs. 2.5, 2.6) (e.g., Paige, 2002; Costard et al., 2002; Kreslavsky and Head, 2004; Kreslavsky et al., 2008) and what the related geomorphic processes might be.

Salts and brines in the martian environment

At the Viking Lander 1 and 2 sites, very high S, high Fe, moderate Mg, and unexpected levels of Cl indicated that the soils were not an admixture of weathering products of known rock types, but rather that other processes, such as the introduction of salts in the form of Mg and Ca sulfates and chlorides (by leaching or addition of volcanic gases), might account for the measurements (e.g., Toulmin et al., 1977; Clark and Baird, 1979; Settle, 1979; Clark and Van Hart, 1981; Clark et al., 1982; Gooding, 1992). Observations at the Pathfinder and Mars Exploration Rover (MER) sites have further emphasized the localized concentration of salts and secondary minerals in the Mars environment (e.g., Larsen et al., 2000; Rieder et al., 2004; Squyres et al., 2004a, 2004b, 2004c, 2006; Haskin et al., 2005; Tosca et al., 2005; Vaniman and Chipera, 2006; Wang et al., 2007). Furthermore, the role of salts in the alteration of the martian surface has been confirmed by analysis of the martian SNC meteorites (Shergottites, Nakhlites, Chassignites, originating from Mars). Assessments of the interaction between martian crustal fluids and parent igneous rocks suggest that the most plausible models for secondary mineral formation (e.g., Fe–Mg–Ca carbonates, anhydrite, gypsum, clays, etc.) involve evaporation of low-temperature brines (e.g., Bridges et al., 2001). Indeed, soils from parts of the MDV contain the same salt phases as those in SNC meteorites from Mars (Gooding et al., 1991; Wentworth et al., 2005).

Brines are also likely to be important on Mars (e.g., Kuzmin and Zabalueva, 1998). Equilibrium ternary phase diagrams indicate that the minimum temperature at which brines could be stable at the surface was 210 K (−63 °C), with a water concentration of ~70 wt% and a high concentration of CaCl (Brass, 1980). Zent and Fanale (1986, 1990) showed that metastable brines could have long lifetimes and that chloride–sulfate brines could be consistent with Viking Lander elemental data. The eutectic brine hypothesis for Mars (Knauth and Burt, 2002) proposes that an original NaCl-rich hydrosphere became evapoconcentrated, pore fluids evolved into complex $CaCl_2$-enriched brines through chemical reaction with the regolith, and continuing freeze-down of Mars left eutectic brines as pore fluids. For comparison, Knauth and Burt (2002) and Burt and Knauth (2003) describe brines in the McMurdo Dry Valleys (Don Juan Pond) with freezing points well below 225 K (−48 °C) and hypothesized that concentrated brines may currently be stable at temperatures between 180 and 210 K (−63 to −93 °C) on Mars. These observations and experiments suggest that salts are present and locally abundant in martian soils. On the basis of the stability of the martian surface, it is likely that salts and brines formed over time as a result of the interaction of surface or subsurface liquid water with basalts, in the presence of the martian atmosphere. Salts and brines on Mars are thus likely to modulate the effects of microclimate conditions (e.g., Clark, 1978, 1979), and influence geomorphic processes (e.g., Malin, 1974), as they do in the MDV. For example, under current conditions on Mars, if the eutectic freezing temperatures of $CaCl_2$-enriched brines were in the range of 180 to 210 K (−63 to −93 °C) (Knauth and Burt, 2002; Burt and Knauth, 2003), at extremely high latitudes they would be frozen, at low latitudes they would have evaporated, but would be stable in mid-latitude bands.

Summary of environmental conditions on Mars

The current nature of the hyperarid, cold-desert environment of Mars has several implications for geomorphological features at various scales.

1. There is a distinctive latitude-dependent variation in mean annual surface temperature (Fig. 2.5a) that is similar to the differences in mean annual temperatures seen in the various MDV microclimate zones (in Fig. 2.1 compare the dashed oval with Mars mean annual temperature at 30°, 50°, and 60° latitudes).
2. The presence of ground ice and the depth to the ice table is also dependent on latitude (Fig. 2.5b, c). These first two factors should map out into potentially significant differences in geomorphology.
3. The current conditions on Mars lead to the prediction that liquid water is not present on the surface or in the near subsurface (see also Lobitz et al., 2001). A major implication of this is that, under present conditions, there can be no active layer, and thus no typical cryoturbation of soils. However, new data on spin axis/orbital parameters and topography permit predictions of the nature of climates in the recent past, and where and when these conditions should have occurred (e.g., Kreslavsky and Head, 2004; Kreslavsky et al., 2008) (Fig. 2.6).
4. Seasonal deposition of CO_2 and H_2O snow and surface ice is strongly latitude dependent. Any geomorphic process linked to this seasonal deposition will also be strongly latitude dependent.
5. Although the current mean surface temperatures everywhere on Mars are typically well below 0 °C (273 K), variations in spin axis/orbital parameters can cause poleward-facing slopes at mid latitudes to reach surface and shallow subsurface

temperatures in excess of 0 °C (273 K) for days to perhaps weeks per year in the recent geological past (e.g., Costard et al., 2002; Paige, 2002; Kreslavsky and Head, 2004; Kreslavsky et al., 2008). This could potentially cause locally enhanced flow of ice, melting of snow, glacial ice and ground ice, and formation of a localized active layer and runoff gullies in topographically favored regions.

6. The potential role of salts and brines in modifying melting temperatures, vapor pressures, and geomorphic responses needs to be taken into consideration.
7. Taken together, these factors make the calculated erosion rates on Mars extremely low (Golombek et al., 1997, 1999), in the range of 0.2 meters per million years (m Ma^{-1}) (Arvidson et al., 1979), very similar to those of the SUZ in the McMurdo Dry Valleys (~0.06–0.3 m Ma^{-1}; Brook et al., 1995; Summerfield et al., 1999).
8. Despite their similarity, there are substantial differences between the MDV and Mars; even at the coldest times of year, the relative humidity in the MDV is equivalent to a water vapor partial pressure ~20 times its probable level on Mars, mean temperatures are 25–50 °C higher than MDV values, soil clay content is generally minor, and the solar ultraviolet does not penetrate as deeply into the atmosphere as it does on Mars (Clark, 1979). In addition, the total atmospheric pressure in the MDV exceeds the triple point of water by a large factor, whereas on Mars the pressure hovers around the triple point of water.

Geomorphic analyses of equilibrium landforms on Earth and Mars

We now summarize the nature of landforms in the MDV and on Mars as a function of size, including (1) macroscale landforms (>250 m in areal extent and/or those features incised in bedrock); (2) mesoscale landforms (unconsolidated deposits from 1 m to ~250 m in areal extent); and (3) microscale landforms (features ≪1 m in size) (Marchant and Head, 2007). One benefit of this scale-dependent organization is that size-specific analyses can be performed for Mars, taking advantage of Viking- and Odyssey-scale image data sets (macroscale), Mars Orbiter Camera (MOC) and High Resolution Imaging Science Experiment (HiRise) image data and Mars Orbiter Laser Altimeter (MOLA) altimetry data (mesoscale), and Viking Lander–Pathfinder–Mars Exploration Rover (MER)-scale image data (microscale). Landforms are additionally grouped according to microclimate zone. We emphasize that although the spatial scale for broad climate zones on Mars (latitude dependent) is vastly greater than that for the MDV microclimate zones, many of the cold-desert landforms and the variability of specific landform types are remarkably similar for both planets. The zones in the MDV are condensed due to the steep thermal gradients that exist in the ~100 km from sea level to the polar plateau (Fig. 2.2).

Macroscale features

McMurdo Dry Valleys: slope asymmetry, drainage-basin asymmetry, and gullies

Coastal thaw zone (CTZ) North-facing slopes that receive relatively high levels of incident solar radiation (e.g., Dana et al., 1998) are shallower (averaging ~20°) than south-facing slopes (~25°), leading to a classic slope asymmetry in the coastal thaw zone (Fig. 2.7). Consistent with this marked valley-side asymmetry is a well-expressed drainage-basin asymmetry (defined here as the ratio of the valley-half width [measured south of the valley thalweg] to the total valley width; an asymmetry factor (AF) of ~50 indicates a symmetrical valley, whereas an AF above or below 50 indicates marked valley-side asymmetry). An AF of ~65 for the CTZ indicates preferred degradation along north-facing walls. In addition to valley-side asymmetry, there is also variation in the size and spacing of gullies. For example, gullies on north-facing slopes in lower Ferrar Valley appear deeper and spaced further apart than those on south-facing slopes (Fig. 2.8).

Melting occurs preferentially on north-facing slopes, although even here the majority of ice loss (~90%) takes place by sublimation (Chinn, 1980, 1981; Frezzotti, 1997; Fountain et al., 1999). The relatively minor meltwater generated may percolate centimeters to tens of centimeters into soils, elevating soil pore pressures sufficiently to induce downslope movement via solifluction, i.e., the slow flow of saturated materials. If channelized, overland flow by meltwater appears sufficient to cut channels 3–5 m deep in loose debris. A saturated 1–2 m wide hyporheic zone (Gooseff et al., 2003a;

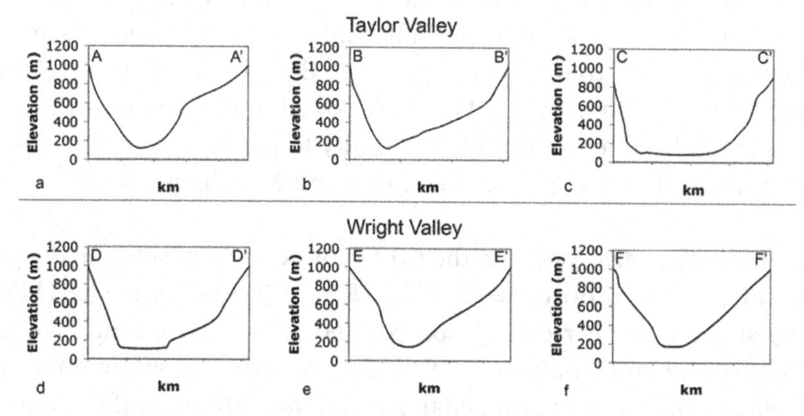

Fig. 2.7. Topographic profiles for the coastal thaw zone and inland mixed zone in the MDV. Locations are shown as lettered lines in Fig. 2.2. In all profiles the north (equator-facing) slopes (those to the right in each profile) are less steep than south (pole-facing) slopes. (From Marchant and Head, 2007.)

Fig. 2.8. Landsat satellite image of lower Ferrar Valley showing gully asymmetry on north-facing (bottom of image) and south-facing (top of image) slopes. See Fig. 2.2 for location and scale. (From Marchant and Head, 2007.)

Ikard et al., 2009) commonly fringes these channels and helps sustain a unique biota of several varieties of cold-adapted nematodes (Gooseff et al., 2003b; Nkem et al., 2006; Northcott et al., 2009). Evaporation of meltwater produces visible salts that coat rock surfaces and intervening soil. Variations in salt crystal size, caused by expansion and contraction upon hydration–dehydration, permits rock breakdown, as salts pry away loosely bound minerals (Huinink et al., 2004). The most common salt species in the CTZ are chlorides and sulfates, reflecting proximity to the Ross Sea (Claridge and Campbell, 1977; Bao et al., 2000, 2008; Bao and Marchant, 2006).

Inland mixed zone (IMZ) Slopes in the IMZ show less valley-side and drainage-basin asymmetry than do those of CTZ (Fig. 2.7). Gullies in the IMZ are relatively shallow, appear closely spaced, and show sharp, knife-like interfluves. They occur commonly on north-facing slopes, and contain three major geomorphic components: alcove, channel, and fan. Although the average air temperature in the IMZ is less than that of the CTZ (Table 2.1), minor snowmelt fringes most snowbanks and snow melts alongside rocks that are heated by solar radiation to temperatures >0 °C (273 K). In addition, snow may be blown from local snowbanks onto the surface of solar-heated rocks; in this way the distribution of meltwater is increased well beyond the immediate margins of snowbanks, although it is localized in the downwind direction. The total meltwater contributed by these processes is minor and does not commonly support solifluction, but appears sufficient to maintain a shallow, discontinuous ice-cemented layer within the upper few centimeters of debris; it may also contribute to shallow subsurface meltwater flow in saline soils (e.g., Lyons et al., 2005; Harris et al., 2007).

Lack of rainfall and associated runoff means that gully streams have little interaction with the broader landscape, and no evidence has been reported that links the sources of gully formation with groundwater flow through an underground aquifer, and penetration to the surface through saline spring sources, as reported in the arctic Axel Heiberg region (e.g., Heldmann et al., 2005). Extensive field evidence indicates that streams and gullies in the MDV form from surface top-down melting of snow and ice due to enhanced summer solar insolation (McKnight et al., 1999; Head et al., 2007a, 2007b) (Fig. 2.9). Because maximum precipitation in the ADV is less than ~10 cm of snow per year, other processes are required to concentrate sources for the meltwater that feeds gully streams. Perennial snowbanks and seasonal wind-blown snow trapped in lows (alcoves, channels, polygon troughs) form significant sources for meltwater that lead to gully formation in the IMZ (e.g., Head et al., 2007a, 2007b; Dickson et al., 2007a; Morgan et al., 2007; Levy et al., 2007) (Fig. 2.10). Surface portions of cold-based glaciers at unique positions related to seasonal insolation intensity and geometry (insolation-induced melting) are another source of water for gullies and streams (Fountain et al., 1999).

Streams flow for less than 20% of the year, some only for a few days; streams show considerable daily, intraseasonal and interannual variation in flow behavior depending on insolation and air temperature (McKnight et al., 1999), and water supply (Dickson et al., 2007a; Morgan et al., 2007; Levy et al., 2007). About half of the stream channels observed in the MDV are currently active; this implies that many of the others date from earlier times (McKnight et al., 1999), and thus may be in the waning stages of formation, or currently inactive. During periods of active channel flow, dark bands form along channel banks, the surface manifestation of a hyporheic zone (where channel water both enters and exits, altering the volume of water in the channel) (Fig. 2.11). Lack of a general groundwater system means that the hyporheic zone is very important for water storage and exchange, and activity can continue there after surface water flow in the channel has ceased (McKnight et al., 1999; Levy et al., 2007).

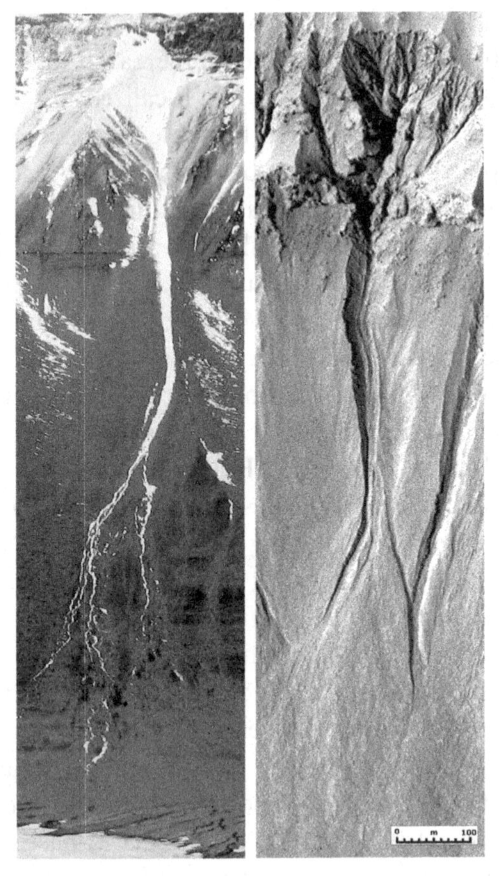

Fig. 2.9. Gullies in the inland mixed zone, MDV and on Mars. (Left) Gully on the northern wall of central Wright Valley, MDV (see Fig. 2.2 for location). Wind-blown and precipitated snow (white material in image) accumulates preferentially in the alcoves and channels of the gullies, and melts during peak summer insolation to form active channels and to erode and deposit sediments in the distal fan. (Right) Gully on the interior of a crater wall on Mars. Gully consists of an alcove at the top, a channel emerging from the alcove and descending down the crater wall and bifurcating on the distal sedimentary fan. Mars gullies are interpreted to represent conditions in the recent past of Mars when climate change supported accumulation of snow in alcoves, melting, and flow and erosion to produce channels and fans. Scales are similar in both images.

Stable upland zone (SUZ) The current extremely cold and arid climate conditions in the SUZ (Table 2.1) prohibit significant meltwater, and thus there is little evidence for ongoing, macroscale geomorphic change. Where present in this zone, gullies are interpreted to be relict and inactive because overlying colluvium is commonly interbedded with near-surface ashfall dated to as

Fig. 2.10. Perspective view of the distal portion of a gully on the north-facing wall of South Fork, upper Wright Valley (see Fig. 2.2 for location). Gully channels descend down the slope and out onto the valley floor, forming a distal fan (foreground). In the middle, white patches represent seasonal snowbanks topographically trapped in the channel from wind-blown snow and undergoing melting to produce water flow and erosion during daily peak solar insolation in the austral summer. The marginal and distal dark regions represent the hyporheic zone (Fig. 2.11). Polygon trough outlines can be seen in the distal hyporheic zone due to the more porous nature of the sand filling the troughs (Fig. 2.14d). The background displays polygonal ground not readily apparent in this perspective view but can be seen in detail in Fig. 2.14a, b. Vertical gray/black line in center is person for scale. The color version of this figure can be found at http://www.cambridge.org/uk/catalogue/catalogue.asp?isbn=9780521889193.

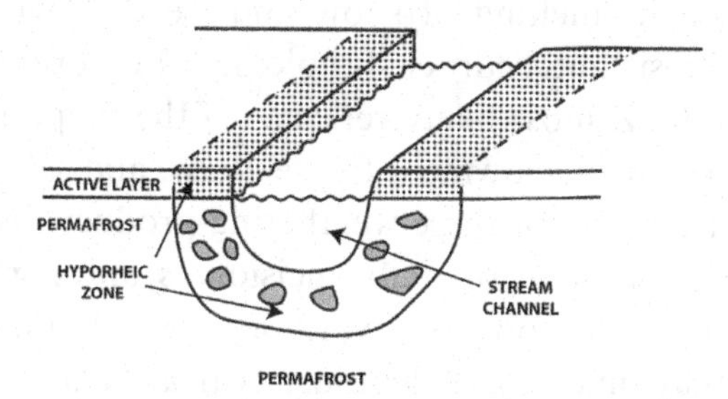

Fig. 2.11. Sketch of the nature of the hyporheic zone associated with MDV streams. Water traveling in the main channel percolates into the stream bed and banks and may reenter the stream along the course. Solutes pass in and out of the hyporheic zones and are often deposited as salts on the channel banks upon desiccation. Nutrients are also transported in and out of the hyporheic zone. (Modified from McKnight et al.,1999.) Compare this sketch with Fig. 2.14b, c, which show active hyporheic zones (dark patches) adjacent to gullies in South Fork, MDV.

much as 11 Ma (Marchant et al., 1993a, 1993b, 1993c, 1996; Marchant and Denton, 1996). Unlike slopes in the CTZ and IMZ, which are incised in relatively uniform igneous and metamorphic rocks of the basement complex, slopes in the SUZ are incised in flat-lying sedimentary rocks and intervening sills of Ferrar Dolerite. Lithology thus plays a significant role in the macroscale geomorphology of the region: cliffs are formed on Ferrar Dolerite and Beacon Heights Orthoquartzite, whereas smooth, rectilinear slopes (~25°–30°) form on weakly cemented sandstones (Altar Mountain Sandstone, New Mountain Sandstone) (Selby, 1971a, 1974; Augustinus and Selby, 1990; Sugden et al., 1995a; Prentice et al., 1998).

Glaciers in the SUZ lose mass almost entirely by sublimation. Where intense reradiation off nearby surfaces elevates local temperatures, the margins of some glaciers and perennial snowbanks may experience melting (e.g., Dana et al., 1998; Fountain et al., 1999). Melting may also occur where a thin cover of wind-blown sand lowers the surface albedo and induces melting of underlying ice and snow. Although this phenomenon occurs in all microclimate zones, it is the only mechanism for relatively significant melting in the SUZ. In all cases, however, the geomorphic impact of this meltwater in the SUZ is negligible. Although meltwater may moisten the upper 5–15 cm of soil, it does not induce widespread solifluction, and evaporation removes most of this water within hours (Kowalewski et al., 2006).

Synthesis The observed variation in valley-side asymmetry, drainage-basin asymmetry, and gully development in the MDV is plausibly related to spatial variations in the melting of snow and ice and, where appropriate, the distribution of resistant bedrock lithologies. For example, the mature gully system of the CTZ most likely reflects (1) the preponderance of rock breakdown associated with snowmelt, freeze–thaw, and salt weathering, and (2) the downslope transport of these weathering products by water, solifluction, and wind. The *duration* of gully incision is another parameter that could be responsible for the observed variation in gully morphology. However, we contend that time is not *the* most critical factor in differentiating gully morphology because many slopes in the IMZ (as well as in the SUZ) have been dated on the basis of $^{40}Ar/^{39}Ar$ analyses of overlying ashfall to ≥7 Ma (Marchant and Denton, 1996). Given the time available for slope evolution in the IMZ, we postulate that gullies would have achieved mature forms if sufficient meltwater had been available (i.e., comparable to that now found in the modern CTZ). The extremely low levels of meltwater produced in the SUZ lead to a preponderance of inherited slopes that originally formed under wetter, and most likely warmer, climate conditions before the onset of cold-polar desert conditions during middle Miocene time (e.g., Marchant et al., 1993a, 1993b; Sugden et al., 1993b, 1995a; Lewis et al., 2007).

Mars: gullies and slope asymmetry

Malin and Edgett (2000, 2001) initially described a class of young features on Mars that they termed gullies, consisting of an alcove, a channel, and a fan (Fig. 2.9). Restricted to middle and high latitude locations, these features were interpreted by Malin and Edgett (2001) to have originated through processes related to the presence of liquid water delivered to the surface by discharge of groundwater from the deeper subsurface. The potential presence of liquid water on the surface of Mars, currently or in the very recent geological past when liquid water is metastable (Hecht, 2002), generated a host of alternative explanations for the gullies (see summary in MEPAG, 2006). Detailed analysis of the conditions under which H_2O could flow as a liquid in the current Mars environment shows a range of conditions under which gully-forming activity is possible (Hecht, 2002; Costard et al., 2002; Heldmann and Mellon, 2004). Recent observations of changes in gully characteristics, interpreted to mean that a few gullies are currently active (Malin et al., 2006), have intensified this discussion. Gullies similar to those seen in the MDV coastal thaw and inland mixed zones are observed on Mars (compare Figs. 2.9 and 2.10), where they show a range in latitude distribution from 30° to 70°, are largely restricted to a band around 45°, and tend to have some preference for pole-facing slopes. Although Malin and Edgett (2000, 2001) initially interpreted gullies to have formed by groundwater release (see also Heldmann and Mellon, 2004), the latitude distribution, orientation, and local geological environment of these features suggest to others that they formed by seasonal melting of snowpack or ground ice to form intermittent streams or debris flows (e.g., Costard et al., 2002; Christensen, 2003; Bridges and Lackner, 2006; Head and Marchant, 2006; Dickson et al., 2007b; Head et al., 2007a, 2007b). Surface melting temperatures are not typically reached for extended periods in the current climate regime (Figs. 2.4, 2.5a, 2.6), and this suggests that these features may date from the very recent geological past (e.g., Costard et al., 2002; Head et al., 2003; Reiss and Jaumann, 2003; Kreslavsky et al., 2008), perhaps extending locally into the present (e.g., Malin et al., 2006).

Latitude-dependent asymmetries in the distribution of steep slopes on Mars have also been observed (Kreslavsky and Head, 2003). For example, the frequency of steep (>20°) slopes on Mars drops more than three orders of magnitude from equatorial to high-latitude regions. This trend is interpreted to be due to the enhanced role of climate-related ice and dust mantling, and ice-assisted creep processes at higher latitude (Kreslavsky and Head, 2003). The importance of slope-related insolation and local microclimate environments is illustrated by the fact that the boundary between preserved

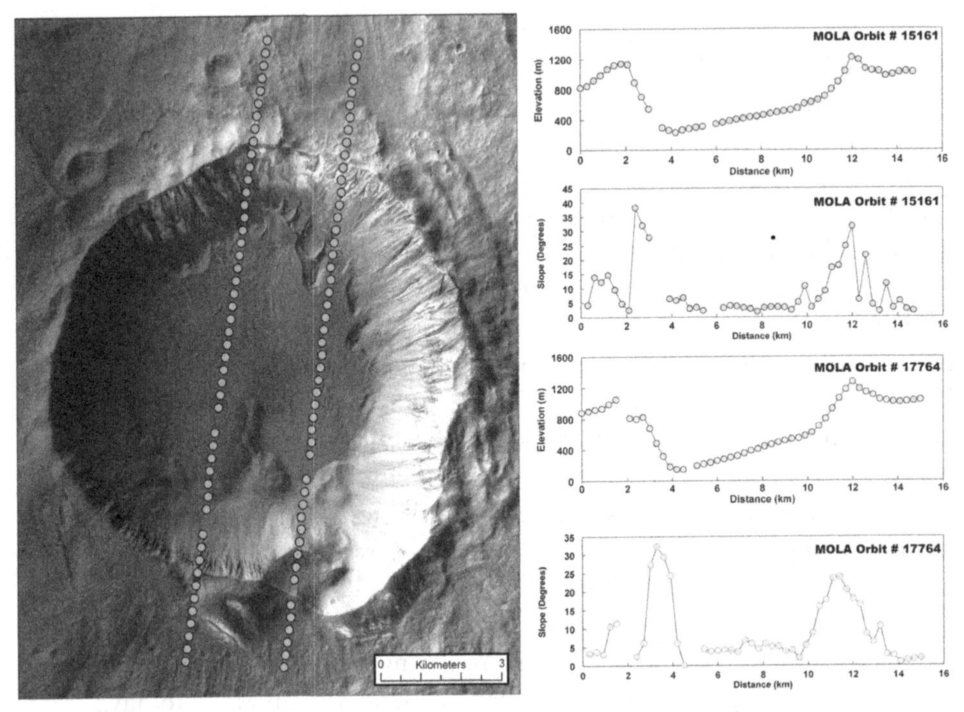

Fig. 2.12. Asymmetrical crater profiles in the 10.5-km diameter unnamed crater in eastern Newton Crater (−155.3E, 40.1S). Two MOLA profiles and derived slopes illustrate the crater interior topography and show that the crater has been modified from its initial fresh-crater morphology. The northern crater wall (top in image) is shallower than the southern wall, and the floor is asymmetrical, tilting from part way up the crater wall to the base of the southern wall. This postformation modification is interpreted to be due to glacial processes during a period of colder climate in recent geological history (see Figs. 2.23–26). Background image is CTX image p02_001842_1397_xi_40S155W.

(equatorial) and modified (higher latitude) steep slopes occurs at higher latitudes for pole-facing slopes and at lower latitudes for equator-facing slopes.

Individual impact craters are assumed to have initially formed generally symmetrical walls and flat floors. Thus, variation on this theme provides important information on modification processes. A specific example of slope and floor asymmetry (Fig. 2.12) in a 10.5-km diameter crater at 155° W and 40° S shows that the north-facing wall slope is steepest (~28–38°), compared with the south-facing wall (~10–32°), and that the floor has a distinctive southward tilt of about 5°. On the basis of the geomorphology of this crater and the presence of features interpreted to have been formed by ice- and glacier-related processes, it is clear that climatic variations have played a major role in crater degradation and that microenvironments in the MDV can be very helpful in the interpretation of features within crater interior microenvironments on Mars (see below).

Mesoscale features

McMurdo Dry Valleys: contraction-crack polygons

In the MDV, contraction-crack polygons form by thermal cracking of ice-rich permafrost and subsequent infilling of cracks with a variety of materials. The vertical cracks in plan view form a variety of polygon shapes (Fig. 2.13). Cracking is favored in regions that experience abrupt seasonal cooling (Berg and Black, 1966; Black, 1973; Mackay, 1977) and where near-surface materials are cohesive. Once cracking has been initiated, various geomorphic processes operating in each microclimate zone yield different classes of contraction-crack polygons (Fig. 2.13).

Coastal thaw zone Ice-wedge polygons are widespread in the CTZ (Fig. 2.13). The seasonal influx of liquid water from wet active layers into open thermal-contraction cracks, along with subsequent growth of ice on freezing, leads to the development of downward-tapering ice wedges outlining raised-rim polygons (Berg and Black, 1966). Polygons average ~10–20 m in diameter in the CTZ. The maximum width of most ice wedges is ~2 m.

Inland mixed zone The dominant forms of contraction-crack polygons in the IMZ are sand-wedge polygons and composite polygons (Fig. 2.13). Sand-wedge polygons develop in a manner analogous to ice-wedge polygons, except that cracks fill with sand rather than with ice (Péwé, 1959; Murton et al., 2000). Sand-wedge polygons in upper Wright Valley (Fig. 2.14) display a variety of forms, and show various relationships with gullies (Fig. 2.14a, b) and with snow trapped in polygon troughs (Fig. 2.14c). Meltwater from these sources means that the sand-filled polygon troughs are often ice cemented. These polygons are broadly characterized by a flat ice-cement table beneath the interior of the polygon, which deepens along a smooth slope toward the thermal-contraction crack (Levy et al., 2006). Retention of open contraction cracks at the ground surface, and the availability of sands to fill these cracks, are two key factors that determine the growth rate of sand-wedge polygons

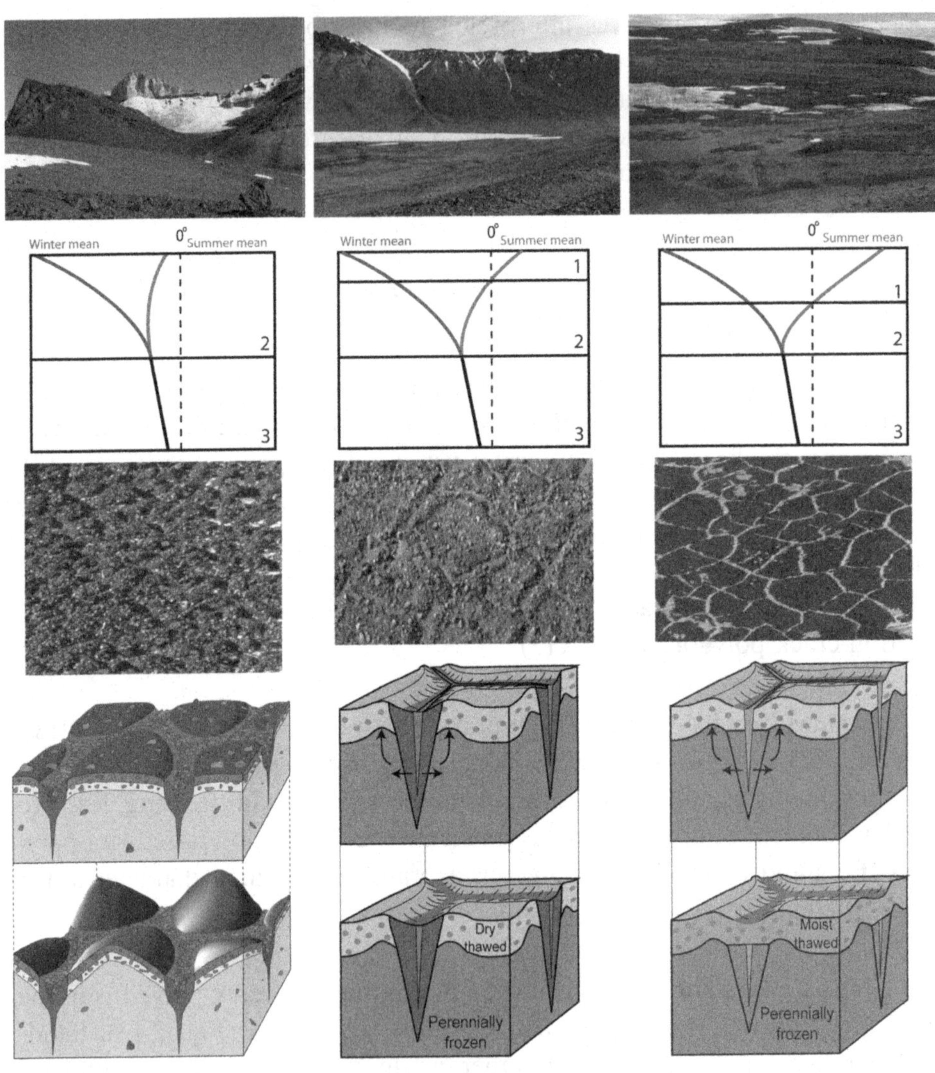

Fig. 2.13. Landscapes, near-surface thermal profiles, and polygons as a function of microclimate zones in the MDV (from Marchant and Head, 2007). (Top row 1) Typical landscapes in each microclimate zone; left to right: stable upland zone (SUZ), inland mixed zone (IMZ), and coastal thaw zone (CTZ). (Second row 2) Schematic vertical thermal profiles for each zone. Dashed line represents 0 °C (273 K) baseline; inclined lines show winter-mean and summer-mean soil temperatures as a function of depth. Numbered soil "horizons" are defined on the basis of temperature profiles. Horizon 1 experiences summer temperatures above 0 °C (273 K). In the case of the CTZ (right), soils are seasonally moist and thus oscillation about 0 °C (273 K) produces a classic active layer (see text for discussion). For the IMZ (center), soils are too dry to produce classic active-layer disturbance, even though summer soil temperatures rise above 0 °C (273 K); instead the IMZ shows a dry active layer. Horizon 1 is not present in the SUZ (left) because mean-summer soil temperatures fail to rise above 0 °C (273 K); this zone thus lacks a traditional active layer. Horizon 2 reflects the depth to which near-surface materials experience

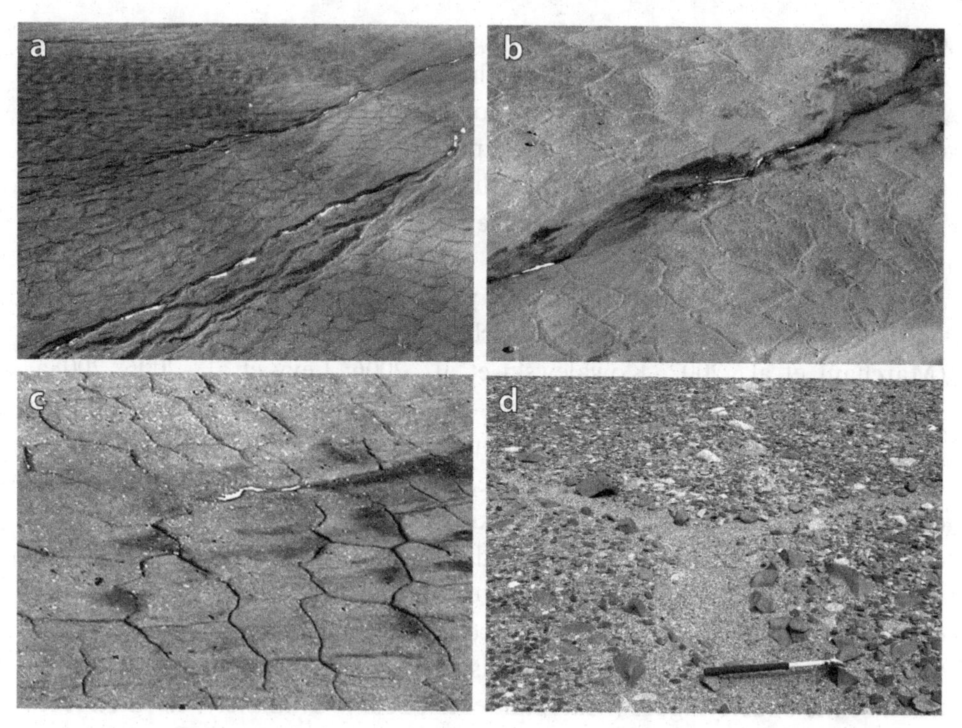

Fig. 2.14. Sand-wedge polygons in South Fork, upper Wright Valley (inland mixed zone; see Fig. 2.2 for location). (a) Southern valley wall showing perspective view of sand-wedge polygons dissected by gully channels. Note that the polygons on the steeper slopes are more irregular in surface topography, possibly linked to downslope movement processes. (b) Details of the superposition relationships between gully and polygonal ground (sand-wedge polygons). Illumination is from the right. Some polygons (left-hand side) appear to have somewhat raised rims, and may be composite polygons, or may be related to downslope movement on slightly steeper slopes; see left-hand side of part (a). Note the bright linear patches in the gully channel; these are trapped wind-blown snowbanks on the channel floor that melt during peak daily insolation, producing meltwater flowing in the stream and contributing to the hyporheic zone (darkened area). (c) Sand-wedge polygons showing elongation preferentially downslope. (d) The junction of sand-filled, sand-wedge polygon troughs (rock pick for scale is ~75 cm).

Caption for Fig. 2.13. (cont.)
seasonal temperature change. Temperature oscillation results in material expansion/contraction and is responsible for the initiation of polygonal terrain; see text. Horizon 3 reflects a zone of uniform temperature increase with depth; the base of the permafrost occurs where temperatures exceed 0 °C. (Third row 3) Left, oblique-aerial view of sublimation-type polygons in SUZ; field of view (FOV) is ~100 m; center, oblique-aerial view of sand-wedge polygons in IMZ; FOV is ~50 m; right, oblique-aerial view of ice-wedge polygons in CTZ; FOV is ~75 m. (Bottom row 4) Block diagrams illustrating the development of sublimation-type polygons (left), sand-wedge polygons (center), and ice-wedge polygons (right). Light gray color in column 1 indicates ice; see text for explanation. The color version of this figure can be found at http://www.cambridge.org/uk/catalogue/catalogue.asp?isbn=9780521889193.

(e.g., Berg and Black, 1966; Marchant et al., 2002). Cracks tend to be wider and remain open longer in soils with cohesive ice- and/or salt-cemented horizons, and thus sand-wedge polygons are most active near the margins (and downwind) of perennial snowbanks that experience minor snowmelt. Composite polygons are those that show wedges with alternating, vertical layers of sand and ice (Ghysels and Heyse, 2006).

Stable upland zone A sublimation-type polygon, a special type of sand-wedge polygon, forms where sediment overlies buried, massive ice in the SUZ (Marchant et al., 2002; Kowalewski et al., 2006; Levy et al., 2006, 2008a, 2008b) (Fig. 2.13). The formation and evolution of sublimation-type polygons is tied to the thermal cracking and sublimation of underlying ice (Fig. 2.13). As cracks form in buried ice, the finest fraction of overlying debris (<2 cm) percolates down into the cracks. Material that is too large (>~2 cm) collects at the buried-ice surface and this process of passive sifting and removal of fine-grained material creates a zone of relatively coarse-grained debris at polygon margins (i.e., above contraction cracks). The relatively high porosity and permeability of this debris causes local enhancement of sublimation of the underlying ice (Marchant et al., 2002). Ultimately, elevated rates of ice sublimation at polygon margins lead to the development of >2–3 m deep troughs that outline conical, sediment-covered mounds of buried ice (Fig. 2.13).

A negative feedback effect prevents runaway ice loss. As sublimation troughs deepen, they become preferred sites for collection of wind-blown snow. The downward flux of vapor and/or minor melt from the base of these snowbanks creates a thin layer of superposed ice that effectively seals the remaining ice from sublimation (Marchant et al., 2002; Kowalewski et al., 2006; Levy et al., 2006). Cohesive plugs of ice-and-salt-cemented sands ultimately form at the top of contraction cracks (at ice level), preventing further infiltration of overlying sediment and reducing the potential for compressive stress and ice deformation at depth (e.g., Marchant et al., 2002).

Synthesis Different types of contraction-crack polygons are strong morphological indicators of conditions prevailing in a particular microclimate zone. Morphologic variation arises from the abundance of liquid water in near-surface soils (Fig. 2.13). Ice-wedge polygons of the CTZ require wet active layers. Sand-wedge polygons of the IMZ (Fig. 2.14) signal soil moisture sufficient to induce widespread thermal cracking but insufficient to fill cracks with ice. Sublimation polygons of the SUZ indicate the location of near-surface buried ice in regions lacking wet active layers (Fig. 2.13). These three types of contraction-crack polygons provide information on the distribution and concentration of subsurface ice. For ice-wedge polygons, the highest concentration of ice occurs in polygon troughs. For sand-wedge polygons, the ice commonly occurs distributed evenly as pore ice (generally <30% ice by volume). For sublimation polygons, excess ice (ice exceeding available pore space and with values $\gg$30% ice by volume) generally occurs $\leq$1 m from the ground surface (Marchant et al., 2002); in sharp contrast to ice-wedge polygons, the lowest concentrations of ice in sublimation polygons are found in the troughs.

Mars: contraction-crack polygons

Several different types of contraction-crack polygons have been observed on Mars. Mangold et al. (2004) and Mangold (2005) classified types of polygons and patterned ground (including slope stripes) into ten different subtypes; their distribution was concentrated above ~55° N and S latitudes and highly correlated with regional ground ice distribution interpreted from Odyssey Neutron Spectrometer data (Fig. 2.5d). At least five subtypes were thought to be controlled by climate zonation because they are located at the same latitudes between 55° and 75° in both hemispheres (Mangold, 2005). Polygons were interpreted to be due to seasonal temperature variations, such as thermal contraction and seasonal thaw, on the basis of comparisons to Earth. Some polygons on Mars are larger than those on Earth by a factor of up to ~5 (Fig. 2.15) and the larger size may be due to deeper propagation of cracks from colder temperature conditions and/or due to larger temperature variations in the past history of Mars. Smaller polygons (50–200 m in width) are found in crater interiors and are relatively homogeneous (Mangold, 2005). The smallest polygons and hummocks observed on Mars (15–40 m in width) are the most abundant polygon type (Fig. 2.15) and are most similar in size to those observed in the MDV and in other periglacial environments on Earth. For these polygons, cracks are usually not observed directly, but thermal contraction and widening by sublimation and desiccation of the surface layer appear to be important in their formation. Mangold (2005) reported that all types of patterned ground on Mars are geologically recent (<10 Ma) but that they did not form simultaneously, finding that some features may correspond to freeze–thaw cycles that occur during periods of higher obliquity (>35°), and that certain subtypes may have been enhanced by specific microenvironments. Kostama et al. (2006) surveyed the northern high latitudes of Mars; polygonal patterned ground there was developed on a thin mantling deposit with a very young surface crater retention age, and

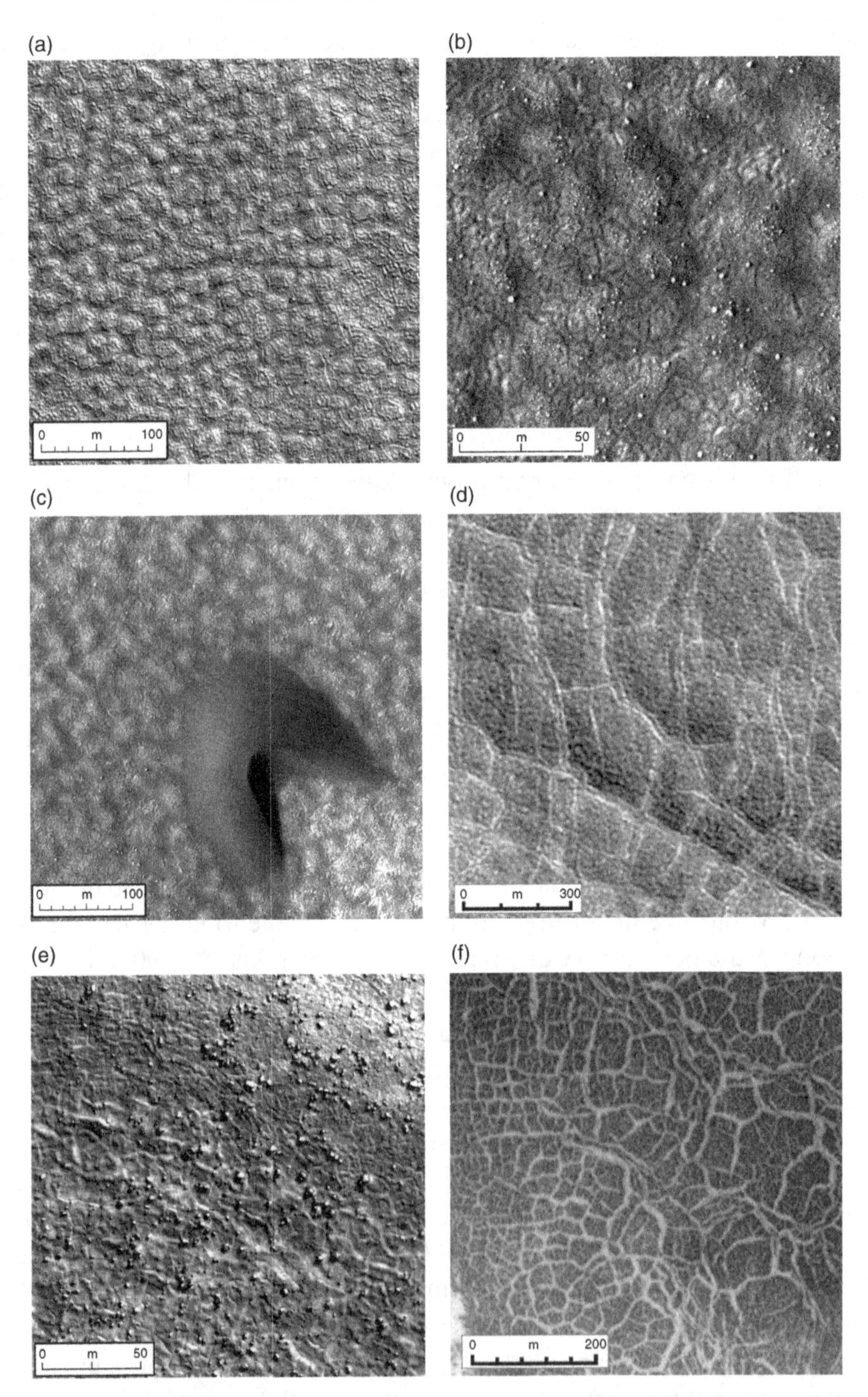

Fig. 2.15. Surface polygons on Mars. (a) Two scales of polygons each resembling "basketball terrain" or sublimation polygons (e.g., Marchant et al., 2002). HiRISE image PSP_001591_2475. (b) Two scales of polygons

the mantle was characterized by patterned ground textures that varied as a function of latitude.

The variety of contraction-crack polygons on Mars thus offers clues to microclimates and climate change on Mars (Marchant and Head, 2007). Many contraction-crack polygons at high latitudes (Fig. 2.15), for example in areas having a shallow ice table (Fig. 2.15a), display the rounded texture, termed "basketball terrain" (Kreslavsky and Head, 2000; Malin and Edgett, 2001; Head et al., 2003; Mangold et al., 2004; Kostama et al., 2006; Levy et al., 2008a, 2008b), similar to the sublimation polygons of the SUZ in the MDV (Marchant et al., 2002) (Fig. 2.13). Other examples (Fig. 2.15b, c) are analogous to sand-wedge polygons. In Fig. 2.15c polygonal troughs ~10 m wide are filled with fine-grained material similar to that forming the barchan dune. In many other places, there is clear evidence of multiple scales of polygons (Fig. 2.15a, d). Many patterns are superposed and have different scales, suggesting recent climate change such as deposition of ice-rich layers during recent ice ages (e.g., Head et al., 2003) and the possible development of high-latitude active layers under conditions of higher spin-axis obliquity (e.g., Mangold, 2005; Kreslavsky et al., 2008) (Fig. 2.6). Such possibilities are further suggested by examples of deformed polygons exhibiting lobate margins at the base of slopes (Fig. 2.15e), suggesting that gelifluction/solifluction has deformed the polygons preferentially downslope, a phenomenon seen in the IMZ in upper Wright Valley (Levy et al., 2007, 2008a, 2008b) (Fig. 2.14). Finally, many high-latitude polygons are characterized by bright bounding troughs (Fig. 2.15f) that may indicate the presence of snow or ice, and possibly ice-wedge polygons.

McMurdo Dry Valleys: viscous-flow features

Solifluction lobes, gelifluction lobes, and debris-covered glaciers "Solifluction" and "gelifluction" are sometimes used interchangeably (e.g., Matsuoka,

Caption for Fig. 2.15. (cont.)
with troughs of larger polygons infilled with sediment (sand-wedge type polygon). Location is bottom left corner of part (c). HiRISE image PSP_001375_2485. (c) Two scales of polygons with larger polygon troughs filled with sediment similar in brightness to that forming the distinctive barchan dune. HiRISE image PSP_001375_2485. (d) Two scales of polygons with the smaller ones resembling "basketball terrain" or sublimation polygons (e.g., Marchant et al., 2002), and the larger ones having sloping margins and filled cracks; 75.16° N, 331.52° W, MOC E21/01593. (e) Polygonal terrain on a slope showing lobes at the base of the slope (lower left) interpreted to be related to downslope movement of ice-rich debris. HiRISE image PSP_001380_2520. (f) Candidate ice wedge or frost/snow filled polygons; 71.74° N, 11.3° W, MOC E09/00249. North is at top in all images.

2001), just as are the terms rock glacier and debris-covered glacier (e.g., Giardino et al., 1987; Whalley and Palmer, 1998). Much of the confusion in nomenclature exists because these features form by a continuum of processes that are all related to the slow movement of hillslope materials. We previously defined solifluction as the slow flow of saturated materials, with or without the presence of nearby ice. Our usage of the term gelifluction differs in that in addition to saturated flow, gelifluction lobes advance by internal deformation of (and perhaps slippage along) buried-ice lenses and/or pore ice. Although rock glaciers and debris-covered glaciers both flow by creep deformation of internal ice, we distinguish rock glaciers as having ice of secondary origin (i.e., not glacial), including ice that forms by the freezing of pore water. In contrast, debris-covered glaciers are those viscous-flow features that have a demonstrable core of buried glacier ice.

Coastal thaw zone In the CTZ, solifluction is the dominant form of viscous flow (Fig. 2.16c). Solifluction lobes are commonly 20–30 cm thick and occur en echelon on slopes as low as 5° (Nichols, 1968; Selby, 1971b). Soil pore pressures, and thus rates of solifluction, are elevated where meltwater is unable to penetrate impermeable subsurface horizons, such as a local ice table. As the summer season progresses, however, shallow ice commonly melts and gives rise to typical, though relatively small, thermorkarst features (shallow depressions, planar slides). The wetting of near-surface debris also produces saline brines (Campbell and Claridge, 1987; Bockheim, 1997; Campbell et al., 1998; Lyons et al., 2005) that remain in a liquid state, facilitating solifluction, even as temperatures drop below the freezing point for pure water.

Inland mixed zone In the IMZ many slopes show well-developed gelifluction lobes (Fig. 2.16b) that flow by downslope movement of saturated debris and internal deformation of pore ice and/or ice lenses at depth. Gelifluction lobes may emanate from steep slopes, though this is not required; they commonly display a series of stacked and/or nested ridges that terminate along coarse-grained, lobate fronts. Gelifluction lobes may exceed the angle of repose and show a relief of 3–5 m.

Stable upland zone Debris-covered glaciers are the most common form of viscous-flow features in the SUZ. Most debris-covered glaciers originate through accumulation of rockfall debris on alpine glaciers and commonly occur downwind from dolerite-capped cliffs (Fig. 2.16a). Debris that falls onto glacier accumulation zones may move englacially before rising toward the surface as overlying ice sublimes. The stratigraphic contact between ice and overlying

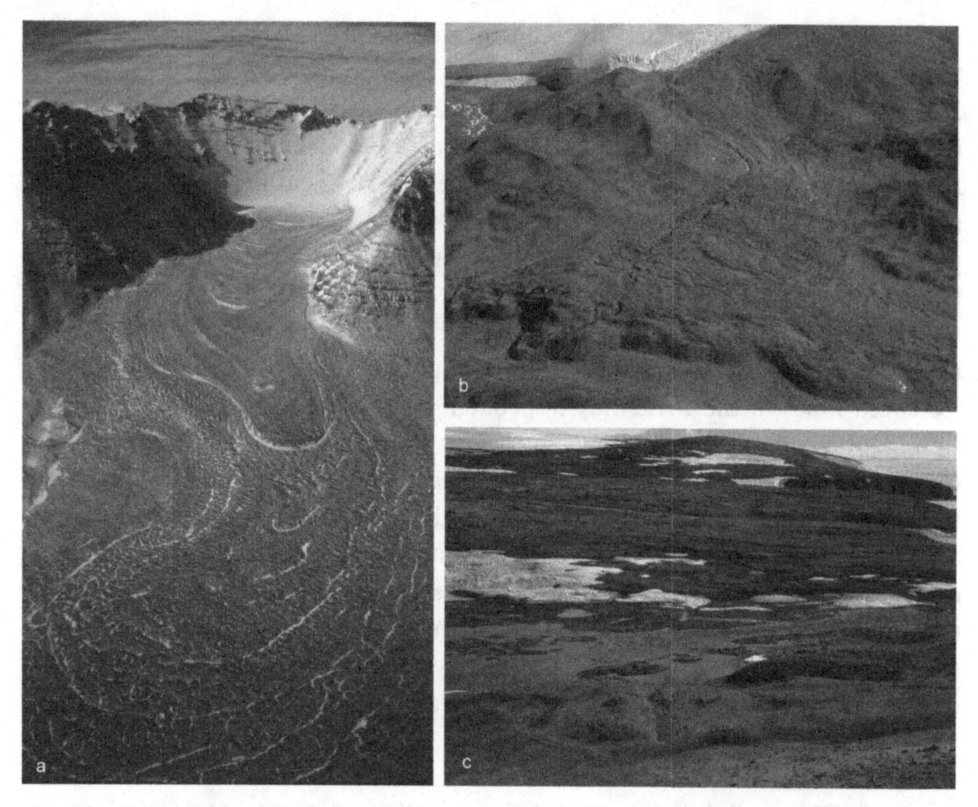

Fig. 2.16. Styles of viscous flow in the McMurdo Dry Valleys. (a) Mullins Valley debris-covered glacier within the SUZ; the small alpine glacier at the valley head transitions down valley into a debris-covered glacier that extends for an additional ~3.5+ km. The surface debris on this glacier originates from rockfall at the valley head, some of which travels englacially before rising to the surface as overlying ice sublimes. The debris, a classic sublimation till, thickens from about 10 cm in the upper portion of the valley to ~50 cm near the base of the photograph; field of view at the base of the photograph is ~0.75 km. (b) A portion of a well-developed ice-cored lobe in IMZ, central Taylor Valley. The front of the lobe in the foreground is up to 5 m high. In places in the upper half of this lobe, ice occurs ~1 m below the ground surface; field of view at the base of the photograph is ~1 km. (c) Minor solifluction lobes/terraces occur just beyond the small meltwater pond and snowbanks in the central portion of this photograph. The lobes are 20 to 25 cm high and are stacked en echelon. Linear features in the immediate foreground are small ice-wedge polygons with slightly raised polygon centers; field of view at the base of the photograph is ~4 m. See Fig. 2.2 for specific locations. (From Marchant and Head, 2007.)

debris is smooth and dry, and mimics the modern ground surface (Marchant et al., 2002). The rate of ice sublimation is dependent on the thickness, porosity, and permeability of overlying debris (sublimation till) (Schaefer et al., 2000; Marchant et al., 2002; Kowalewski et al., 2006; Levy et al., 2006).

Concentric surface ridges are seen in most debris-covered glaciers in the SUZ and are characteristic of subsurface ice flow (Fig. 2.16a; Levy et al., 2006). Horizontal ice-surface velocities are generally <~40 mm per year (Rignot et al., 2002; Shean et al., 2007b), a rate considerably less than that for comparably sized ice-cored lobes in the IMZ (~3 cm per year; Hassinger and Mayewski, 1983). Ice temperature, and its control on ice rheology, provides one explanation for this difference in flow rate: all other things being equal, warm ice flows faster than cold ice (Patterson, 2001). The temperature of ice in most debris-covered glaciers in the SUZ is $\leq -25\,^{\circ}C$ (248 K) (Kowalewski et al., 2006 and unpublished data), whereas that in ice-cored lobes in the IMZ may approach $\sim -18\,^{\circ}C$. Debris-covered glaciers that exhibit topographic hollows in former snow and ice accumulation areas (e.g., beheaded glaciers) occur in many places in the SUZ. The hollows may reflect a change in climate leading to a marked reduction in ice accumulation and/or to a reduction in the rate of rockfall; the latter may lead to greater expanses of exposed glacier ice and rapid ice-surface lowering via sublimation (e.g., Marchant and Head, 2004). Sugden et al. (1995b) argue that some stagnant glacier ice in the SUZ has survived for at least 8.1 Ma beneath a dry, debris layer ~50 cm thick (see also van der Wateren and Hindmarsh, 1995; Kowalewski et al., 2006).

Synthesis The availability and state of water determines the style of viscous-flow feature developed in each microclimate zone. If water saturates near-surface soils, as it does in the CTZ (aided in places by the freezing point depression associated with brines), then flow is best accommodated by solifluction. If water freezes at depth to form excess pore ice, then gelifluction lobes may dominate, as they may do in the IMZ. In the SUZ, there is insufficient meltwater to promote solifluction and/or produce excess subsurface ice, and debris-covered glaciers dominate. The downward flux of vapor into soils has been postulated as a process that could account for significant accumulation of near-surface ice in the SUZ and on Mars (e.g., Squyres et al., 1992; Helbert et al., 2007), but in a test of alternative, nonglacial source mechanisms for subsurface ice in the SUZ, Kowalewski et al. (2006) showed that the downward flux of vapor into soils is currently incapable of producing requisite pore-ice volumes that would sustain viscous flow.

Mars: viscous-flow features

Lobate, viscous-flow features and debris aprons (Fig. 2.17) similar to those seen in the MDV (compare to Fig. 2.16a) have been observed on Mars

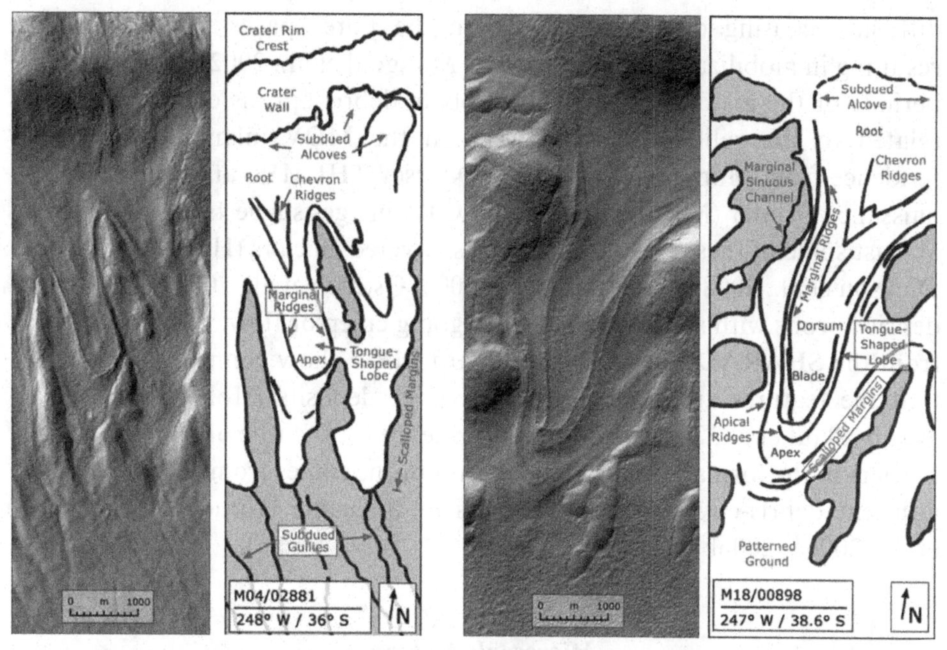

Fig. 2.17. Viscous flow features at mid latitudes on Mars. (Left) Image and sketch map of viscous-flow features on the interior of an impact crater wall. Note the tongue-shaped lobes, the marginal ridges, and the alcoves. Subdued gullies emerge from the distal portions of the depressions occupied by the tongue-shaped lobes. MOC image M04/02881, 248°W, 36°S. North is at the top of the image and illumination is from the northwest. (Right) Image and sketch map of viscous-flow features on the interior of an impact crater wall. Note the details of the tongue-shaped lobe, including multiple marginal ridges, multiple apical ridges, and the distal patterned ground. MOC image M18/00898, 247° W, 38.6° S. North is at the top of the image and illumination is from the northwest. (From Marchant and Head, 2003, 2007.)

and both types of features display a distribution limited in latitude. Morphologies similar to debris-covered-glaciers cluster in latitude bands at about 45° (e.g., Milliken et al., 2003) on Mars and commonly occur in microenvironments on crater interior walls, where topography favors accumulation and preservation of ice and snow, where slopes are typically steeper, and where slope orientation may favor higher surface insolation and temperatures sufficient to cause flow (Fig. 2.17) (e.g., Hecht, 2002; Milliken et al., 2003; Kreslavsky and Head, 2004; Kreslavsky et al., 2008). Between 30° and 60° latitude, larger lobate debris aprons commonly occur and are concentrated around the base of massifs that are shedding debris (e.g., Squyres et al., 1992; Pierce and Crown, 2003; Head et al., 2005). These features are interpreted by many to be due to atmospheric water vapor diffusion causing deposition of ice in the pore spaces of debris fans, resulting in mobilization and flow (e.g., Mangold et al., 2002) during certain periods of the history of Mars. The lobate debris aprons could be closely related to the gelifluction lobes seen in the MDV inland mixed zone. Evidence from Mars Express HRSC, Odyssey THEMIS, and Mars Reconnaissance Orbiter (MRO) HiRISE and CTX images suggests, however, that at least some of these features are debris-covered glaciers (Head et al., 2005; Dickson et al., 2008; Kress and Head, 2008; Ostrach et al., 2008) rather than debris aprons with secondary ice undergoing creep or flow. New radar data from the SHARAD instrument on board MRO show compelling evidence that the features are cored by extensive ice deposits (Holt et al., 2008a, 2008b; Plaut et al., 2009). Additional evidence for the role of debris-covered glaciers in the formation of lobate debris aprons comes from the presence of adjacent debris-covered valley glaciers at the same latitudes (e.g., Head et al., 2006a, 2006b).

Microscale features

McMurdo Dry Valleys: tafoni, weathering pits, desert pavements, and duricrusts

Coastal thaw zone Salt encrustations (Nichols, 1968; Gibson et al., 1983; Campbell and Claridge, 1987; Hall, 1991) are seen on many rocks at the surface in the CTZ, and often in other zones, and are produced by evaporation of near-surface brines and saline meltwater (Fig. 2.18). Mineral grains are prised loose by the growth and expansion of salts, particularly on coarse-grained rocks, and this leads to the development of widespread grus (loose collections of mineral grains). Salt weathering in the CTZ also leads to the development of tafoni through cavernous weathering of coarse-grained rocks (Conca and Astor, 1987). Typical salt species in the CTZ include gypsum ($CaSO_4 \cdot 2H_2O$), halite (NaCl), mirabilite ($Na_2SO4 \cdot 10H_2O$), and nitratine ($NaNO_3$). The long-term effect of salt weathering in the CTZ is to smooth bedrock slopes over 10 m baselines, although considerable relief is produced at meter and submeter scales (Conca and Astor, 1987).

Inland mixed zone Climate conditions of the IMZ foster the development of widespread desert pavements, with wind-faceted cobbles (ventifacts) and intervening lags of coarse-grained sand and gravel. The spacing of ventifacts varies approximately with soil age: surfaces with interlocking ventifacts and little intervening sand are generally older than are sandy surfaces with widely spaced ventifacts (Selby, 1977).

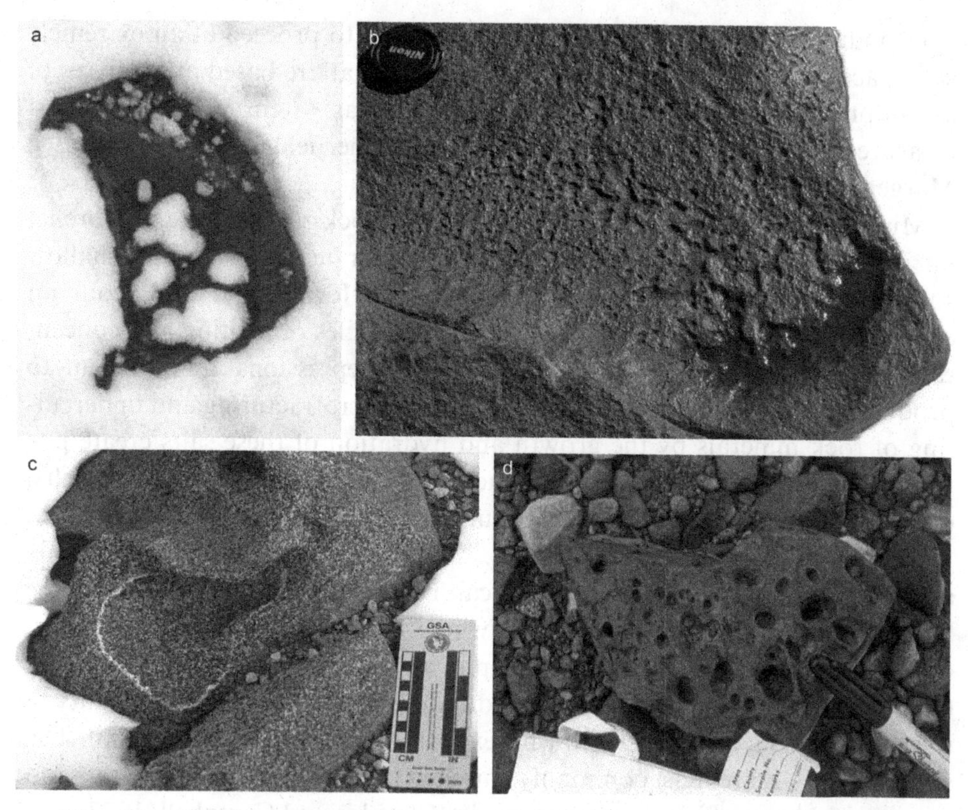

Fig. 2.18. Modification of rock-surface textures by salt weathering in the MDV. (a) Snow preferentially collects in surface lows and pits on rocks; here, recent snowfall is trapped within cm-scale pits cut into low-albedo dolerite in the SUZ; the largest snow-filled pit is ~1.5 cm in diameter. (b) Solar radiation on the surface of low-albedo dolerite in the SUZ causes snow in the pits to melt; subsequent evaporation creates brines that ultimately form salt-encrusted pits. The surface slope of the rock in part (b) dips to the lower right corner, and meltwater accumulates in the large surface pit. The rill-like forms are produced as brines repeatedly overfill small pits and trickle downslope. (c) The maximum meltwater level for one pit on another dolerite boulder appears as a white (salt-encrusted) line (e.g., "bathtub ring"); this relatively coarse-grained dolerite shows greater susceptibility to salt weathering than the relatively fine-grained dolerite rocks in parts (a), (b), and (d); hence the pit here is larger and deeper. (d) Ultimately, snowmelt, evaporation, and crystal growth (including hydration/dehydration cycles) produce deep pits, even for this fine-grained dolerite, that tend to increase in width and depth linearly with age (e.g., Staiger et al., 2006). (From Marchant and Head, 2007.) See text for details.

Stable upland zone Rocks in the SUZ are not subjected to episodes of burial and exposure, as commonly occurs in areas with traditional, wet active layers (e.g., Hallet and Waddington, 1991). Therefore, rock surface textures are almost exclusively a function of wind erosion and salt weathering. The very dry conditions of the SUZ cause salt weathering to proceed at an extremely slow pace. Rates of surface erosion in the SUZ that are based on analyses of *in situ* produced cosmogenic nuclides are as low as $\sim$6 cm Ma^{-1}, the lowest measured on Earth (Brook et al., 1995; Summerfield et al., 1998, 1999; Margerison et al., 2005; Staiger et al., 2006).

Minor snowmelt that forms on solar-heated rocks can initiate microrelief on rock surfaces (Fig. 2.18a, b). The meltwater tends to occupy shallow surface depressions, which originally may have formed by wind scour on exposed rock surfaces (Selby, 1977). Repeated cycles of melting and concentration of ions through water evaporation in depressions are sufficient to produce visible salt encrustations (Fig. 2.18c). Microfracturing and undercutting of rock minerals by the growth and hydration of these salts create pits that, through positive feedback, attract more snow, producing more salts; ever deeper and wider pits can be produced by this process (Allen and Conca, 1991; Parsons et al., 2005; Staiger et al., 2006) (Fig. 2.18d). A network of pits and microrills on the surface of rocks can be produced by this process in the SUZ (Fig. 2.18b, d). Microrills form where rock surfaces are inclined such that saline meltwater spills out from pits and flows down the rock surface (Fig. 2.18b). The average depth and width of weathering pits at one locality in the SUZ shows a linear increase with exposure age (Staiger et al., 2006). The salts in the SUZ are most commonly enriched in nitrates, reflecting derivation from snow blown off the ice-sheet margin (Claridge and Campbell, 1977; Bao and Marchant, 2006).

Rocks that are fringed with thermally cleaved and spalled fragments are known as puzzle rocks and are also common in the SUZ. For rocks composed of low albedo Ferrar Dolerite ($\sim$0.07; Campbell et al., 1997b), the transition from solid boulders to partially disintegrated puzzle rocks in the SUZ likely requires 3 Ma to 5 Ma (Marchant et al., 1993a; Staiger et al., 2006).

Snowmelt infiltrating into the upper few centimeters of silty soils in the SUZ helps create a fragile, salt-cemented duricrust. This fragile crust is held together by salt crystals that bind detrital grains and is susceptible to breakdown upon wetting. Wetted crusts may yield saline solutions that locally enhance downslope movement of near-surface soil–water mixtures (millimeter to centimeter scale thickness). In the absence of significant surface water, as is commonly the case in the SUZ, the salts are enduring and tend to retard downslope movement and aeolian deflation. Furthermore, salts in duricrusts may slow the flux of water vapor into and out of underlying soil owing to their chemical properties and physical ability to reduce near-surface porosity and permeability (Marchant et al., 2002; Kowalewski et al., 2006).

Synthesis The relatively wet climate conditions of the CTZ foster the development of widespread near-surface brines. Successive hydration and dehydration cycles lead to the formation of multiple salt species and relatively rapid rates of salt weathering. These rates of rock disintegration in the CTZ appear to outpace the development of well-formed ventifacts and small, rock-weathering pits on most rock surfaces. The rate of degradation via salt weathering decreases inland. Measured rates for bedrock erosion in the SUZ are an order of magnitude lower than that measured near the coast (Summerfield et al., 1999). Nonetheless, the effects of salt weathering are visible in the SUZ as thin duricrusts and pitted-rock surfaces; well-formed pits appear only on rocks in the SUZ that have been exposed at the ground surface for $>\sim 1$ Ma (Parsons et al., 2005; Staiger et al., 2006). Duricrusts tend to be most extensive where wind-blown snow commonly melts on the surface of solar-heated rocks. This meltwater may infiltrate the upper few centimeters of soil, evaporate, and over time leave behind appreciable salts (e.g., Claridge and Campbell, 1977; Bockheim, 1997; Bao and Marchant, 2006). As it binds adjacent sand grains, the salt mixture creates a thin duricrust from 1 to 2 cm thick. A simple but important point to emphasize is that surface and near-surface brines in the MDV may remain in a liquid state at temperatures $<0\,°C$. In extreme situations, such as the case for Don Juan Pond, upper Wright Valley, surface water remains below the solidus even as temperatures drop below $-40\,°C$ to $-50\,°C$ (e.g., Marion, 1997; Takamatsu et al., 1998; Healy et al., 2006). In addition to influencing geomorphic change, such a situation could impact microbial activity (Siegel et al., 1979; Torii et al., 1989).

Mars: microscale topography

There are differences at the microscale between the rocks and surface features found at different latitudes on Mars, such as at the high-latitude Viking 2 landing site (47.7° N; Mutch et al., 1976a, 1977) and those at lower latitudes (Viking 1, Pathfinder, and Mars Exploration Rovers, all below 23° latitude; Mutch et al., 1976b; Binder et al., 1977; Golombek et al., 1999; Squyres et al., 2004a, 2004b, 2004c). Sediment-filled contraction-crack polygons observed at the high-latitude Viking 2 site are similar to those seen in the MDV inland mixed zone (Fig. 2.19) (Mutch et al., 1976a, 1977), and are also seen elsewhere in this latitude band in orbital MOC high-resolution images (e.g., Malin and Edgett, 2001; Mangold et al., 2004). Heavily pitted rocks are also observed at the Viking 2 site (Fig. 2.20). The pits were previously interpreted to be formed by vesiculation or differential weathering of clasts or phenocrysts (e.g., Mutch et al., 1976a). Allen and Conca (1991) suggest, however, that these

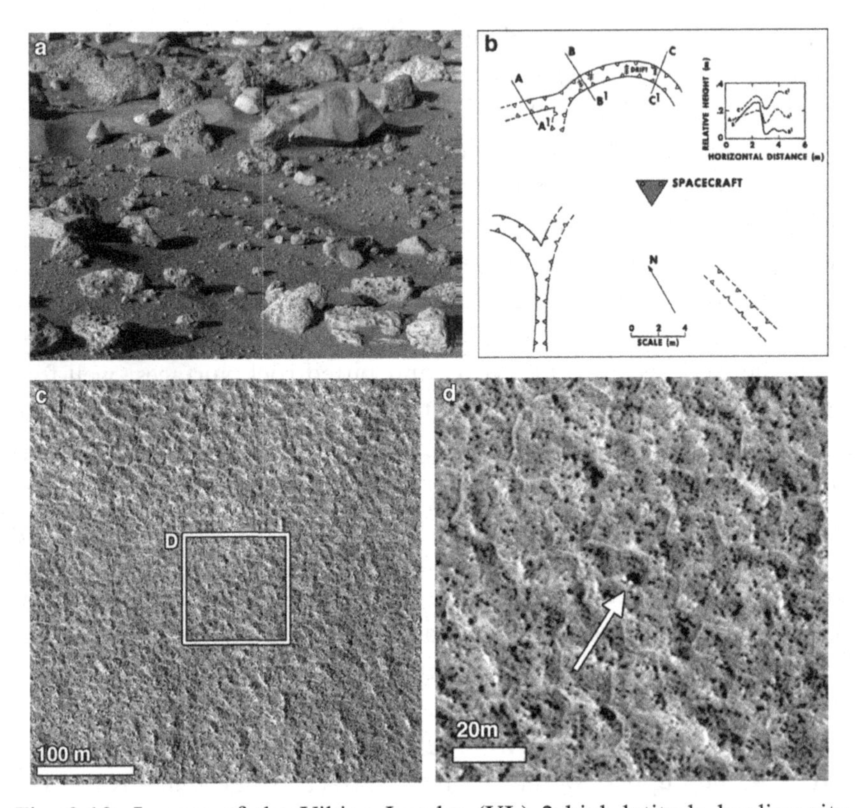

Fig. 2.19. Images of the Viking Lander (VL) 2 high-latitude landing site (47.97° N, 229.59° W) showing evidence for sand-wedge and ice-wedge polygons. (a) In the near field of the site, about 8 m north of the spacecraft, a trough ~1 m across and ~10 cm deep trends from the upper left to lower right and is filled with small drifts of fine-grained sediment. The trough is asymmetrical; the northern wall is relatively steep and high and a prominent upraised rim is seen along the northern edge. Blocks are relatively rare in the troughs, and rocks that are observed there are partly buried, in contrast to many in the adjacent terrain. Camera event 21A024. (b) The general distribution of troughs mapped in the vicinity of VL 2, and topographic profiles across the northern trough (inset). Abrupt termination of troughs near the spacecraft are due to obscuration of the camera view by the spacecraft. The trough shown in part (a) is the one to the north of the spacecraft. Profiles show the presence of a prominent raised rim on the northern edge, and more subtle differences in elevation suggest that the southern rim may also be raised. The presence of several trough junctions suggests a polygonal network. Similar features were not observed at the Viking Lander 1 site, at a more southerly latitude (22.48° N, 47.82° W) (from Mutch et al., 1976a, 1977). The combination of raised rims and sediment fill suggests that these features are analogous to sand-wedge and/or ice-wedge polygons in the MDV (e.g., Fig. 2.13). The depth to the current ice table at these latitudes (Fig. 2.5b), the amount of water ice in the substrate (Fig. 2.5c), and the annual temperature changes at the site all indicate that thermal cycling could produce contraction-crack polygons, which became traps for fine-grained

Fig. 2.20. Images of the Viking Lander (VL) 2 high-latitude landing site (47.97° N, 229.59° W) showing evidence for heavily pitted rock surfaces. (a) Pitted rocks in the vicinity of the lander footpad; the rock at the top of the footpad is ~35 cm across (portion of 22A001). (b) Pitted rocks near the edge of a trough (top of image; see Fig. 2.23 for context) (portion of 21A024). Pits are ~1–2 cm in diameter. (c) Frost and snow deposition at the VL 2 site in northern winter (image P21873). (d) Enlargement of area in the center of part (c) showing the location of snow and frost deposition in lows and small drifts between rocks and ripples, and in pits and troughs in the rocks. Seasonal snow and frost deposition (Jones et al., 1979; Wall, 1981) may have led to the formation of local meltwater on heated rock surfaces on Mars and focused chemical alteration to produce pits (Head and Kreslavsky, 2006), similar to the situation on Earth (see Fig. 2.18; see Parsons et al., 2005; Staiger et al., 2006). (e) Duricrusts similar to those seen in the MDV were also observed at the Viking Lander 2 site on Mars; the descent engines blew away surface sediment between rocks in several places, revealing a cracked duricrust layer (from Marchant and Head, 2007).

Caption for Fig. 2.19. (cont.)
sediment. (c) Polygon networks in the region surrounding the VL 2 site are observed from orbit in MOC images (MSSS.com release MOC2–1802). Box indicates location of part (d), which shows the position of the VL 2 lander (arrow) and the generally polygonally textured landscape surrounding the site. Compare with surface images in part (a) and polygon scale in part (b) in the vicinity of the VL 2 lander. (From Marchant and Head, 2007.)

features may form by processes much more similar to those that formed the pits in the MDV (see also Parsons et al., 2005; Staiger et al., 2006). Deposition of frost and snow on rocks and on intervening ground was observed during winter at the Viking 2 site (Fig. 2.20; compare with Fig. 2.18), but was not seen at lower latitudes. These observations suggest that the deposition of snow on rocks with a thermal inertia sufficient to retain heat and cause localized melting of snowfall (Head and Kreslavsky, 2006) could have been a factor in the formation of the observed pitting, as appears to be the case in the stable upland zone and parts of the inland mixed zone of the MDV (e.g., Allen and Conca, 1991). Duricrusts similar to those seen in the MDV were also observed at the Viking Lander 1 site on Mars (Fig. 2.20e).

Geomorphic evidence for climate change

Changes in the spatial distribution of equilibrium landforms over time, or the presence of relict equilibrium landforms, provide a morphological basis for interpreting past and ongoing climate change. In the MDV, a slight climatic warming might be registered as a landward shift of the CTZ at the expense of the IMZ (e.g., Marchant and Denton, 1996). Such a shift would likely increase the area of active-layer cryoturbation. Wind-blown snow in gullies of the IMZ might undergo greater melting, promote solifluction and salt weathering, and possibly lead to the development of wide, low-density gullies with rounded interfluves such as those that today occur in the CTZ. In addition, composite- and sand-wedge polygons of the IMZ would likely give way to ice-wedge polygons, and soils would show an overall increase in moisture content (Swanger and Marchant, 2007). On Mars, climate change might also be registered as latitudinal shifts in specific landforms. We emphasize that although the spatial scale for broad climate zones on Mars (latitude dependent) is vastly greater than that for the MDV microclimate zones, many of the cold-desert landforms and the variability of specific landform types is remarkably similar for both planets. Detailed knowledge of the processes derived from the MDV microclimates can thus be applied to Mars and will be helpful in deconvolving the signal of climatic zonation and climate change there.

Geomorphic evidence for recent climate change in the McMurdo Dry Valleys

Extensive field inspection of the IMZ reveals geomorphic evidence consistent with ongoing climatic warming. While the majority of landforms in the IMZ appear in equilibrium with local environmental conditions, some ice-cored lobes show evidence for considerable stream dissection, with channels bisecting lobes and producing locally extensive fans of stratified debris (Fig. 2.21). It is

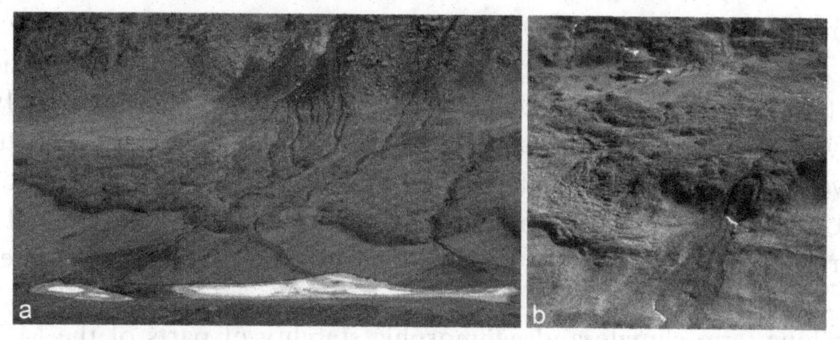

Fig. 2.21. Modern stream dissection of gelifluction lobes in the IMZ (see Fig. 2.2 for locations). (a) Meltwater channels derived from snowmelt higher on the valley wall (snowbanks not shown) cut up to 5 m into prominent lobes. Buried ice is exposed in places along channel walls. The channels feed fans that occur just beyond the front of these prominent gelifluction lobes. Excess water fills saline ponds at the distal margin of the fans. The darker colors on the fans show water-filled channels where surrounding sands are moistened and/or saturated (e.g., hyporheic zones). The photograph was taken on December 12, 2004; field of view at base is ~0.65 km. (b) A braided stream cuts through the front of a gelifluction lobe in Taylor Valley. Channel incision has cut through the lobe and removed a portion of a lateral moraine from Taylor Glacier (faint horizontal line to the right of the braided stream). Photograph taken on January 10, 1998; field of view at base is ~0.45 km. Wind-blown sands cap the upper part of gelifluction lobes in each image. (From Marchant and Head, 2007.)

unlikely, given the magnitude of stream dissection, that these ice-cored lobes could have sustained this level of incision for >1 Ma, the likely age of sediment in some of these lobes (Hartman, 1998). Instead, we argue that these lobes are no longer in equilibrium with current microclimate conditions, and the superposition relationships may be a harbinger of local climate warming in the MDV. The stream dissection that crosses these lobes does not appear related to potential insolation changes that might arise as lobes descend to lower elevations along valley walls, because the dissection occurs over a range of elevations and slope gradients. We argue that the ice-cored lobes and superposed channels reflect a palimpsest landscape, showing the migration of dominant geomorphic processes of the CTZ to the IMZ. This conclusion is consistent with observations for rising lake levels in the MDV since 1900 (Chinn, 1993; Priscu, 1998, and references therein; see also Lyons et al., 1997).

Is the warming apparent in other microclimate zones? The presence of ancient, *in situ* ashfall on slopes and elsewhere in the SUZ indicates that the relatively warm and wet microclimate conditions of the CTZ, which foster active layer cryoturbation and widespread solifluction, have not advanced into the SUZ for at least ~13 Ma (Marchant et al., 1996; Marchant and

Denton, 1996; Lewis et al., 2007, 2008). The geomorphic stability of the SUZ likely arises from the lasting presence of a robust polar East Antarctic Ice Sheet (without extensive surface melting zones; Marchant and Denton, 1996). East–west thermal gradients across the MDV may have steepened during relatively warm periods (i.e., the middle Pliocene), but climate conditions in the SUZ have most probably remained little modified over the last ~13 Ma (Beyer et al., 1999; Staiger et al., 2006 and references therein; Lewis et al., 2007, 2008).

The long-term climate and geomorphic stability of parts of the MDV is extremely rare on Earth. This stability lies in stark contrast to measured, large-scale changes in climate and landscapes that have occurred in arctic regions throughout the Pliocene and Quaternary Periods (for example, Baffin Island, Miller et al., 2005; Kleman et al., 2001). Consequently most macro-scale (and perhaps mesoscale) landforms in the Arctic reflect the culmination of alternating geomorphic processes operating under a variety of climates (Bradley, 1999) and true equilibrium landforms there may rarely occur.

Geomorphic evidence for recent climate change on Mars

Mars has undergone major changes in climate during its geological history, as well as in the recent geological past (e.g., Head et al., 2003 and references therein). The most recent changes have been linked to large variations in spin axis/orbital parameters (eccentricity and obliquity; Laskar et al., 2004), with indications of major lateral changes in ice deposition and stability, and "ice age" deposits extending down to 30° latitude. Mapping of the distribution of different geomorphologic features and their stratigraphic relations support this view (e.g., Kreslavsky and Head, 2000, 2003; Mustard et al., 2001; Milliken et al., 2003), and individual geomorphologic features can be used to establish the sequence and sign of climate change.

For example, geomorphologic features on crater interior walls in mid latitudes on Mars commonly reveal a sequence of events that illustrates the nature of changing climate. Marchant and Head (2007) described the interior wall of a mid-latitude impact crater in which these types of changes are observed in geomorphic features (e.g., sublimation polygons, crenulate concentric ridges, alcoves, talus cones) (Fig. 2.22). These features are typical of many seen on crater walls at mid latitudes on Mars (e.g., Malin and Edgett, 2000, 2001; Berman et al., 2005). Marchant and Head (2007) showed that their stratigraphic relationships provided evidence for changing climate. In this scenario, the climate was sufficiently different in the relatively recent geologic past (e.g., Fig. 2.6) to cause the accumulation of snow and ice in the

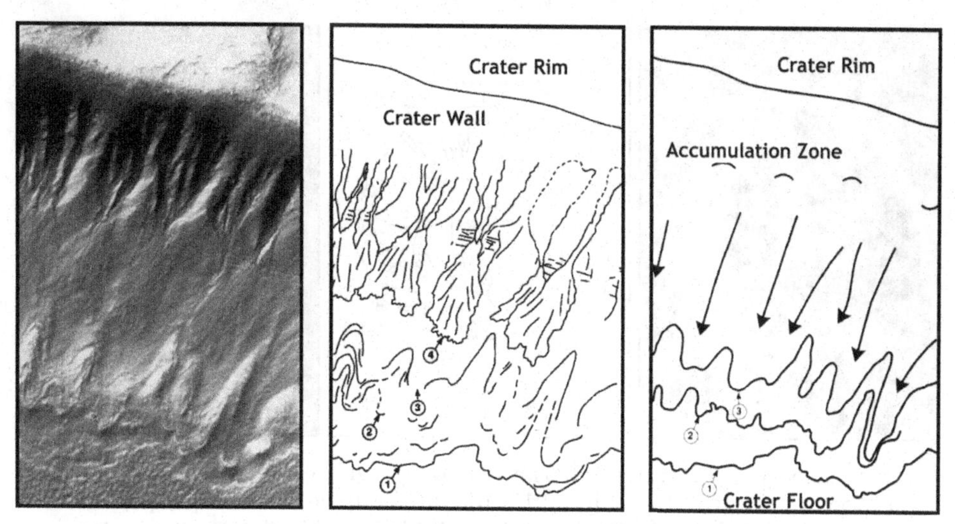

Fig. 2.22. Sequential development of climate-dependent landforms on Mars on the interior wall of an impact crater. (Left) MOC image of a crater rim crest (top), wall (middle), and floor (bottom) showing the sequence of modification of the interior of a crater wall by glacial and mass wasting processes. (Middle) Underlying the talus cones (4) is a sequence of units ranging from the upper alcoves, to debris-covered glacier remnants (1–3), to sublimation polygons and till on the crater floor. (Right) These relationships suggest the former presence of a microclimate zone that included ice and snow accumulation, debris-covered glaciers, and sublimation and polygon formation on the floor. Climate change caused retreat and ice loss, the formation of moraines, and dehydration of the climate resulted in the advent of superposed dry talus cones (middle, 4). These types of relationships can be used to establish the nature of past microclimate zones and the sign (direction) of climate change. (From Marchant and Head, 2007.)

alcoves, leading to the formation of debris-covered glaciers (e.g., Fig. 2.17) that descended toward the crater floor (e.g., Milliken et al., 2003; Head et al., 2008). As the climate changed to conditions less favorable for snow and ice deposition, the debris-covered glaciers retreated, leaving a series of moraines; the higher concentrations of ice in the regions proximal to the accumulation zones (below the alcoves) underwent sublimation, beheading the debris-covered glaciers and leaving the lobate depressions. The microclimate environment entered a new phase dominated by relatively dry talus cone development (for example, the type of change from high obliquity to lower obliquity shown in Fig. 2.6), and alcove and crater-wall debris, weakened by the previous glacial action, mass wasted down the crater wall to form talus cones, filling the lows previously occupied by the debris-covered glaciers.

Similar evidence for climate change on Mars is observed in the interior of a 10.5-km diameter crater in Newton Crater (Fig. 2.23). The distinctive wall

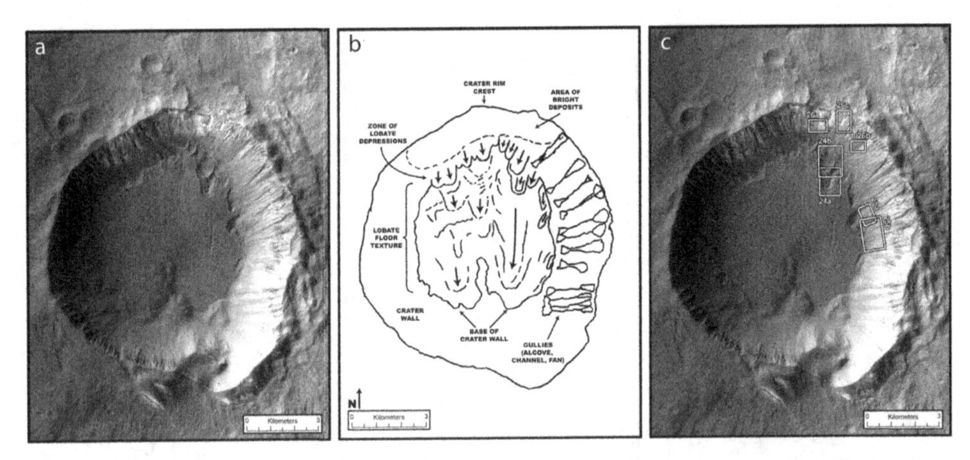

Fig. 2.23. Impact crater (10.5 km diameter) in eastern Newton Crater (−155.3E, 40.1S). See Fig. 2.12 for topographic and slope profiles. (a) CTX image p02_001842_1397_xi_40S155W. North is at the top. (b) Sketch map showing main geologic features and structures. (c) Index map showing location of image enlargements in Figs. 2.24–26. Portions of HiRISE image PSP-002094_1435.

and floor asymmetry (Fig. 2.12) is due to the same types of processes and sequence (compare to Fig. 2.22) and includes:

1. Glacial phase. The beheaded glacier-shaped lobes and moraines at the base of the wall slope are interpreted to have formed from accumulation of ice on the northern wall and downslope glacial flow. These contributed debris-covered ice to the crater floor to create the lobate floor texture (Figs. 2.23b, 2.24) and the southward floor slope (Fig. 2.12).
2. Transitional phase. Conditions changed such that there was an insufficient supply of snow and ice to maintain glacial flow. This caused sublimation of existing ice in the accumulation zone, beheading of the glacial lobes (Fig. 2.23), formation of sublimation till, and deformation of debris-covered ice at the base of the slopes to produce the terraces and ridges along the base of the slope (Figs. 2.24b, 2.25). Thermal cycling formed contraction-crack polygons in the sublimation till (Fig. 2.26). As the snow and ice sublimated and began exposing bedrock and alcoves along the northern wall, talus aprons and gullies began to form (Fig. 2.25). The stratigraphic relationships between gully fans and terrace/ridge formation (Fig. 2.25) clearly indicate the transitional nature of this period.
3. Gully phase. The most recent period is marked by gully formation. Gullies begin in alcoves (Figs. 2.23, 2.26), channels extend down the slopes (Fig. 2.25), and debris fans accumulate at the base of the slopes (Fig. 2.25). The most recent gully activity clearly cuts the polygonal terrain (Fig. 2.26b), and is superposed on the deformed fans (Fig. 2.25a, b). The current presence of snow and ice in the alcoves (e.g., Figs. 2.23, 2.26c) suggests that a source of meltwater may exist for very recent gully activity (e.g., Costard et al., 2002; Malin et al., 2006; Head et al., 2007, 2008).

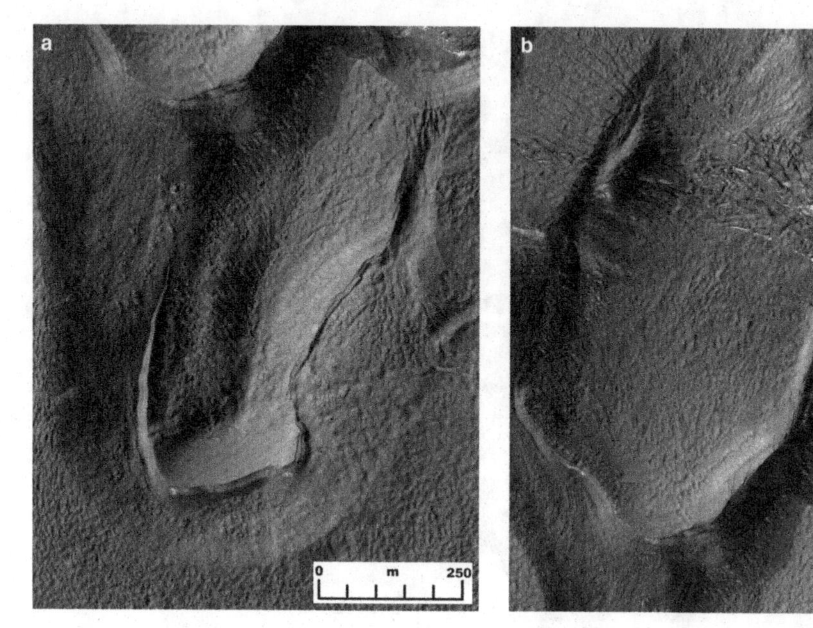

Fig. 2.24. (a), (b) Lobate depressions along the base of the northern wall of the 10.5-km diameter crater. These depressions are interpreted to be the beheaded remnants of lobate glaciers extending downslope from an accumulation zone on the crater wall (top) and contributing to the debris-covered lobate deposit on the crater floor (bottom). Portions of HiRISE image PSP-001842_1395. See Fig. 2.23c for location map and Fig. 2.12 for topography and slopes.

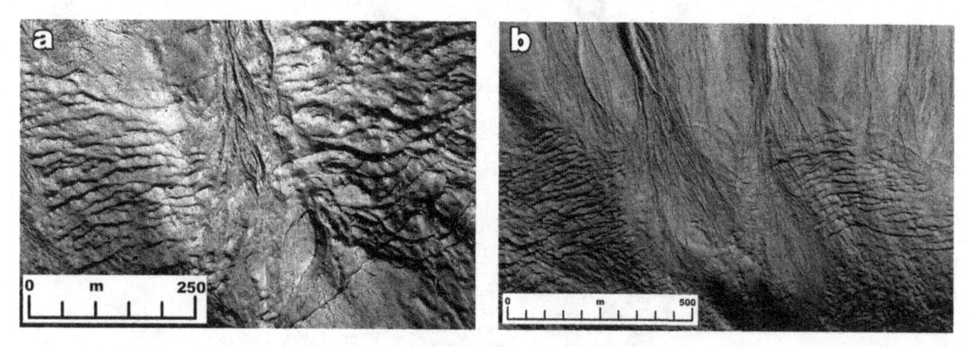

Fig. 2.25. Deformed deposits forming terraces and ridges near the base of the slope along the northeastern wall of the 10.5-km diameter crater. The most recent channels and fans clearly postdate the terraces and ridges; center of (a) and left and center of (b). Some older channels and fans, however, are clearly cut by the ridges; left- and right-hand sides of (b). Portions of HiRISE image PSP-001842_1395. See Fig. 2.23c for location map and Fig. 2.12 for topography and slopes.

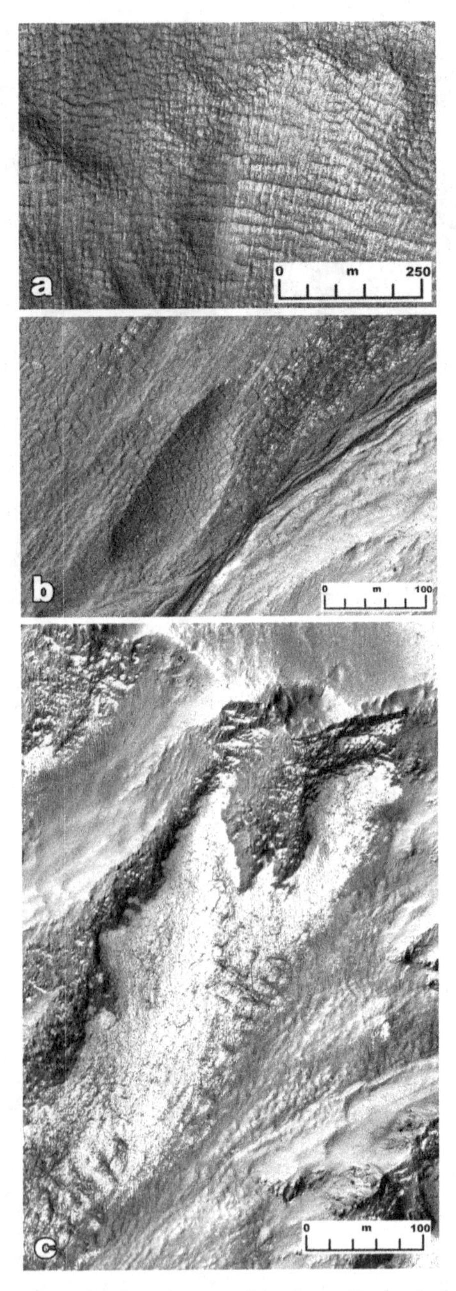

Fig. 2.26. Polygonal ground and snow/ice deposits in the upper parts of the northern crater wall of the 10.5-km diameter crater. (a) Polygonal ground formed in the potential accumulation zone for past glaciers. Seasonal ice deposits modify the albedo in the right-hand side of the image. (b) Oval depression in the polygonal terrain lends support to the interpretation of ice downwasting and sublimation till formation. Note the gully channel deposits (lower right) cutting and superposed on the polygonal terrain. (c) Alcove in

In summary, in a manner analogous to evidence for climate changes discussed for the MDV, changing and superposed landforms can be used to infer the sign of climate change on Mars. Mapping of similar microclimate zones and their superposition over the critical latitude-dependent transition zones on Mars holds promise for deconvolving the details of geologically recent climate change there (Fig. 2.17) (e.g., Head et al., 2003; Kreslavsky et al., 2008). Indeed, the upcoming Phoenix Mission to the northern high latitudes of Mars (Smith et al., 2007) will provide important new information on the climate conditions and the microscale and mesoscale geomorphology there.

Climate change in the earlier history of Mars

These examples provide abundant evidence for the use of antarctic McMurdo Dry Valley geomorphic process analogs in interpreting the nature of recent climate change on Mars. The stability of the martian crust and lithosphere, combined with the extremely low erosion rates, mean that the geologic record of climate change throughout the history of Mars is available for study as well (Fig. 2.27), and that MDV analogs will be important in the interpretation of this record. For example, the MDV record has been important in the interpretation of fan-shaped deposits on the northwest flanks of the Tharsis Montes volcanoes as tropical mountain glaciers; the enigmatic ridge terrain there (Fig. 2.3b) has been linked to cold-based glacial drop moraines similar to those seen in the MDV (Fig. 2.3a) (e.g., Head and Marchant, 2003; Shean et al., 2005, 2007a). Similarly, MDV analogs have been applied to the interpretation of Early Amazonian valley networks formed from melting snowpack at the summits of volcanic edifices (Fassett and Head, 2006, 2007), and to features interpreted as glacial lobes at the margins of Hesperian-aged outflow channel sources (Head et al., 2004). Additional documentation and analysis of the MDV microenvironments and their interactions, together with new high-resolution data from Mars, will help in the understanding of the

Caption for Fig. 2.26. (cont.)
the upper part of the northern crater wall. Note the polygonal ground on the floor of the alcove and the superposed snow and ice deposits covering the polygons and filling in local lows. The presence of snow and ice deposits under present-day conditions in the areas interpreted to be glacial accumulation zones lends support to the glacial interpretation. Increased accumulations in these areas under earlier climate regimes would lead to ice accumulation and flow. Portions of HiRISE image PSP-001842_1395. See Fig. 2.23c for location map and Fig. 2.12 for topography and slopes.

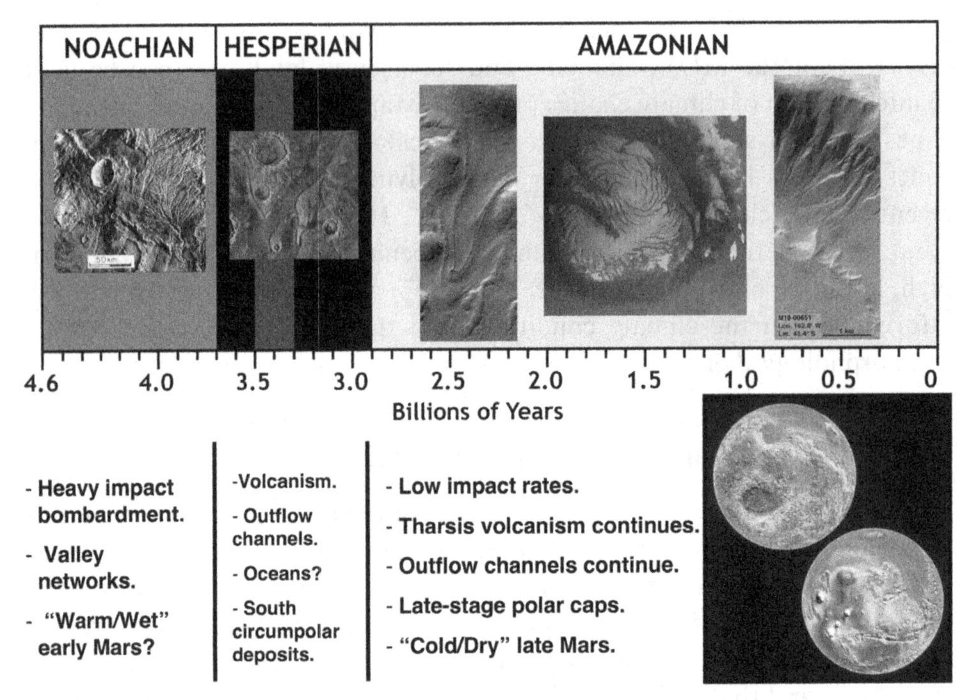

Fig. 2.27. Major events during the geological history of Mars. In the Noachian, valley networks are thought to represent pluvial activity and a possibly warm and wet early Mars. Subsequently, changing climate conditions in the Hesperian caused the formation of a global cryosphere and the sequestration of groundwater, with the occasional catastrophic release forming outflow channels, standing bodies of water, and potential episodic climate change. The Amazonian has apparently been characterized by a cold and dry Mars much like today, but significant variations in spin axis and orbital parameters (e.g., Laskar et al., 2004) has led to changing climate conditions during this period, including tropical mountain glaciers (e.g., Head and Marchant, 2003). MDV analogs will be useful in interpreting the entire climate history of Mars and will help inform investigators about conditions that might have been conducive to the origin and maintenance of a biota.

cold, hyperarid climate of Mars and the analysis of climate change there throughout its history.

Antarctica as an astrobiological analog for Mars

Extreme environmental conditions in the McMurdo Dry Valleys, such as those described above in the three microclimate zones, limit the presence of life in this harsh environment. Multicellular organisms representative of

Fig. 2.28. Algal mats in the vicinity of annual snow patches and gullies. (a) Desiccated algal mats downslope from a small snow patch. Note the cracking, upturning, and breaking of the margins of the mats; broken fragments are often transported by the wind. (b) Water from melting snow patch reviving algal mats. Note the dampening and wetting of the algal mat and the folded and broken textures. Mats are a few millimeters thick. South Fork of the Upper Wright Valley, MDV.

higher trophic levels tend to be low in diversity and abundance. Nematodes are among the largest animals in the MDV ecosystems (Gooseff et al., 2003b; Nkem et al., 2006). The biota that grow in the lakes, soils, and glacial meltwater streams are primarily cyanobacteria, eukaryotic algae, heterotrophic and chemoautotrophic bacteria, and mosses (McKnight et al., 1999). Despite this generally bleak outlook, life has adapted to many of the local microenvironments described above. For example, gullies and their associated streams can be "oases" of life. Diverse microbial communities are observed in many streams, including cyanobacteria, chlorophytes, diatoms, nematodes, rotifers, and tardigrades. Algal communities grow as mats in streams and can survive long periods (decades) of desiccation and extreme cold in a cryptobiotic state (McKnight et al., 1999). Abundant algal mats exist even in those late-stage channels wetted only by annual snow patch melting (Fig. 2.10) (Head et al., 2007b). Between periods of wetting, algal mats desiccate, curl up and crack (Fig. 2.28), and then are transported by wind to other parts of the stream and related microenvironments. Melting of small amounts of seasonally accumulated snow can dampen, wet, and revive the algal mats (Fig. 2.28). In addition, recent work has shown that microbes may be preserved in ancient ice for millions of years (Bidle et al., 2007; Gilichinsky et al., 2007). Thus, the further definition of microclimate zones, geomorphic processes and features, and associated microenvironment will help to provide a better understanding of the range of habitability in the McMurdo Dry Valleys and how these might be manifested on Mars.

Summary and conclusions

The hyperarid, cold-polar desert of the McMurdo Dry Valleys (MDV) provides an important terrestrial analog for Mars. The cold and dry MDV climate and the associated suite of landforms closely resemble those occurring on the surface of Mars at several different scales. Extreme hyperaridity on Mars and in the MDV underlines the importance of salts and brines on soil development, phase transitions from liquid water to water ice, and in turn, on process geomorphology, landscape evolution, and biology at a range of scales. The three MDV microclimate zones (coastal thaw, inland mixed, and stable upland) are defined on the basis of climate parameters (atmospheric temperature, soil moisture, and relative humidity). These zones are hypersensitive to climate change: minor differences can lead to large differences in the distribution and morphology of features at the macroscale (e.g., slopes and gullies); mesoscale (e.g., polygons, viscous-flow features, and debris-covered glaciers); and microscale (e.g., rock-weathering processes/features, including wind erosion, salt weathering, and surface pitting). Similar types of landform analyses have been applied to Mars where microclimates and equilibrium landforms analogous to those in the MDV occur in a variety of local environments, in different latitudinal bands, and in units of different ages. Documentation of the nature and evolution of microclimate zones and associated geomorphic processes in the MDV is providing a quantitative framework for assessing the evolution of climate on Mars, and the range of habitats that might have been conducive to the formation and evolution of life there.

Equilibrium landforms form in balance with environmental conditions within fixed microclimate zones. Past and/or ongoing shifts in climate zonation are indicated by landforms that today appear in disequilibrium with local microclimate conditions in the MDV, providing a record of the sign and magnitude of climate change there. Analysis of the record of recent climate change on Mars is under way (e.g., Laskar et al., 2002; Head and Marchant, 2003; Levrard et al., 2004; Milkovich and Head, 2005; Tanaka, 2005; Forget et al., 2006; Kreslavsky et al., 2008) and will continue over the coming decades as automated spacecraft and eventually humans explore the planet, and begin to understand how to extrapolate the recent record to earlier periods of the climate history of Mars (e.g., Bibring et al., 2006; Carr, 2006). In parallel, the exploration and analysis of the MDV microclimate zones described here, and their comparison with the emerging details of similar zones on Mars, will provide important insight into the current climate on Mars, the nature and distribution of martian microclimate zones, and the geomorphological evidence for climate change in the past history of Mars (Fig. 2.1), including those climates most favorable for life.

The modes of occurrence of the limited and unusual biota in the MDV provide terrestrial laboratories for the study of possible environments for life on Mars. The range of microenvironments in the MDV are hypersensitive to climate variability, and their stability and change provide important indications of climate history, potential stress on the biota, and ways in which a biota might have accommodated to the evolution from early Mars to the harsh surface environment of today.

Acknowledgments

We gratefully acknowledge the financial assistance of the National Science Foundation, Office of Polar Programs, through Grant NSF ANT-0338291 to DRM and JWH, and of the National Aeronautics and Space Administration, Mars Fundamental Research Program, through Grant NNX06AE32G to DRM, the NASA Applied Information Systems Research Program, through Grant NNGO5GA61G to JWH, and the Mars Data Analysis Program, through Grant NNG04GJ99G to JWH. We thank Chris McKay and William Boynton for very helpful reviews. We gratefully acknowledge the Mars Orbiter Camera, THEMIS, and HiRISE teams for their tireless efforts in data acquisition and rapid data verification and delivery to the scientific community. Thanks are also extended to Jay Dickson, Doug Kowalewski, Kate Swanger, and Anne Côté for help in manuscript preparation.

References

Allen, C. C. and Conca, J. L. (1991). Weathering of basaltic rocks under cold, arid conditions: Antarctica and Mars. In *Proceedings of 21st Lunar Planetary Science Conference*, Houston, TX, pp. 711–717.

Anderson, D. M., Gatto, L. W., and Ugolini, F. C. (1972). An Antarctic analog of martian permafrost terrain. *Antarctic Journal of the United States*, **7**, 114–116.

Arvidson, R. E., Guinness, E., and Lee, S. (1979). Differential aeolian redistribution rates on Mars. *Nature*, **278**, 533–535.

Atkins, C. B., Barrett, P. J., and Hicock, S. R. (2002). Cold glaciers erode and deposit: evidence from Allan Hills, Antarctica. *Geology*, **30**, 659–662.

Augustinus, P. C. and Selby, M. J. (1990). Rock slope development in McMurdo Oasis, Antarctica, and implications for interpretations of glacial history. *Geografiska Annaler*, **72**(A), 55–62.

Baker, V. R. (2001). Water and the martian landscape. *Nature*, **412**, 228–236.

Bao, H. and Marchant, D. R. (2006). Quantifying sulfate components and their variations in soils of the McMurdo Dry Valleys, Antarctica. *Journal of Geophysical Research*, **111**, doi: 10.1029/2005JD006669.

Bao, H., Campbell, D. A., Bockheim, J. G., and Thiemens, M. H. (2000). Origin of sulfate in Antarctic Dry Valley soils as deduced from anomalous ^{17}O compositions. *Nature*, **407**, 499–502.

Bao, H., Barnes, J. D., Sharp, Z. D., and Marchant, D. R. (2008). Two chloride sources in soils of the McMurdo Dry Valleys, Antarctica. *Journal of Geophysical Research*, **113**(D3) D03301, doi: 10.1029/2007JD008703.

Berg, T. E. and Black, R. F. (1966). Preliminary measurements of growth of non-sorted polygons, Victoria Land, Antarctica. In *Antarctic Soils and Soil Forming Processes*, ed. J. C. F. Tedrow. American Geophysical Union Antarctic Research Series 8. Washington, D.C.: AGU, pp. 61–108.

Berman, D. C., Hartmann, W. K., Crown, D. A., and Baker, V. R. (2005). The role of arcuate ridges and gullies in the degradation of craters in the Newton Basin region of Mars. *Icarus*, **178**(2), 465–486.

Beyer, L., Bockheim, J. G., Campbell, I. B., and Claridge, G. G. C. (1999). Review: genesis, properties, and sensitivity of Antarctic gelisols. *Antarctic Science*, **11**, 387–398.

Bibring, J.-B., Langevin, Y., Mustard, J. F., and the OMEGA Team (2006). Global mineralogical and aqueous Mars history derived from OMEGA/Mars Express data. *Science*, **312**, 400–404.

Bidle, K. D., SangHoon, L., Marchant, D. R., and Falkowski, P. G. (2007). Fossil genes and microbes in the oldest ice on Earth. *Proceedings of the National Academy of Sciences*, **104**(33), 13 455–13 460.

Binder, A. B., Arvidson, R. E., Guinness, E. A., et al. (1977). The geology of the Viking 1 landing site. *Journal of Geophysical Research*, **82**, 4439–4451.

Black, R. F. (1973). Growth of patterned ground in Victoria Land, Antarctica. In: *Permafrost Second International Conference*, National Academy of Sciences, Yakutsk, Siberia, pp. 193–203.

Bockheim, J. G. (1997). Properties and classification of cold desert soils from Antarctica. *Soil Society of America Journal*, **61**(1), 224–231.

Bockheim, J. G. (2002). Landform and soil development in the McMurdo Dry Valleys: a regional synthesis. *Arctic Antarctic and Alpine Research*, **34**, 308–317.

Bockheim, J. G. (2003). *University of Wisconsin Antarctic Soils Database*. Boulder, CO: National Snow and Ice Data Center/World Data Center for Glaciology. Digital media.

Bockheim, J. G. and McLeod, M. (2006). Soil formation in Wright Valley, Antarctica since the late Neogene. *Geoderma*, **137**, 109–116.

Borgstrom, I. (1999). Basal ice temperatures during late Weichselian deglaciation: comparison of landform assemblages in west-central Sweden. *Annals of Glaciology*, **28**, 9–15.

Boynton, W. V. and 24 colleagues (2002). Distribution of hydrogen in the near-surface of Mars: evidence for subsurface ice deposits. *Science*, **296**, 81–85.

Bradley, R. S. (1999). *Paleoclimatology: Reconstructing Climates of the Quaternary*. International Geophysics Series, 64. London: Academic Press, 613 pp.

Brass, G. W. (1980). Stability of brines on Mars. *Icarus*, **42**, 20–28.

Bridges, J. C., Catling, D. C., Saxton, J. M., et al. (2001). Alteration assemblages in martian meteorites: implications for near-surface processes. *Space Science Review*, **96**, 365–392.

Bridges, N. T. and Lackner, C. N. (2006). Northern hemisphere Martian gullies and mantled terrain: implications for near-surface water migration in Mars' recent past. *Journal of Geophysical Research, Planets*, **111**, doi: 10.1029/2006JE002702.

Brook, E. J., Kurz, M. D. Ackert, Jr., R. P., et al. (1993). Chronology of Taylor Glacier advance in Arena Valley, Antarctica, using in situ cosmogenic ^{3}He and ^{10}Be. *Quaternary Research*, **39**,11–23.

Brook, E. J., Brown, E. T., Kurz, M. D., et al. (1995). Constraints on age, erosion, and uplift of Neogene glacial deposits in the Transantarctic Mountains determined from in situ cosmogenic ^{10}Berillium and ^{26}Aluminum. *Geology*, **23**, 1063–1066.

Brook, M. S., Kirkbride, M. P., and Bock, B. W. (2006). Quantified time scale for glacial valley cross-profile evolution in alpine mountains. *Geology*, **34**, 637–640.

Burt, D. M. and Knauth, L. P. (2003). Electrically conducting, Ca-rich brines, rather than water, expected in the Martian subsurface. *Journal of Geophysical Research*, **108**, doi: 10.1029/2002JE001862.

Campbell, I. B. and Claridge, G. G. C. (1969). A classification of frigic soils: the zonal soils of the Antarctic continent. *Soil Science*, **107**, 75–85.

Campbell, I. B. and Claridge, G. G. C. (1987). *Antarctica: Soils, Weathering Processes, and Environment*. Developments in Soil Science 16. New York: Elsevier, 368 pp.

Campbell, I. B. and Claridge, G. G. C. (2006). Permafrost properties, patterns and processes in the Transantarctic Mountains region. *Permafrost and Periglacial Processes*, **17**, 215–232.

Campbell, I. B., Claridge, G. G. C., Balks, M. R., and Campbell, D. I. (1997a). Moisture content in soils of the McMurdo Sound and Dry Valley region of Antarctica. In *Ecosystem Processes in Antarctic Ice-free Landscapes*, ed. W. B. Lyons, C. Howard-Williams, and I. Hawes. Rotterdam, Netherlands: A. A. Balkema, pp. 61–76.

Campbell, D. I., MacCulloch, R. J. L., and Campbell, I. B. (1997b). Thermal regimes of some soils in the McMurdo Sound and Dry Valley region. In *Ecosystem Processes in Antarctic Ice-free Landscapes*, ed. W. B. Lyons, C. Howard-Williams, and I. Hawes. Rotterdam, Netherlands: A. A. Balkema, pp. 45–56.

Campbell, I. B, Claridge, G. G. C., Campbell, D. I, and Balks, M. R. (1998). The soil environment of the McMurdo Dry Valleys, Antarctica. In *Ecosystem Dynamics in a Polar Desert: The McMurdo Dry Valleys, Antarctica*, ed. J. C. Priscu. Antarctic Research Series 72. Washington, D.C.: American Geophysical Union, pp. 297–322.

Carr, M. H. (1981). *The Surface of Mars*. New Haven, CT: Yale University Press.

Carr, M. H. (1996). *Water on Mars*. New York: Oxford University Press.

Carr, M. H. (2006). *The Surface of Mars*. New York: Cambridge University Press.

Chinn, T. J. (1980). Glacier balances in the Dry valleys area, Victoria Land, Antarctica. In *Proceedings of the Riederlap Workshop*. IAHS-AISH Publication 126, pp. 237–247.

Chinn, T. J. (1981). Hydrology and climate in the Ross Sea area. *Journal of the Royal Society of New Zealand*, **11**(4), 373–386.

Chinn, T. J. (1993). Physical hydrology of the Dry Valley lakes. In *Physical and Biogeochemical Processes in Antarctic Lakes*, ed. W. J. Green and E. I. Freidmann. Antarctic Research Series 59. Washington, D.C.: American Geophysical Union, pp. 1–51.

Christensen, P. R. (2003). Formation of recent martian gullies through melting of extensive water-rich snow deposits. *Nature*, **422**, 45–48.

Christensen, P. R. and 25 colleagues (2001). Mars Global Surveyor Thermal Emission Spectrometer experiment: investigation description and surface science results. *Journal of Geophysical Research*, **106**(E10), 23 823–23 872.

Claridge, G. G. C. and Campbell, I. B. (1968). Soils of the Shackleton Glacier, Queen Maude Range, Antarctica. *New Zealand Journal of Science*, **11**, 171–218.

Claridge, G. G. C. and Campbell, I. B. (1977). The salts in Antarctic soils, their distribution and relationship to soil processes. *Soil Science*, **28**, 377–384.

Claridge, G. G. C. and Campbell, I. B. (2005). Weathering processes in arid cryosols. In *Cryosols: Permafrost Affected Soils*, ed. J. Kimble. Berlin: Springer, pp. 447–458.

Clark B. C. (1978). Implications of abundant hygroscopic materials in the martian regolith. *Icarus*, **34**, 645–665.

Clark, B. C. (1979). Chemical and physical microenvironments at the Viking landing sites. *Journal of Molecular Evolution*, **14**, 13–31.

Clark, B. C. and Baird, A. K. (1979). Is the martian lithosphere sulfur rich? *Journal of Geophysical Research*, **84**, 8395–8403.

Clark, B. C. and Van Hart, D. C. (1981). The salts of Mars. *Icarus*, **45**, 370–378.

Clark, B. C., Baird, A. K., Weldon, R. J., et al. (1982). Chemical composition of martian fines. *Journal of Geophysical Research*, **87**, 10 059–10 067.

Conca, J. L. and Astor, A. M. (1987). Capillary moisture flow and the origin of cavernous weathering in dolerites of Bull Pass, Antarctica. *Geology*, **15**, 151–154.

Costard, F., Forget, F., Mangold, N., and Peulvast, J. -P. (2002). Formation of recent martian debris flows by melting of near-surface ground ice at high obliquity. *Science*, **295**, 110–113.

Craddock, R. A. and Howard, A. D. (2002). The case for rainfall on a warm, wet early Mars. *Journal of Geophysical Research*, **107**(E11), doi: 10.1029/2001JE001505.

Cuffey, K. M., Conway, H., Gades, A. M., et al. (2000). Entrainment at cold glacier beds. *Geology*, **28**, 351–354.

Dana, G. L., Wharton, R. A., and Dubayah, R. (1998). Solar radiation in the McMurdo dry valleys, Antarctica. In *Ecosystem Dynamics in a Polar Desert: The McMurdo Dry Valleys, Antarctica*, ed. J. C. Priscu. Antarctic Research Series 72. Washington, D.C.: American Geophysical Union, pp. 39–64.

Davis, N. (2001). *Permafrost: A Guide to Frozen Ground in Transition*. Fairbanks, AK: University of Alaska Press.

Denton, G. H. and Marchant, D. R. (2000). The geologic basis for a reconstruction of a grounded ice sheet in McMurdo Sound, Antarctica, at the last glacial maximum. *Geografiska Annaler*, **82**(A), 167–211.

Denton, G. H., Sugden, D. E., Marchant, D. R., Hall, B. L., and Wilch, T. I. (1993). East Antarctic Ice Sheet sensitivity to Pliocene climatic change from a Dry Valleys perspective. *Geografiska Annaler*, **75A**, 155–204.

Dickinson, W. W. and Rosen, M. R. (2003). Antarctic permafrost: an analogue for water and diagenetic minerals on Mars. *Geology*, **31**, 199–202, doi: 10.1130/0091–7613.

Dickson, J. L., Head, J. W., Marchant, D. R., Morgan, G. A., and Levy, J. S. (2007a). Recent gully activity on Mars: clues from late-stage water flow in gully systems and channels in the Antarctic Dry Valleys. *Lunar Planetary Science Conference*, **38**, abstract 1678.

Dickson, J. L., Head, J. W., and Kreslavsky, M. (2007b). Martian gullies in the southern mid-latitudes of Mars: evidence for climate-controlled formation of young fluvial features. *Icarus*, **188**, 315–323.

Dickson, J. L., Head, J. W., and Marchant, D. R. (2008). Late Amazonian glaciation at the dichotomy boundary on Mars: evidence for glacial thickness maxima and multiple glacial phases. *Geology*, **36**, 411–414.

Doran, P. T., McKay, C. P., Clow, G. D., et al. (2002). Valley floor climate observations from the McMurdo dry valleys, Antarctica, 1986–2000. *Journal of Geophysical Research*, **107**(D24), doi: 10.1029/2001JD002045.

Elliot, D. E. and Fleming, T. H. (2004). Occurrence and dispersal of magmas in the Jurassic Ferrar large igneous province, Antarctica. *Gondwana Research*, **7**(1), 223–237.

Fabel, D., Stroeven, A. P., Harbor, J., et al. (2002). Landscape preservation under Fennoscandian ice sheets determined from in situ produced ^{10}Be and ^{26}Al. *Earth and Planetary Science Letters*, **201**(2), 397–406.

Farmer, C. B. (1976). Liquid water on Mars. *Icarus*, **28**, 279–289.

Farmer, C. B. and Doms, P. E. (1979). Global seasonal variation of water vapor on Mars and the implications of permafrost. *Journal of Geophysical Research*, **84**, 2881–2888.

Fassett, C. I. and Head, J. W. (2006). Valleys on Hecates Tholus Mars: origin by basal melting of summit snowpack. *Planetary and Space Science*, **54**, 370–378, doi: 10.1016/j.pss.2005.12.011.

Fassett, C. I. and Head, J. W. (2007). Valley formation on martian volcanoes in the Hesperian: evidence for melting of summit snowpack, caldera lake formation, drainage and erosion on Ceraunius Tholus, Mars. *Icarus*, **189**, 118–135, doi: 10.1016/j.icarus.2006.12.021.

Feldman, W. C. and 12 colleagues (2002). Global distribution of neutrons from Mars: results from Mars Odyssey. *Science*, **297**, 75–78.

Forget, F., Hourdin, F., Fournier, R., et al. (1999). Improved general circulation models of the Martian atmosphere from the surface to above 80 km. *Journal of Geophysical Research*, **104**(E10), 24 155–24 176.

Forget, F., Haberle, R. M., Montmessin, F., Levrard, B., and Head, J. W. (2006). Formation of glaciers on Mars by atmospheric precipitation at high obliquity. *Science*, **311**, 368–371.

Fountain, A. G., Lewis, K. J., and Doran, P. T. (1999). Spatial climatic variation and its control on glacier equilibrium line altitude in Taylor Valley, Antarctica. *Global and Planetary Change*, **22**, 1–10.

Frezotti, M. (1997). Ice front fluctuation, iceberg calving flux and mass balance of Victoria Land glaciers. *Antarctic Science*, **9**, 61–73.

Garvin, J. B., Head, J. W., Marchant, D. R., and Kreslavsky, M. A. (2006). High-latitude cold-based glacial deposits on Mars: multiple superposed drop moraines in a crater interior at 70 °N latitude. *Meteoritics and Planetary Science*, **41**, 1659–1674.

Ghysels, G. and Heyse, I. (2006). Composite-wedge pseudomorphs in Flanders, Belgium. *Permafrost and Periglacial Processes*, **17**, 145–161.

Giardino, J. R., Shroder, J. F., and Vitek, J. D. (1987). *Rock Glaciers*. London: Allen and Unwin.

Gibson, E. K., Wentworth, S. T., and McKay, D. S. (1983). Chemical weathering and diagenesis of a cold desert soil from Wright Valley, Antarctica: an analog for Martian weathering processes. *Journal of Geophysical Research, Supplement* **88**, A812–A918.

Gilichinsky, D. A., Wilson, G. S., Friedmann, E. I., et al. (2007). Microbial populations in Antarctic permafrost: biodiversity, state, age, and implication for astrobiology. *Astrobiology*, **7**(2), 275–311. doi: 10.1089/ast.2006.0012.

Golombek, M. P. and 13 colleagues (1997). Overview of the Mars Pathfinder Mission and assessment of landing site predictions. *Science*, **278**, 1743.

Golombek, M. P. and 22 colleagues (1999). Overview of the Mars Pathfinder Mission: launch through landing, surface operations, data sets, and science results. *Journal of Geophysical Research*, **104**(E4), 8523–8554.

Gooding, J. L. (1992). Soil mineralogy and chemistry on Mars: possible clues from salts and clays in SNC meteorites. *Icarus*, **99**, 28–41.

Gooding, J. L., Wentworth, S. J., and Zolensky, M. E. (1991). Aqueous alteration of the Nakhla meteorite. *Meteoritics*, **26**, 135–143.

Gooseff, M. N., McKnight, D. M., Runkel, R. L., and Vaughn, B. H. (2003a). Determining long time-scale hyporheic zone flow paths in Antarctic streams. *Hydrological Processes*, **17**(9), 1691–1710.

Gooseff, M. N., Barrett, J. E., Doran, P. T., et al. (2003b). Snow-patch influence on soil biogeochemical processes and invertebrate distribution in the McMurdo Dry Valleys, Antarctica. *Arctic Antarctic and Alpine Research*, **35**, 92–100.

Haberle, R. M., McKay, C. P., Schaeffer, J., et al. (2001). On the possibility of liquid water on present-day Mars. *Journal of Geophysical Research*, **106**(E10), 23 317–23 326.

Hall, B. L., Denton, G. H., Lux, D. R., and Bockheim, J. G. (1993). Late Tertiary Antarctic paleoclimate and ice-sheet dynamics inferred from surficial deposits in Wright Valley. *Geografiska Annaler*, **75**(A), 239–268.

Hall, K. J. (1991). Mechanical weathering in the Antarctic: a maritime perspective. In *Periglacial Geomorphology*, ed. J. C. Dixon and A. D. Abrahams. Chichester, UK: John Wiley and Sons, pp. 103–123.

Hallet, B. and Waddington, D. E. (1991). Buoyancy forces induced by freeze-thaw in the active layer: implications for diapirism and soil circulation. In *Periglacial Geomorphology*, ed. J. C. Dixon and A. D. Abrahams. Chichester, UK: John Wiley and Sons, pp. 251–279.

Hallet, B., Hunter, L., and Bogen, J. (1996). Rates of erosion and sediment evacuation by glaciers: a review of field data and their implications. *Global and Planetary Change*, **12**, 213–235.

Harris, K. J., Carey, A. E., Lyons, W. B., Welch, K. A., and Fountain, A. G. (2007). Solute and isotope geochemistry of subsurface ice melt seeps in Taylor Valley, Antarctica. *Geological Society of America Bulletin*, **119**, 548–555.

Hartman, B. N. (1998). Miocene paleoclimate and ice sheet dynamics as recorded in central Taylor Valley, Antarctica. Unpublished M.S. thesis. Boston University, MA.

Haskin, L. A. and 29 colleagues (2005). Water alteration of rocks and soils on Mars at the Spirit rover site in Gusev crater. *Nature*, **436**, 66–69.

Hassinger, J. M. and Mayewski, P. A. (1983). Morphology and dynamics of rock glaciers in southern Victoria Land, Antarctica. *Arctic and Alpine Research*, **15**, 351–368.

Head, J. W. and Kreslavsky, M. A. (2006). Formation of weathering pits on rock surfaces in the Antarctic Dry Valleys and on Mars. Paper presented at the 44th Brown/Vernadsky Microsymposium, Moscow, Russia, abstract m44–25.

Head, J. W. and Marchant, D. R. (2003). Cold-based mountain glaciers on Mars: Western Arsia Mons. *Geology*, **31**(7), 641–644.

Head, J. W. and Marchant, D. R. (2006). Gullies and saline ponds in the cold hyper-arid desert of the Antarctic Dry Valleys: clues to interpreting climate conditions on Mars. Paper presented at the 44th Brown/Vernadsky Microsymposium, Moscow, Russia, Abstract m44–26.

Head, J. W., Mustard, J. F., Kreslavsky, M. A., Milliken, R. E., and Marchant, D. R. (2003). Recent ice ages on Mars. *Nature*, **426**, 797–802.

Head, J. W., Marchant, D. R., and Ghatan, G. J. (2004). Glacial deposits on the rim of a Hesperian-Amazonian outflow channel source trough: Mangala Valles, Mars. *Geophysical Research Letters*, L10701, doi: 10.1029/2004GL020294.

Head, J. W. and 13 colleagues (2005). Tropical to mid-latitude snow and ice accumulation, flow and glaciation on Mars. *Nature*, **434**, 346–351.

Head, J. W., Marchant, D. R., Agnew, M. C., Fassett, C. I., and Kreslavsky, M. A (2006a). Extensive valley glacier deposits in the northern mid-latitudes of Mars: evidence for Late Amazonian obliquity-driven climate change. *Earth and Planetary Science Letters*, **241**, 663–671.

Head, J. W., Nahm, A. L., Marchant, D. R., Neukum, G., and the HRSC Co-Investigator Team (2006b). Modification of the dichotomy boundary on Mars by Amazonian mid-latitude regional glaciation. *Geophysical Research Letters*, **33**(8), doi: 2005GL024360.

Head, J. W., Marchant, D. R., Dickson, J. L., Levy, J. S., and Morgan, G. A. (2007a). Mars gully analogs in the Antarctic Dry Valleys: geological setting and processes. *Lunar Planetary Science Conference*, **38**, Abstract 1617.

Head, J. W., Marchant, D. R., Dickson, J., Levy, J., and Morgan, G. (2007b). Transient streams and gullies in the Antarctic Dry Valleys: geological setting, processes and analogs to Mars. Online Proceedings of the 10th ISAES X, ed. A. K. Cooper, C. R. Raymond, et al. USGS Open-File Report 2007-1047, Extended Abstract 1763, 1–4.

Head, J. W., Marchant, D. R., and Kreslavsky, M. A. (2008). Formation of gullies on Mars: link to recent climate history and insolation microenvironments implicate surface water flow origin. *Proceedings of the National Academy of Sciences*, **105**, 13 258–13 263, doi 10.1073 pnas.0803760105.

Healy, M., Webster-Brown, J. G., Brown, K. L., and Lane, V. (2006). Chemistry and stratification of Antarctic meltwater ponds. II. Inland ponds in the McMurdo Dry Valleys, Victoria Land. *Antarctic Science*, **18**, 525–533.

Hecht, M. H. (2002). Metastability of liquid water on Mars. *Icarus*, **156**, 373–386.

Helbert, J., Head, J. W., and Kreslavsky, M. (2007). A global physical and morphological survey of candidate ice-rich environments and deposits on Mars. *Lunar Planetary Science Conference*, **38**, Abstract 1279.

Heldmann, J. L. and Mellon, M. T. (2004). Observations of martian gullies and constraints on potential formation mechanisms. *Icarus*, **168**, 285–304.

Heldmann, J. L., Toon, O. B., Pollard, W. H., et al. (2005b). Formation of Martian gullies by the action of liquid water flowing under current Martian environmental conditions. *Journal of Geophysical Research*, **110**, E05004, doi: 10.1029/2004JE002261.

Holt, J. W., Safaeinili, A., Plaut, J. J., et al., and the SHARAD Team (2008a). Radar sounding evidence for ice within lobate debris aprons near Hellas basin, mid-southern latitudes on Mars, *Lunar Planetary Science Conference*, **39**, Abstract 2441.

Holt, J. W., Safaeinili, A., Plaut, J. J., et al. (2008b). Radar sounding evidence for buried glaciers in the southern mid-latitudes of Mars. *Science*, **322**, 1235–1238.

Huinink, H. P., Pel, L., and Kopinga, K. (2004). Stimulating the growth of tafoni. *Earth Surface Process Landforms*, **29**(10), 1225–1233.

Ikard, S. J, Gooseff, M. N., Barrett, J. E., and Takacs-Vesbach, C. (2009). Thermal characterization of active layer across a soil moisture gradient in the McMurdo Dry Valleys, Antarctica. *Permafrost and Periglacial Processes*, **20**, 27–39.

Jakosky, B. M. and Mellon, M. T. (2001). High-resolution thermal intertia mapping of Mars: sites of exobiological interest. *Journal of Geophysical Research*, **106**(E10), 23 887–23 908.

Jones, K. L., Bragg, S. L., Wall, S. D., Carlston, C. E., and Pidek, D. G. (1979). One Mars year: Viking lander imaging observations. *Science*, **204**, 799–806.

Kleman, J. and Hattestrand, C. (1999). Frozen-bed Fennoscandian and Laurentide ice sheets during the last glacial maximum. *Nature*, **402**, 63–66.

Kleman, J., Marchant, D. R., and Borgstrom, I. (2001). Late glacial ice dynamics on southern Baffin Island and in Hudson Strait. *Arctic Antarctic and Alpine Research*, **33**(3), 249–257.

Knauth, L. P. and Burt, D. M. (2002). Eutectic brine seeps on Mars: origin and possible relation to young seepage features. *Icarus*, **158**, 267–271.

Koppes, M. and Hallet, B. (2006). Erosion rates during rapid deglaciation in Icy Bay, Alaska. *Journal of Geophysical Research*, **111**, doi: 10.1029/2005JF000349.

Kostama, V. -P., Kreslavsky, M. A., and Head, J. W. (2006). Recent high-latitude icy mantle in the northern plains of Mars: characteristics and ages of emplacement. *Geophysical Research Letters*, **33**(11), doi: 10.1029/2006GL025946.

Kowalewski, D. E., Marchant, D. R., Levy, J. S., and Head, J. W. III. (2006). Quantifying low rates of summertime sublimation for buried glacier ice in Beacon Valley, Antarctica. *Antarctic Science*, **18**(3), 421–428.

Kreslavsky, M. A. and Head, J. W. (1999). Kilometer-scale slopes on Mars and their correlation with geologic units: initial results from Mars Orbiter Laser Altimeter (MOLA) data. *Journal of Geophysical Research*, **104**(E9), 21 911–21 924.

Kreslavsky, M. A. and Head, J. W. (2000). Kilometer-scale roughness of Mars: results from MOLA data analysis. *Journal of Geophysical Research*, **105**(E11), 26 695–26 712.

Kreslavsky, M. A. and Head, J. W. (2002a). Mars: nature and evolution of young latitude-dependent water-ice-rich mantle. *Geophysical Research Letters*, **29**(15), doi: 10.1029/2002GL015392.

Kreslavsky, M. A. and Head, J. W. (2002b). The fate of outflow channel effluents in the Northern Lowlands of Mars: the Vastitas Borealis Formation as a sublimation residue from frozen ponded bodies of water. *Journal of Geophysical Research*, **107**(E12), 5121, doi: 10.1029/2001JE001831.

Kreslavsky, M. A. and Head, J. W. (2003). North-south topographic slope asymmetry on Mars: evidence for insolation-related erosion at high obliquity. *Geophysical Research Letters*, **30**(15), doi: 10.1029/2003GL017795.

Kreslavsky, M. A. and Head, J. W. (2004). Periods of active permafrost layer formation in the recent geological history of Mars. *Lunar Planetary Science Conference*, **35**, Abstract 1201.

Kreslavsky, M. A. and Head, J. W. (2005). Mars at very low obliquity: atmospheric collapse and the fate of volatiles. *Geophysical Research Letters*, **32**, doi: 10.1029/2005GL022645.

Kreslavsky, M. A., Head, III, J. W., and Marchant, D. R. (2008). Periods of active permafrost layer formation during the geological history of Mars: implications for circum-polar and mid-latitude surface processes. *Planetary and Space Sciences*, **56**, 289–302, doi:10.1016/j.pss.2006.02.010.

Kress, A. M. and Head, J. W. (2008). Ring-mold craters in lineated valley fill and lobate debris aprons on Mars: evidence for subsurface glacial ice. *Geophysical Research Letters*, **35**, L23206, doi:10.1029/2008GL035501.

Kuzmin, R. O. and Zabalueva, E. V. (1998). On salt solutions of the martian cryolithosphere. *Solar System Research*, **32**, 187–197.

Lachenbruch, A. H. (1962). Mechanics of thermal contraction cracks and ice-wedge polygons in permafrost. *Geological Society of America Special Paper*, **70**, 1–69.

Lancaster, N. (2002). Flux of eolian sediment in the McMurdo Dry Valleys, Antarctica: a preliminary assessment. *Arctic Antarctic and Alpine Research*, **34**(3), 318–323.

Langbein, W. B. and Schumm, S. A. (1958). Yield of sediment in relation to mean annual precipitation. *AGU Transactions*, **39**, 1023–1036.

Larsen, K. W., Arvidson, R. E., Jolliff, B. L., and Clark, B. C. (2000). Correspondence and least-squares analysis of soil and rock compositions for the Viking Lander 1 and Pathfinder landing sites. *Journal of Geophysical Research*, **105**, 29 207–29 221.

Laskar, J., Levrard, B., and Mustard, J. F. (2002). Orbital forcing of the martian polar layered deposits. *Nature*, **419**, 375–377.

Laskar, J., Gastineau, M., Joutel, F., et al. (2004). Long term evolution and chaotic diffusion of the insolation quantities of Mars. *Icarus*, **170**, 343–364.

Levrard, B., Forget, F., Montmessin, F., and Laskar, J. (2004). Recent ice-rich deposits formed at high latitudes on Mars by sublimation of unstable equatorial ice during low obliquity. *Nature*, **431**, 1072–1075.

Levy, J. S., Marchant, D. R., and Head, III, J. W. (2006). Distribution and origin of patterned ground on Mullins Valley debris-covered glacier, Antarctica: the roles of ice flow and sublimation. *Antarctic Science*, **18**(3), 385–397.

Levy, J., Head, J. W., Marchant, D. R., Morgan, G. A., and Dickson, J. L. (2007). Gully surface and shallow subsurface structure in the South Fork of Wright Valley, Antarctic Dry Valleys: implications for gully activity on Mars. *Lunar Planetary Science Conference*, **38**, Abstract 1728.

Levy, J. S., Head, J. W., and Marchant, D. R. (2008a). The role of thermal contraction crack polygons in cold-desert fluvial systems. *Antarctic Science*, **20**, 565–579, doi: 10.1017/S0954102008001375.

Levy, J. S., Head, J. W., Marchant, D. R., and Kowalewski, D. E. (2008b). Identification of sublimation-type thermal contraction crack polygons at the proposed NASA landing site: implications for substrate properties and climate-driven morphological evolution. *Geophysical Research Letters*, **35**, L04202, doi: 10.1029/2007GL032813.

Lewis, A. R., Marchant, D. R., Ashworth, A. C., Hemming, S. R., and Machlus, M. (2007). Major middle Miocene global climate change: evidence from East Antarctica and the Transantarctic Mountains. *Geological Society of America Bulletin*, **119**(11/12), 1449–1461, doi: 10.1130B26134.1.

Lewis, A. R., Marchant, D. R., Ashworth, A. C., et al. (2008). Mid-Miocene cooling and the extinction of tundra in continental Antarctica. *Proceedings of the National Academy of Sciences*, **105**(31), 10 676–10 689, doi: 10.1073/ pnas.0802501105.

Lobitz, B., Wood, B. L., Averner, M. M., and McKay, C. P. (2001). Use of spacecraft data to derive regions on Mars where liquid water would be stable. *Proceedings of the National Academy of Sciences*, **98**(5), 2132–2137, doi: 10.1073/ pnas.031581098.

Lyons, W. B., Howard-Williams, C., and Hawes, I., eds. (1997). *Ecosystem Processes in Antarctic Ice-Free Landscapes*. Rotterdam, Netherlands: A.A. Balkema

Lyons, W. B., Welch, K. A., Carey, A. E., et al. (2005). Groundwater seeps in Taylor Valley, Antarctica: an example of a subsurface melt event. *Annals of Glaciology*, **40**, 200–206.

Mackay, J. R. (1977). The widths of ice wedges. *Geological Survey of Canada Professional Paper*, **77–1A**, 43–44.

Mahaney, W. C., Dohm, J. M., Baker, V. R., et al. (2001). Morphogenesis of Antarctic Paleosols: Martian analogue. *Icarus*, **154**(1), 113–130.

Malin, M. C. (1974). Salt weathering on Mars. *Journal of Geophysical Research*, **79**, 3888–3894.

Malin, M. C. (1984). Abrasion rate observations in Victoria Valley, Antarctica: 340-day experiment. *Antarctic Journal of the United States*, **19**(5), 14–16.

Malin, M. C. (1987). Abrasion in ice-free areas of southern Victoria Land, Antarctica. *Antarctic Journal of the United States*, **21**(5), 38–39.

Malin, M. C. and Edgett, K. S. (2000). Evidence for recent groundwater seepage and surface runoff on Mars. *Science*, **288**, 2330–2335.

Malin, M. C. and Edgett, K. S. (2001). Mars Global Surveyor Mars Orbiter Camera: interplanetary cruise through primary mission. *Journal of Geophysical Research*, **106**(E10), 23 429–23 570.

Malin, M. C., Edgett, K. S., Posiolova, L. V., McColley, S. M., and Noe Dobrea, E. Z. (2006). Present-day impact cratering rate and contemporary gully activity on Mars. *Science*, **314**, 1573–1577.

Mangold, N. (2005). High latitude patterned grounds on Mars: classification, distribution and climatic control. *Icarus*, **174**, 336–359.

Mangold, N., Allemand, P., Duval, P., Geraud, Y., and Thomas, P. (2002). Experimental and theoretical deformation of ice-rock mixtures: implications on rheology and ice content of Martian permafrost. *Planetary and Space Science*, **50**(4), 385–401.

Mangold, N., Maurice, S., Feldman, W. C., Costard, F., and Forget, F. (2004). Spatial relationships between patterned ground and ground ice detected by the Neutron Spectrometer on Mars. *Journal of Geophysical Research*, **109**(E8), doi: 10.1029/2004JE002235.

Marchant, D. R. and Denton, G. H. (1996). Miocene and Pliocene paleoclimate of the Dry Valleys region, southern Victoria Land: a geomorphological approach. *Marine Micropaleontology*, **27**, 253–271.

Marchant, D. R. and Head, J. W. (2003). Origin of sublimation polygons in the Antarctic western Dry Valleys region: implications for patterned ground development on Mars. *EOS* (fall Suppl.), **84**(46), Abstract C12C–06.

Marchant, D. R. and Head, III, J. W. (2004). Microclimates zones of the Dry Valleys of Antarctica: implications for landscape evolution and climate change on Mars. *Lunar Planetary Science Conference*, **35**, Abstract 1405.

Marchant, D. R. and Head, J. W. (2007). Antarctic Dry Valleys: microclimate zonation, variable geomorphic processes, and implications for assessing climate change on Mars. *Icarus*, **192**(1), 187–222, doi: 10.1016/j.icarus.2007.06.018.

Marchant, D. R., Denton, G. H., and Swisher, III, C. C. (1993a). Miocene-Pliocene-Pleistocene glacial history of Arena Valley, Quartermain Mountains, Antarctica. *Geografiska Annaler*, **75A**, 269–302.

Marchant, D. R., Denton, G. H., and Sugden, D. E. (1993b). Miocene glacial stratigraphy and landscape evolution of the western Asgard Range, Antarctica. *Geografiska Annaler*, **75A**, 303–330.

Marchant, D. R., Swisher, III, C. C., Lux, D. R., West, Jr., D. P., and Denton, G. H. (1993c). Pliocene paleoclimate and East Antarctic ice-sheet history from surficial ash deposits. *Science*, **260**, 667–670.

Marchant, D. R., Denton, G. H., Bockheim, J. G., Wilson, S. C., and Kerr, A. R. (1994). Quaternary ice-level changes of upper Taylor Glacier, Antarctica: implications for paleoclimate and ice-sheet dynamics. *Boreas*, **23**, 29–42.

Marchant, D. R., Denton, G. H., Swisher, III, C. C., and Potter, Jr., N. (1996). Late Cenozoic Antarctic paleoclimate reconstructed from volcanic ashes in the Dry Valleys region, south Victoria Land. *Geological Society of America Bulletin*, **108**(2), 181–194.

Marchant, D. R., Lewis, A., Phillips, W. C., et al. (2002). Formation of patterned-ground and sublimation till over Miocene glacier ice in Beacon Valley, Antarctica. *Geological Society of America Bulletin*, **114**, 718–730.

Margerison, H. R., Phillips, W. M., Stuart, F. M., and Sugden, D. E. (2005). Cosmogenic He-3 concentrations in ancient flood deposits from the Coombs Hills, northern Dry Valleys, East Antarctica: interpreting exposure ages and erosion rates. *Earth and Planetary Science Letters*, **230**, 163–175.

Marion, G. M. (1997). A theoretical evaluation of mineral stability in Don Juan Pond, Wright Valley, Victoria Land. *Antarctic Science*, **9**, 92–99.

Marshall, G. J. and Turner, J. (1997). Katabatic wind propagation over the western Ross Sea observed using ERS-1 scatterometer data. *Antarctic Science*, **9**, 221–226.

Matsuoka, N. (2001). Solifluction rates, processes, and landforms: a global review. *Earth-science Reviews*, **44**, 107–134.

McKay, C. P. (2009). Snow recurrence sets the depth of dry permafrost at high elevations in the McMurdo Dry Valleys of Antarctica. *Antarctic Science*, **21**, 89–94.

McKay, C. P., Mellon, M. T., and Friedmann, E. I. (1998). Soil temperatures and stability of ice-cemented ground in the McMurdo Dry Valleys, Antarctica. *Antarctic Science*, **10**, 31–38.

McKnight, D. M., Niyogi, D. K., Alger, A. S., et al. (1999). Dry valley streams in Antarctica: ecosystems waiting for water. *BioScience*, **49**, 985–995, doi: 10.2307/1313732.

Mellon, M. T. (1997). Small-scale polygonal features on Mars: seasonal thermal contraction cracks in permafrost. *Journal of Geophysical Research*, **102**, 25 617–25 628.

Mellon, M. T. (2003). Theory of ground ice on Mars and implications to the neutron leakage flux. *Lunar Planetary Science Conference*, **34**, Abstract 1916.

Mellon, M. T. and Jakosky, B. M. (1995). The distribution and behavior of martian ground ice during past and present epochs. *Journal of Geophysical Research*, **100**, 11 781–11 799.

MEPAG Next Decade Science Analysis Group (2008). Science priorities for Mars sample return. *Astrobiology*, **8**, doi: 10.1089/ast.2008.0759.

Milkovich, S. M. and Head, J. W. (2005). North polar cap of Mars: polar layered deposit characterization and identification of a fundamental climate signal. *Journal of Geophysical Research*, **110**(E1), doi: 10.1029/2004JE002349.

Milkovich, S. M., Head, J. W., Neukum, G., and the HRSC Co-Investigator Team (2008). Stratigraphic analysis of the northern polar layered deposits of Mars: implications for recent climate history. *Planetary and Space Science*, **56**, 266–288, doi: 10.1016/j.pss.2007.08.004.

Miller, G. H., Wolfe, A. P., Briner, J. P., Sauer, P. E., and Nesje, A. (2005). Holocene glaciation and climate evolution of Baffin Island, Arctic Canada. *Quaternary Science Reviews*, **24**, 1703–1721.

Milliken, R. E., Mustard, J. F., and Goldsby, D. L. (2003). Viscous flow features on the surface of Mars: observations from high-resolution Mars Orbiter Camera (MOC) images. *Journal of Geophysical Research*, **108**(E6), doi: 10.1029/2002JE002005.

Mitrofanov, I. and 11 colleagues (2002). Maps of subsurface hydrogen from the high energy neutron detector, Mars Odyssey. *Science*, **297**, 78–81.

Morgan, G., Head, J. W., Marchant, D. R., Dickson, J. L., and Levy, J. S. (2007). Gully formation on Mars: testing the snowpack hypothesis from analysis of analogs in the Antarctic Dry Valleys. *Lunar Planetary Science Conference*, **38**, Abstract 1656.

Murton, J. B., Worsley, P., and Gozdzik, J. (2000). Sand veins and wedges in cold aeolian environments. *Quaternary Science Reviews*, **19**, 899–922.

Mustard, J. F., Cooper, C. D., and Rifkin, M. K. (2001). Evidence for recent climate change on Mars from the identification of youthful near-surface ground ice. *Nature*, **412**, 411–414.

Mutch, T. A., Grenander, S. U., Jones, K. L., et al. (1976a). The surface of Mars: the view from the Viking 2 lander. *Science*, **194**, 1277–1283.

Mutch, T. A., Patterson, W. R., Binder, A. B., et al. (1976b). The surface of Mars: the view from the Viking 1 Lander. *Science*, **193**, 791–801.

Mutch, T. A., Arvidson, R. E., Guinness, E. A., Binder, A. B., and Morris, E. C. (1977). The geology of the Viking Lander 2 site. *Journal of Geophysical Research*, **82**, 4452–4467.

Nichols, R. L. (1968). Coastal geomorphology, McMurdo Sound, Antarctica. *Journal of Glaciology*, **7**, 449–478.

Nkem, J. N., Virginia, R. A., Barrett, J. E., Wall, D. H., and Li, G. (2006). Salt tolerance and survival thresholds for two species of Antarctic soil nematodes. *Polar Biology*, **28**(8), 643–651.

Northcott, M. L., Gooseff, M. N., Barrett, J. E., et al. (2009). Hydrologic characteristics of lake- and stream-side riparian wetted margins in the McMurdo Dry Valleys, Antarctica. *Hydrological Processes*, **23**, 1255–1267.

Nylen, T., Fountain, A. G., and Doran, P. (2004). Climatology of katabatic winds in the McMurdo Dry Valleys, southern Victoria Land, Antarctica. *Journal of Geophysical Research*, **109**, doi: 10.10292–2003JD003937.

Ostrach, L. R., Head, J. W., and Kress, A. M. (2008). Ring-mold craters (RMC) in lobate debris aprons (LDA) in the Deuteronilus Mensae region of Mars: evidence for shallow subsurface glacial ice in lobate debris aprons. *Lunar Planetary Science Conference*, **39**, Abstract 2422.

Paige, D. A. (2002). Near-surface liquid water on Mars. *Lunar Planetary Science Conference*, **33**, Abstract 2049.

Parsons, R., Head, J. W., and Marchant, D. R. (2005). Weathering pits in the Antarctic Dry Valleys: insolation-induced heating and melting and applications to Mars. *Lunar Planetary Science Conference*, **36**, Abstract 1138.

Patterson, W. S. B. (2001). *The Physics of Glaciers*. London: Butterworth-Heinemann.

Péwé, T. L. (1959). Sand-wedge polygons (tesselations) in the McMurdo Sound region, Antarctica: a progress report. *American Journal of Science*, **257**, 545–552.

Pierce, T. L. and Crown, D. A. (2003). Morphologic and topographic analyses of debris aprons in the eastern Hellas region, Mars. *Icarus*, **163**, 46–65.

Plaut, J. J., Safaeinili, A., and Holt, J. W., et al. (2009). Radar evidence for ice in lobate debris aprons in the mid-northern latitudes of Mars. *Geophysical Research Letters*, **36**, L02203, doi: 10.1029/2008GL036379.

Prentice, M. L., Kleman, J. K., and Stroeven, A. P. (1998). The composite glacial erosional landscape of the northern McMurdo Dry Valleys: implications for Antarctic Tertiary glacial history. In *Ecosystem Dynamics in a Polar Desert: The McMurdo Dry Valleys, Antarctica*, ed. J. C. Priscu. Antarctic Research Series 72. Washington, D.C.: American Geophysical Union, pp. 1–38.

Priscu, J. C., ed. (1998). *Ecosystem Dynamics in a Polar Desert: The McMurdo Dry Valleys, Antarctica.* AGU Antarctic Research Series 72. Washington, D.C.: American Geophysical Union, 370 pp.

Reiss, D. and Jaumann, R. (2003). Recent debris flows on Mars: seasonal observations of the Russell Crater dune field. *Geophysical Research Letters*, **30**(6), doi: 10.1029/2002GL016704.

Richardson, M. I. and Mischna, M. A. (2005). Long-term evolution of transient liquid water on Mars. *Journal of Geophysical Research*, **110**(E3), doi: 10.1029/2004JE002367.

Rieder, R. and 14 colleagues (2004). Chemistry of rocks and soils at Meridiani Planum from the alpha particle X-ray Spectrometer. *Science*, **306**, 1746–1749.

Rignot, E., Hallet, B., and Fountain, A. (2002). Rock glacier surface motion in Beacon Valley, Antarctica, from synthetic-aperture radar interferometry. *Geophysical Research Letters*, **29**(12), doi: 10.1029/2001GL013494.

Schaefer, J. M., Bauer, H., Denton, G. H., et al. (2000). The oldest ice on Earth in Beacon Valley, Antarctica: new evidence from surface exposure dating. *Earth and Planetary Science Letters*, **179**, 91–99.

Schumm, S. A. (1965). Quaternary paleohydrology. In *The Quaternary of the United States*, ed. H. E. Wright and D. G. Frey. Princeton, NJ: Princeton University Press, p. 922.

Schumm, S. A. and Lichty, R. W. (1965). Time, space and causality in geomorphology. *American Journal of Science*, **263**, 110–119.

Schwerdtfeger, W. (1984). *Weather and Climate of the Antarctic*. Developments in Atmospheric Science 15. Amsterdam, Netherlands: Elsevier, 262 pp.

Selby, M. J. (1971a). Slopes and their development in an ice-free, arid area of Antarctica. *Geografiska Annaler*, **53**(A), 235–245.

Selby, M. J. (1971b). Some solifluction surfaces and terraces in the ice-free valleys of Victoria Land, Antarctica. *New Zealand Journal of Geology and Geophysics*, **14**(3), 469–476.

Selby, M. J. (1974). Slope evolution in an Antarctic oasis. *New Zealand Geographer*, **30**, 18–34.

Selby, M. J. (1977). Transverse erosional marks on ventifacts from Antarctica. *New Zealand Journal of Geology and Geophysics*, **20**(5), 949–969.

Settle, M. (1979). Formation and deposition of volcanic sulfate aerosols on Mars. *Journal of Geophysical Research*, **84**, 8343–8354.

Shean, D. E., Head, J. W., and Marchant, D. R. (2005). Origin and evolution of a cold-based tropical mountain glacier on Mars: the Pavonis Mons fan-shaped deposit. *Journal of Geophysical Research*, **110**, E05001, doi: 10.1029/2004JE002360.

Shean, D. E., Head, III, J. W., Fastook, J. L., and Marchant, D. R. (2007a). Recent glaciation at high elevations on Arsia Mons, Mars: implications for the formation and evolution of large tropical mountain glaciers. *Journal of Geophysical Research*, **112**, E03004, doi: 10.1029/2006JE002761.

Shean, D. E., Head, J. W., and Marchant, D. R. (2007b). Shallow seismic surveys and ice thickness estimates of the Mullins Valley debris-covered glacier,

McMurdo Dry Valleys, Antarctica. *Antarctic Science*, **19**, 485–496, doi: 10.1017/S0954102007000624.

Siegel, B. Z., McMurty, G., Siegel, S. M., Chen, J., and Larock, P. (1979). Life in the calcium chloride environment of Don Juan Pond, Antarctica. *Nature*, **280**, 828–829.

Sletten, R. S., Hallet, B., and Fletcher, R. C. (2003). Resurfacing time of terrestrial surfaces by the formation and maturation of polygonally patterned ground. *Journal of Geophysical Research*, **108**, doi: 10.1029/2002JE001914.

Smith, D. E. and 23 colleagues (2001). Mars Orbiter Laser Altimeter: experiment summary after the first year of global mapping of Mars. *Journal of Geophysical Research*, **106**, 23 689–23 722.

Smith, P. H. and the Phoenix Science Team (2007). The Phoenix mission. *In Seventh International Conference on Mars*, Abstract 3180.

Spotila, J. A, Buscher, J. T, Meigs, A. J., and Reiners, P. W. (2004). Long-term glacial erosion of active mountain belts: example of the Chugach-St. Elias Range, Alaska. *Geology*, **32**, 501–504.

Squyres, S. W., Clifford, S. M., Kuzmin, R. O., Zimbelman, J. R., and Costard, F. M. (1992). Ice in the martian regolith. In *Mars*, ed. H. H. Kieffer, B. M. Jakosky, C. W. Snyder, and M. S. Matthews. Tucson, AZ: University of Arizona Press, pp. 523–554.

Squyres, S. W. and 49 colleagues (2004a). The Spirit rover's Athena science investigation at Gusev Crater, Mars. *Science*, **305**, 794–799.

Squyres, S. W. and 49 colleagues (2004b). The Opportunity rover's Athena science investigation at Meridiani Planum, Mars. *Science*, **306**, 1698–1703.

Squyres, S. W. and 17 colleagues (2004c). In-situ evidence for an ancient aqueous environment at Meridiani Planum, Mars. *Science*, **306**, 1709–1714.

Squyres, S. W. and 17 colleagues (2006). Two years at Meridiani Planum: results from the Opportunity rover. *Science*, **313**, 1403–1407.

Staiger, J. W., Marchant, D. R., Schaefer, J. M., et al. (2006). Plio-Pleistocene history of Ferrar Glacier, Antarctica: implications for climate and ice sheet stability. *Earth and Planetary Science Letters*, **243**, 489–503.

Sugden, D. E., Denton, G. H., and Marchant, D. R. (1995a). Landscape evolution of the Dry Valleys, Transantarctic Mountains: tectonic implications. *Journal of Geophysical Research*, **100**(B7), 9949–9967.

Sugden, D. E., Marchant, D. R., Potter, Jr., N., et al. (1995b). Miocene glacier ice in Beacon Valley, Antarctica. *Nature*, **376**, 412–416.

Summerfield, M. A., Stuart, F. M., Cockburn, H. A. P., et al. (1998). Long-term rates of denudation in the Dry Valleys region of the Transantarctic Mountains, southern Victoria Land based on in-situ produced cosmogenic Ne-21. *Geomorphology*, **27**, 113–129.

Summerfield, M. A., Sugden, D. E., Denton, G. H., et al. (1999). Cosmogenic isotope data support previous evidence of extremely low rates of denudation in the Dry Valleys region, southern Victoria Land, Antarctica. *Geological Society of London Special Publication*, **162**, 255–267.

Swanger, K. M. and Marchant, D. R. (2007). Sensitivity of ice-cemented Antarctic soils to greenhouse-induced thawing: are terrestrial archives at risk? *Earth and Planetary Science Letters*, **259**, 347–359.

Takamatsu, N., Kato, N., and Matsumoto, G. I. (1998). The origin of salts in water bodies of the McMurdo Dry Valleys. *Antarctic Science*, **10**, 439–448.

Tanaka, K. L. (2005). Geology and insolation-driven climatic history of Amazonian north polar materials on Mars. *Nature*, **437**, 991–994.

Tillman, J. E. (1988). Mars global atmospheric oscillations: annually synchronized, transient normal mode oscillations and the triggering of global dust storms. *Journal of Geophysical Research*, **93**, 9433–9451.

Torii, T., Nakaya, S., Matsubaya, O., et al. (1989). Chemical characteristics of pond waters in the Labyrinth of southern Victoria Land, Antarctica. *Hydrobiologia*, **172**, 255–264.

Tosca, N. J., McLennan, S. M., Clark, B. C., et al. (2005). Geochemical modeling of evaporation processes on Mars: insight from the sedimentary record at Meridiani Planum. *Earth and Planetary Science Letters*, **240**, 122–148.

Toulmin, P., Baird, A. K., Clark, B. C., et al. (1977). Geochemical and mineralogical interpretation of the Viking inorganic chemical results. *Journal of Geophysical Research*, **82**, 4625–4634.

van der Wateren, D. and Hindmarsh, R. (1995). Stabilists strike again. *Nature*, **376**, 389–391.

Vaniman, D. T. and Chipera, S. J. (2006). Transformation of Mg- and Ca-sulfate hydrates in Mars regolith. *American Mineralogist*, **91**, 1628–1642.

Wall, S. D. (1981). Analysis of condensates formed at the Viking 2 lander site: the first winter. *Icarus*, **47**, 173–183.

Wang, A. and 11 colleagues (2007). Sulfate-rich soils exposed by Spirit rover at multiple locations in Guseve Crater on Mars. In *Seventh International Conference on Mars*, Abstract 3348.

Wentworth, S. K. Gibson, E. K. Velbel, M. A., and McKay, D. S. (2005). Antarctic Dry Valleys and indigenous weathering in Mars meteorites: implications for water and life on Mars. *Icarus*, **174**, 383–395.

Whalley, W. B. and Palmer, C. F. (1998). A glacial interpretation for the origin and formation of the Marinet Rock Glacier, Alpes Maritimes, France. *Geografiska Annaler*, **80**, 221–236.

Wilson, L. (1969). Les relations entre les processus geomorphologique et le climat moderne comme méthode de paléoclimatologie. *Revue De Géographie Physique et de Geologie Dynamique*, **11**, 309–314.

Yershov, E. D. (1998). *General Geocryology*. Studies in Polar Research. Cambridge, UK: Cambridge University Press, 580 pp.

Zent, A. P. and Fanale, F. P. (1986). Possible Mars brines: equilibrium and kinetic considerations. *Journal of Geophysical Research*, **91**, 439–445.

Zent, A. P. and Fanale, F. P. (1990). Possible martian brines: radar observations and models. *Journal of Geophysical Research*, **95**, 14 531–14 542.

Zent, A. P., Fanale, F. P., Salvail, J. R., and Postawko, S. E. (1986). Distribution and state of H_2O in the high-latitude shallow subsurface of Mars. *Icarus*, **67**, 19–36.

Zurek, R. (1992). Comparative aspects of the climate of Mars: an introduction to the current atmosphere. In *Mars*, ed. H. H. Kieffer, B. M. Jakosky, C. W. Snyder, and M. S. Matthews. Tucson, AZ: University of Arizona Press, pp. 799–817.

Zurek, R. W., Barnes, J. R., Haberle, R. M., et al. (1992). Dynamics of the atmosphere of Mars. In *Mars*, ed. H. H. Kieffer, B. M. Jakosky, C. W. Snyder, and M. S. Matthews. Tucson, AZ: University of Arizona Press, pp. 835–933.

3

The legacy of aqueous environments on soils of the McMurdo Dry Valleys: contexts for future exploration of martian soils

J. E. BARRETT, MICHAEL A. POAGE,
MICHAEL N. GOOSEFF, AND CRISTINA TAKACS-VESBACH

Introduction

The McMurdo Dry Valleys are the largest and one of the most southernly exposed terrestrial antarctic environments (Ugolini and Bockheim, 2008) and have been a prominent analog environment for speculations about surface processes (Mahaney et al., 2001; Dickenson and Rosen, 2003; Marchant and Head, 2007) and potential biology (McKay, 1997; Wynn-Williams and Edwards, 2000) on Mars. The extremes in cold and aridity, the paucity of visually conspicuous life forms, and the undisturbed conditions of the McMurdo Dry Valleys make this region an obvious candidate for such comparisons. Recent discoveries of evidence demonstrating past and perhaps present availability of liquid water on the martian surface detected by the Mars Global Surveyor (Malin and Edgett, 2000; Baker, 2001) and the Spirit and Opportunity rovers (Squires et al., 2004a; Haskin et al., 2005) have extended the foundation of these comparisons beyond similarities in climate to surface geomorphology, geochemistry, and mineralogy (Chevrier et al., 2006; Marchant and Head, 2007; Amundson et al., 2008).

Water is the primary limitation to geochemical weathering and biological activity in the McMurdo Dry Valleys of Antarctica and other cold desert ecosystems where availability and movement of liquid water is limited by low temperatures (Kennedy, 1993; Convey et al., 2003; Barrett et al., 2008). This limitation of liquid water results in slow weathering and highly constrained biological activity contributing to relatively stable geochemical conditions in surface environments. Thus, in the McMurdo Dry Valleys, the legacy of paleo-aquatic environments is preserved in contemporary patterns

Life in Antarctic Deserts and Other Cold Dry Environments: Astrobiological Analogs, ed. Peter T. Doran, W. Berry Lyons and Diane M. McKnight. Published by Cambridge University Press.

of soil geochemistry. Examination of contemporary spatial variation in surface geochemistry points to mechanisms responsible for these prominent patterns in dry valley landscapes. For example, stable isotope ratios of entrained organic matter and surface geochemistry are evidence of ancient lake shore environments (Burkins et al., 2000; Lyons et al., 2000; Foley et al., 2006).

In this chapter we review research describing these patterns and discuss possible mechanisms behind their formation. We briefly assess the similarities and dissimilarities in the McMurdo Dry Valleys and martian soil environments and describe the contemporary and paleo-influences of surface water on soil geochemistry and distribution and activity of biota. The objectives of this chapter are to illustrate how spatial variation in surface soil chemistry of the dry valleys can be used to interpret landscape history and to provide context for the interpretation of ancient water bodies on Mars and their role in shaping the current extant landforms and surface geochemistry of the Red Planet.

Comparisons of the antarctic and martian surface environments

The McMurdo Dry Valleys (Fig. 3.1) are a series of generally east to west oriented valleys located between the Ross Sea and the polar plateau in Southern Victoria Land, Antarctica. The valleys are dry (i.e., not covered by ice sheets) because the Trans-Antarctic Mountains restrict the flow of the East Antarctic Ice Sheet to a few terminal lobes that penetrate to the valleys; numerous alpine glaciers are also present in the mountains, some flowing to valley floors (Fountain et al., 1999). The seasonal melt of these glaciers provides the primary source of liquid water in this environment and contributes to ephemeral streams and ice-covered lakes in the valleys (Lewis et al., 1999; Bomblies et al., 2001). This scarcity of liquid water and resulting aridity of soils provides a basis of comparison between antarctic dry valley environments and some locations on Mars where cold temperatures result in similar constraints, despite the presence of frozen water in surface deposits in polar regions and in ground ice in circumpolar latitudes (Titus et al., 2003; Mellon et al., 2004).

Climatic and atmospheric conditions

In the McMurdo Dry Valleys mean daily air temperatures average −17 °C, with winter minimum temperatures often below −40 °C, summer maximums up to 10 °C (Doran et al., 2002), and surface soil temperatures as high as

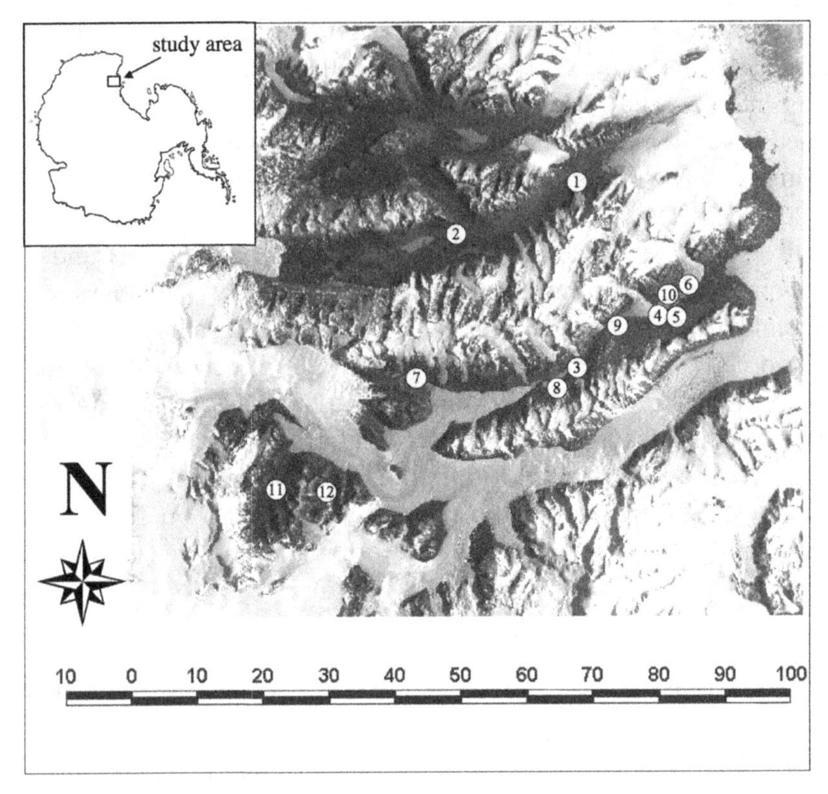

Fig. 3.1. Map of study sites in the McMurdo Dry Valleys, Antarctica: 1, Lower Onyx River; 2, Upper Onyx River; 3, Priscu Stream; 4, Green Creek; 5, Upper Delta Stream and Lower Delta Stream; 6, Lost Seal Stream; 7, Lake Joyce; 8, Lake Bonney; 9, Lake Hoare; 10, Lake Fryxell; and 11, Beacon, and 12, Arena Valleys in the Quartermain Mountains.

15 °C (Barrett et al., 2008). Regionally, average annual temperatures range from −18 °C in coastal zones to −24 °C in high elevation soils of the dry valleys (Mahaney et al., 2001; Aislabie et al., 2006). Mean diurnal surface temperatures on Mars vary from lows of −123 °C at the poles to −33 °C at the warmest locations (north-facing slopes) during midsummer in the southern hemisphere (Leovy, 2001). Daily maxima can reach as high as 27 °C during summer at mid-southern latitudes, but only for the surface soils where thermal inertia is low; above freezing temperatures are likely limited to the top centimeter (Carr, 2006). Thus, martian soil environments exhibit climate conditions most comparable to the most climatically extreme soils in alpine regions of the McMurdo Dry Valleys, such as those found on Mount Fleming and in the Quartermain Mountains, for example Beacon Valley (Mahaney et al., 2001; Marchant and Head, 2007).

A key difference between the terrestrial and martian surface environments is the total pressure of the atmosphere. The dry valleys experience typical atmospheric pressures for Earth (~1 atm) that are more than two orders of magnitude greater than the typical surface pressures on Mars. Thus, physical conditions permit the melting of ice to form liquid water in the McMurdo Dry Valleys during the brief austral summer, while on the contemporary martian surface this is not possible because the combination of low atmospheric pressures (~0.006 atm) and temperatures do not permit the stable presence of liquid water. Under these conditions, ice heated in excess of the melting point would boil away rapidly. A key conclusion from this is that the pressure on Mars must have been higher in the past to stabilize liquid water, though temperatures may not necessarily have been higher. As in Antarctica, diel peaks of temperatures above zero with seasonal means lower than zero can generate liquid water if the pressure is adequate to stabilize it.

Low levels of free oxygen (O_2) in the martian atmosphere (Carr, 2006) contribute to significant differences between the martian and antarctic surface environments, contributing to very distinct conditions in which minerals, weather, and biological processes could potentially operate. Extensive oxidation of surface minerals is evidenced by the iconic red color of iron minerals on the martian surface. Much of the martian surface is deeply covered by a fine dust composed of nanophase Fe^{3+}-bearing oxide minerals, especially hematite and goethite (Bell et al., 2000; Morris et al., 2004; Yen et al., 2005; Lichtenberg et al., 2007) which bear evidence of *in situ* weathering in aqueous environments (Squyres et al., 2004b), comparable to the oxidation fronts in soil profiles of dry valley soils occurring on dolerite-rich tills in the western dry valley region, most notably in the Quartermain Mountains (Campbell and Claridge, 1987; Bockheim, 2007).

The absence of atmospheric O_2 does not necessarily preclude the presence of autotrophs (photo- or chemo-) or heterotrophs, provided an availability of other electron acceptors (e.g., NO_3^-, SO_4^{2-}, Fe^{3+}) in martian soils (e.g., Westall, 2005; Weber et al., 2006). Indeed, Earth's biota existed for millions of years in the absence of free oxygen and many biochemical pathways crucial to microbial metabolism and global biogeochemical cycling of elements remain strictly anaerobic under current conditions (Lazcano and Miller, 1999; Nisbet and Sleep, 2001; Kasting and Siefert, 2002).

Absence of O_2 prohibits the development of an ozone layer in the martian atmosphere permitting full penetration of ultraviolet (UV) radiation (e.g., Kasting, 1993). In addition, the absence of a martian magnetosphere results in constant bombardment by cosmic rays and the solar wind (Lundin et al., 2004). This photochemically reactive environment on the martian surface may contribute to the presence of strong oxidants and supposed "super oxides" in the surface soils (Klein, 1978; Benner et al., 2000; Quinn et al., 2005). These compounds, including hydrogen peroxide and perchlorates, are toxic to most Earth biota, suggesting that martian soil biota would likely be confined to protected subsurface environments. Alternatively, some researchers have speculated that these oxides could play a similar role as molecular oxygen in metabolic processes (Navarro-Gonzalez et al., 2003).

Such conditions are not without historic parallels on Earth; the origin and radiation of diverse biochemical pathways are thought to have preceded the development of the ozone layer following the evolution of oxygen-producing photosynthesis by cyanobacteria-like organisms *c.* 2 billion years ago (Margulis and Sagan, 1986; Lazcano and Miller, 1994). Currently, the antarctic environment is subject to strong penetration of UVB radiation (Madronich et al., 1998). Resident biota exhibit a number of adaptations including pigmentation (Wynn-Williams and Edwards, 2000) and avoidance strategies by living at soil depths where UV radiation cannot penetrate; some organisms such as nematodes and fungi are found in higher densities at 5–10 cm depth relative to the soil surface (Powers et al., 1995; Tosi et al., 2005). Any organisms that evolved in paleo-martian environments when water was likely present (e.g., Noachian Epoch) may have shared analogous adaptations to high radiation exposure (Westall, 2005).

Hydrology

Typical contemporary atmospheric pressures (average of 5.6 mbar) and temperatures are too low to permit the existence of stable pools of liquid water on the martian surface (Haberle et al., 2001). Satellite and geomorphologic evidence does suggest transient occurrences of contemporary liquid water (Malin and Edgett, 2000; Baker 2001; Heldmann et al., 2007), perhaps associated with geothermal heating and permafrost failures in confined aquifers (Mellon and Phillips, 2001), but surface conditions prohibit the availability of liquid water for biologically relevant durations (Carr, 2006). However, the martian landscape does preserve extensive evidence of historic occurrences of liquid water including geomorphological features and surface mineralogy suggestive of outwash gullies, rivers, and evaporative zones associated with lakes and seas (Klingelhöfer et al., 2004; Rieder et al., 2004; Squyres et al., 2004a, 2004b; Haskin et al., 2005; McLennan et al., 2005; Ming et al., 2006; Zuber, 2007).

Although liquid water is unstable, water ice is present even in the contemporary martian environment as polar ice caps (Titus et al., 2003;

Hvidberg, 2005) and ground ice (Mellon et al., 2004; Kuzmin, 2005; Mitrofanov, 2005; Sizemore and Mellon, 2006). The martian soil is thought to be underlain by continuous permafrost at high latitudes, and at latitudes as low as 60° in the southern hemisphere (Mellon et al., 2004; Mitrofanov, 2005). Even in the relatively recent geological past on Mars (<500 Ka), when orbital and axial variation generated higher than current insolation, and where geography (latitude, slope, aspect, etc.) was favorable to large sensible energy fluxes, climate conditions may have permitted the existence of liquid water and allowed active chemical weathering and geomorphological development (Mellon et al., 2004). Moreover, the presence of salts, weathering products, and other solutes may have facilitated the presence of liquid water at even lower temperatures because of the eutectic properties of brines (Mellon and Phillips, 2001; Kuzmin, 2005).

In the McMurdo Dry Valleys, water is also scarce across much of the terrestrial landscape. Together with organic matter, liquid water strongly limits the spatial distribution and activity of soil organisms (Treonis et al., 1999, 2000; Barrett et al., 2008; Zeglin et al., 2009; Takacs-Vesbach et al., this volume, Chapter 8). Liquid water is unavailable for most of the year, even in nearshore environments because temperatures exceed 0 °C for only brief periods (10–75 days) during the austral summer (Doran et al., 2002; Barrett et al., 2008), though liquid water may be available in brine films at even lower temperatures (Mahaney et al., 2001). In most soils more than ~10 m from streams or lakes, precipitation and ground ice are the only sources of water (Gooseff et al., 2003, 2007; Lyons et al., 2005). Annual precipitation inferred from snow pits is less than 10 cm water equivalent (Witherow et al., 2006), with sublimation losses dominating ablation rates, further limiting the availability of liquid water (Chinn, 1993; Ng et al., 2005).

While movement of soil water is thought to occur primarily in the vapor phase in the coldest regions of the McMurdo Dry Valleys most comparable to martian environments (Mahaney et al., 2001; Dickinson and Rosen, 2003; Hagedorn et al., 2007), there is limited migration of snowmelt in surface soils, particularly during midsummer (Fountain et al., 1999; Campbell, 2003; Gooseff et al., 2003). Thus, little liquid water is available to drive geochemical weathering and biological activity in low-elevation soils for days to weeks during the austral summer, though hydrological dynamics exhibit high degrees of temporal variability over seasonal and interannual scales (Treonis et al., 2000; Gooseff et al., 2003; Barrett et al., 2008). Few examples of *in situ* measurements are currently available, although estimates suggest that under conditions of elevated liquid water availability, weathering rates can exceed those in temperate environments (Lyons et al., 1997), and increased biological activity, in the form of soil respiration and growing populations of microfauna, have been observed under such conditions (Parsons et al., 2003; Barrett et al., 2008). During particularly warm events groundwater seeps are widespread in the McMurdo Dry Valleys (Lyons et al., 2005; Harris et al., 2007; Barrett et al., 2008), suggesting subsurface deposits of massive ice are important sources of liquid water.

Soils and geomorphology

Arid soils underlain by dry permafrost (<1% soil water by weight) are the most extensive landform of the dry valleys occupying 61% of glacier ice-free surfaces below 1000 m elevation (Burkins et al., 2001; Bockheim et al., 2007). On Mars, gamma ray and neutron mapping by the Mars Odyssey orbiter suggests the majority of low- and mid-latitude soils share similar ranges of water content (Mitrofanov, 2005). Soils of the dry valleys have formed from a variety of parent materials, including sandstone, granite, diorite, dolerite, and basalt, originating primarily in the glacial till derived from several cycles of glaciation (Hall et al., 2000). The surface of Mars appears to be composed primarily of basalt based upon orbital observations (Bandfield et al., 2000; Christensen et al., 2001), surface explorations (McSween et al., 2004; Squyres et al., 2004a), and examination of martian meteorites discovered on Earth (Morris et al., 2000; Wentworth et al., 2005; Chevrier and Mathe, 2007).

Contributions of basalt composition rocks to dry valley soils is of particular interest here since soils forming on basaltic parent material may share similar mineralogy with martian environments (Chevrier et al., 2006). The Ferrar Dolerite, present throughout the dry valleys (Campbell and Claridge, 1987, Bedard et al., 2007), and volcanic clasts, present in the eastern regions of the dry valleys, are examples of these mafic parent materials, which can comprise a high proportion of Taylor Valley soils by mass (Hall et al., 2000). In the McMurdo Dry Valley region influences of basaltic parent material include higher total soil phosphorus content (Bate et al., 2008) and higher rates of net primary productivity in neighboring aquatic ecosystems (Barrett et al., 2007).

Dry valley soils are broadly classified as gelisols based upon their solute content and permafrost dynamics (Bockheim, 1997). Taylor Valley soils occurring in coastal areas and near contemporary lake edges are characterized by limited horizon development and shallow profiles with dry permafrost or ice-cemented layers occurring generally within 30 cm of the surface (Campbell and Claridge, 1987; Bockheim, 1997). Pervasive patterned ground formations (ice-wedge and sand-wedge polygons) in the dry valleys indicate active permafrost processes (Bockheim, 2002; Sletten et al., 2003; Marchant and Head, 2007; Fig 3.2a). Ground ice can be broadly classified as pore ice cemented sediments, modern or remnant buried glacial ice, and epigenetic ice in contraction cracks of ice wedges patterned ground. Pore ice is the most common form of subsurface ice in the valleys (Bockheim, 2002). Buried massive ice, some of which is glacial in origin, is also commonly found in the valleys (Bockheim and Hall, 2002) from coastal margins to regions farther inland (Sugden et al., 1995).

Soils of the McMurdo Dry Valleys are among the coldest and driest terrestrial environments on Earth and exhibit orders of magnitude variation in soil salinity and soluble ion concentrations (Bockheim, 1997; Poage et al., 2008). The geochemistry of dry valley soils is fundamentally linked to each valley's landscape history, integrating the effects of climate, soil hydrology, glacial activity, lake inundation, atmospheric deposition, and surface age (Barrett et al., 2006, 2007; Foley et al., 2006). For example, longer surface exposure ages result in higher soil salinities due to accumulation of atmospheric inputs (Bockheim, 1997; Michalski et al., 2005), while the geochemistry of low-elevation environments is influenced by local hydrology and evaporative wicking in lake-margin sediments (Barrett et al., 2007; Gooseff et al., 2007; Northcott et al., 2009). Soils are typically neutral to alkaline in pH, and often have high concentrations of soluble salts due to low precipitation and negligible leaching losses (Bockheim, 1997; Barrett et al., 2006). The extreme aridity of both Mars and the McMurdo Dry Valleys has focused attention on the importance of salts and their role in phase states and transitions among solid, vapor, and liquid phases of water.

Organic matter content of these soils is low, with a large proportion attributed to the deposition of ancient and contemporary lake sediments (Burkins et al., 2000; Barrett et al., 2006). Martian soils may be similarly marked by such legacy influences of paleo-aqueous environments (e.g., Ming et al., 2006). Contemporary biological communities in the McMurdo Dry Valleys are characterized by low biomass, and are composed of cyanobacteria, algae, bacteria, fungi, protozoans, and a very limited diversity of metazoan invertebrates (Adams et al., 2006; Aislabie et al., 2006; Connell et al., 2006; Wood et al., 2008), though geochemical and fossil evidence preserve a history of previously rich vascular flora (Cuneo et al., 2003).

Martian soils formed under climatic conditions of extreme cold and aridity similar to the McMurdo Dry Valleys, with surface features indicating similar influences of aeolian (Howard, 2000; Lancaster, 2002), aqueous (Ming et al., 2006), and periglacial processes on soil development (Marchant and Head, 2007). Evidence from the Mars Exploration rovers supports the existence of ubiquitous aeolian sediment consisting of low density, fine grained "bright dust" of uniform chemical composition (Gellert et al., 2004; Morris et al., 2004;

(a)

(b)

Fig. 3.2. (cont.)

Haskin et al., 2005), together with millimeter-sized grains and larger lithic fragments in a desert pavement surface (Yen et al., 2005). Moreover, influence of salts and brine films on the phase transitions of water in martian soils may be analogous to geochemical processes in the McMurdo Dry Valleys (Mahaney et al., 2001). For example, increased levels of sulfur, chlorine, and bromine are consistent with mobilization of soluble salts and deposition of evaporates by thins films of liquid water (Rieder et al., 2004; Squyres et al., 2004a, 2004b; Haskin et al., 2005; McLennan et al., 2005; Yen et al., 2005; Ming et al., 2006; Chevrier and Mathe, 2007; Amundson et al., 2008). Recent data provided by the Mars Phoenix Lander indicates that soils share a similar range of pH values and comparable ion composition to dry valley soils (S. Kounaves, unpublished data).

More is probably known about martian geomorphology than other physical or chemical properties of the surface environment because of the availability of high quality imagery of martian landforms collected by a succession of orbital spacecraft over the past 37 years. Geomorphologic features also provide some of the most compelling evidence for ancient and perhaps contemporary influences of liquid water on surface processes of Mars (Malin and Edgett, 2000). A variety of patterned ground formations (Fig. 3.2b) associated with cryogenic processes similar to those in the McMurdo Dry Valleys have been identified in mid to high latitudes of both hemispheres on Mars, testifying to the presence of ground ice near the soil surface (Seibert and Kargel, 2001; Mangold, 2005; Sizemore and Mellon, 2006; Marchant and Head, 2007). These features form from thermal contraction and expansion following seasonal temperature variation. Small (15–40 m diameter) contraction-crack polygons and hummocks (similar to those found in the McMurdo Dry Valleys; Marchant et al., 2002; Barrett et al., 2004) are the most abundant patterned ground type (Marchant and Head, 2007). Many of these polygonal features are thought to be of relatively recent origin, <10 Ma, because they occur on deposits devoid of impact craters (Mangold, 2005).

In addition to the influence of the cryosphere on martian surface processes, multiple lines of evidence from analysis of surface features support the current or former existence of martian surface water as lakes, seas, and gully-forming outwash streams (Malin and Edgett, 2000; Baker, 2001; Zuber, 2007). Evidence for paleolakes on the martian surface abounds in features

Caption for Fig. 3.2. (cont.)
Patterned ground formations in: (a) Taylor Valley, Antarctica, latitude 77.35° N, longitude 163.90° E (photo credit: J. E. Barrett) and (b) in "Green Valley" on Mars (latitude 68.22° N, lon,gitude 234.25° E (image courtesy of the NASA Phoenix Mars Lander Project, www.nasa.gov/mission_pages/phoenix/images/).

resembling lake deposits present in many of the large impact craters (Cabrol and Grin, 1999; Baker, 2001). Lake terraces and perched deltas indicate that the lakes may have persisted for 10^3-10^4 years (Ori et al., 2000), probably as perennially ice-covered lakes (Squyres, 1989; Baker, 2001). Moreover, large areas of lowlands in the northern hemisphere may have been inundated by an ancient martian ocean ~2 Ga ago based upon analysis of possible paleoshorelines identified from satellite images (Baker et al., 1991; Head et al., 1999; Carr and Head, 2003; Perron et al., 2007). In contrast, debris-flow gullies superimposed on uncratered landforms consisting of aeolian deposits and contraction-crack polygons are evidence of relatively recent surface water within the past several million years (Malin and Edgett, 2000). Thus, aqueous environments are likely to have been present at multiple times over the history of Mars and contributed to geomorphology and soil development and geochemistry.

Geochemical legacies of aqueous environments in the McMurdo Dry Valleys

In the McMurdo Dry Valleys, major landscape features such as glaciers, meltwater streams, lakes, and soils are hydrologically linked over multiple timescales with the dynamics of local and regional hydrology dominantly driven by climatic variability (e.g., Lyons et al., 2000; Doran et al., 2008). Soil development and surface geochemistry is fundamentally linked to the complex hydrology of the dry valleys; thus, patterns in surface soil geochemistry can be used in the interpretation landscape history and provide insights into hydrologic and geomorphologic processes. If past surface conditions on Mars included similar thermal conditions but higher atmospheric pressures, then the behavior of water and its influence on landscape processes and geochemistry may have exhibited similar influences as observed in the McMurdo Dry Valleys. Thus understanding the mechanisms underlying landscape development and surface geochemistry in the dry valleys may help develop hypotheses for understanding surface processes on Mars.

In the McMurdo Dry Valleys surface soil geochemistry is influenced by many components of landscape history including till provenance and exposure age (Bockheim, 1997; Barrett et al., 2007; Bate et al., 2008), as well as inundation and recession of lake waters (Burkins et al., 2000; Lyons et al., 2000; Foley et al., 2006). With respect to the influence of landscape history and geomorphology on contemporary surface features and geochemistry, Taylor Valley is the best understood of the McMurdo Dry Valleys. The Dry Valley Drilling Project (McGinnis, 1981), the extensive description of glacial sediments by G. Denton and colleagues (e.g., Denton et al., 1989; Denton and Hall, 2000; Hendy et al., 2000), and the work of the McMurdo Dry Valleys Long Term Ecological Research (MCM-LTER) Program in describing the influence of contemporary and paleoclimate on ecological and hydrological processes (Lyons et al., 2000, 2005; Foreman et al., 2004; Barrett et al., 2007) have all contributed significantly to this understanding.

Low elevation soils in Taylor Valley have formed on two major tills: the Taylor II drift and the Ross Sea drift. The Taylor II drift was deposited as the Taylor Glacier advanced eastward into Taylor Valley between 70 000 and 120 000 years before present (BP) (Higgins et al., 2000). As the older of the two major tills, soils and lake ecosystems forming on Taylor II have higher solute content than soils developing on the younger Ross Sea drift, which was deposited between 8000 and 24 000 years BP in the area that now comprises the Lake Fryxell and eastern portion of the Lake Hoare basins (Hall and Denton, 2000; Hall et al., 2000). Ice from the Ross Sea advanced westward into Taylor Valley, damming Glacial Lake Washburn and initiating a lake-ice conveyor that produced a complex array of mounds and ridges of lacustrine sediments on the valley floor (Hall et al., 2000). High stands of Glacial Lake Washburn inundated Taylor Valley perhaps as high as 300 m a.s.l. 40 000 years BP (Denton et al., 1989) resulting in entrainment of lake organic matter in low elevation tills (Burkins et al., 2000). By about 7000 years BP the West Antarctic Ice Sheet receded and Lake Washburn drained, leaving smaller lakes impounded in valley basins. Subsequent drawdown during a climatic drying period reduced the lakes greatly by ~1000 years BP (Lyons et al., 2000; Poreda et al., 2004). Since that time Taylor Valley lakes have refilled to their present state and stratification (Lyons et al., 2000), though in recent years lake levels have exhibited wide fluctuations (Barrett et al., 2008; Doran et al., 2008).

The history of these lake inundations and recessions are recorded in lacustrine sediment mounds (Hall et al., 2000; Higgins et al., 2000), entrained lake sediments in perched deltas (Denton et al., 1989; Hall and Denton, 2000), the stable isotopic composition of soil organic matter (Burkins et al., 2000), and the distribution of carbonate minerals in the soils (Foley et al., 2006). These geochemical patterns overlie existing chemical variation in underlying tills, and are in turn influenced by contemporary hydrological and, in some cases, biological processes (Barrett et al., 2007; Gooseff et al., 2007; Bate et al., 2008; Northcott et al., 2009; Zeglin et al., 2009). In this section we present data illustrating the contemporary influences of stream and lake waters on adjacent sediments and soils and interpret broader scale patterns of soil geochemistry in light of these insights.

Contemporary influences of water on surface geochemistry

In order to assess the influence of water on soil geochemistry, we present data collected from sampling transects established along a gradient of soil moisture, adjacent to streams and lakes in the McMurdo Dry Valleys. Soils (0–10 cm depth) were collected from these transects in the austral summer of December 2004 during a period of active streamflow and lake-moat meltout when open water conditions were present (Doran et al., 2008). These transects (Fig. 3.3) included positions 0.2 m from open water in stream and lake margins (first position), together with transect positions on the nearshore and uphill sides of the visual edge of the wetted front (third and fourth positions), and a transect position midway between open water and the visual edge of the wetted front (second positions). Salt deposits are typically visible within a few centimeters of the wetted fronts adjacent to stream and lake environments. We use Analysis of Variance (ANOVA) to partition variability in soil properties across the margin transects and between lake and stream environments (Tables 3.1 and 3.2). Fig. 3.4 illustrates the influence of hydrological margins on soil biology and biochemical properties for fixed-distance transects extending from open water to dry soils near three lake and stream environments (e.g., Treonis et al., 1999; Barrett et al., 2002; Ayres et al., 2007).

The most conspicuous influence of aquatic environments on neighboring sediment and soils is the wicking of water by capillary action into drier environments (Gooseff et al., 2007). Water content varied unimodally across hydrological margins adjacent to both streams and lakes in the McMurdo Dry Valleys (Table 3.1), with water content of sediment and soils decreasing as a function of distance from liquid water (Fig. 3.3a). Lake margins are wetter on average than stream environments, though distance from liquid water accounted for most of the explainable variance across all the environments studied (Table 3.1).

These results are consistent with the visual observations of the wetted front which extended up to the third transect positions, 4–5 m on average, from open water (Fig. 3.3a). Water content in samples collected from the fourth transect position in stream and lake margins was typically below 5% soil water by weight. Treonis et al. (2000) suggested that the threshold for biological activity of antarctic metazoan invertebrates is 4–5% soil moisture content; thus these hydrological margins may delineate zones of potential biological activity by multicellular organisms in the McMurdo Dry Valleys.

Nearshore environments (within 20 cm of liquid water) had significantly different pH in lake (9.0) versus stream (8.3) sediments (Table 3.1), but were

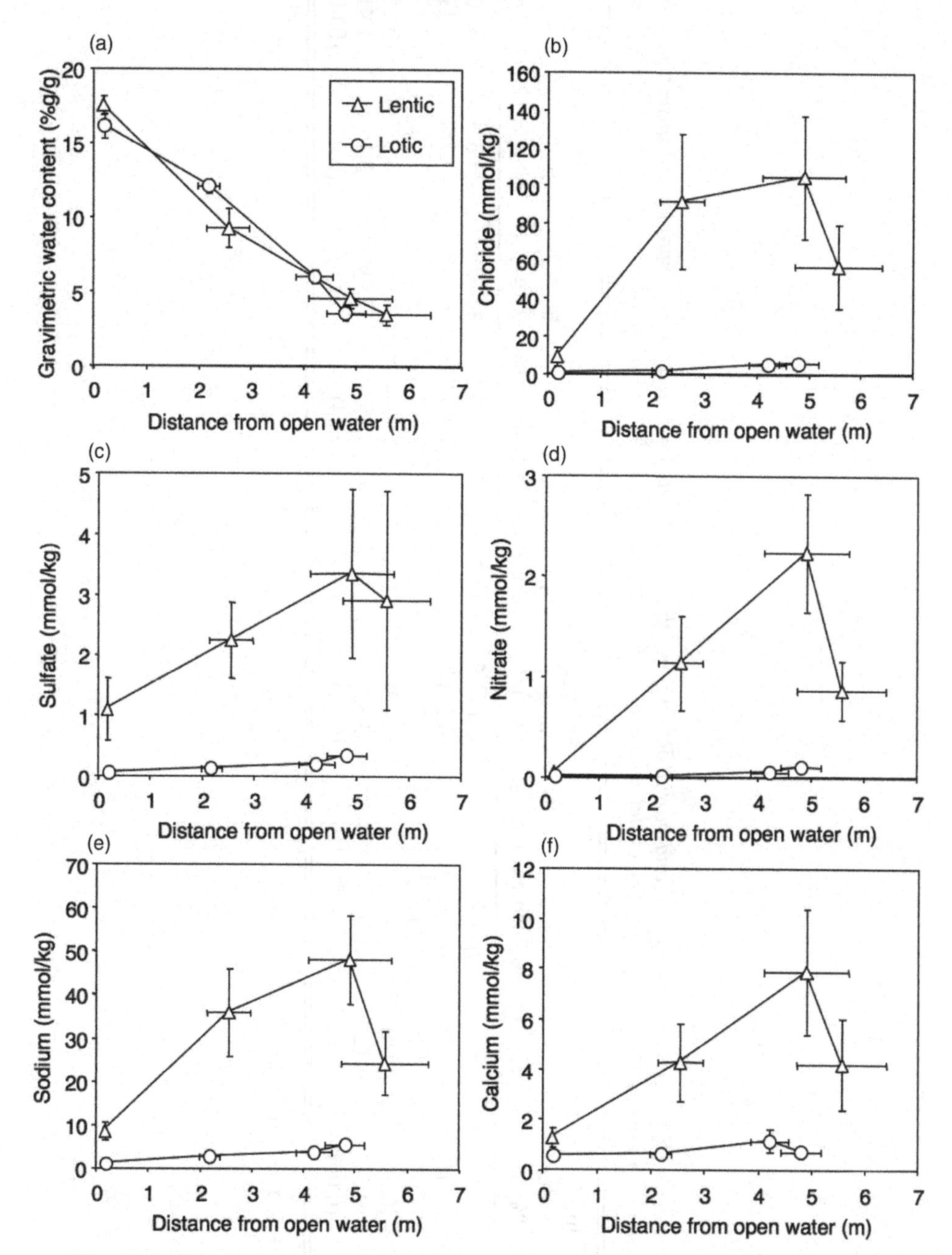

Fig. 3.3. Contemporary influences of stream and lake waters on nearshore surface water content and major ion concentrations in 11 hydrological margins of the McMurdo Dry Valleys. Significance of landform (stream versus lakes) and transects on soil properties are shown in Tables 3.1 and 3.2.

Table 3.1. *ANOVA summaries (*F *statistic with* P *values, and partial* R^2 *in parenthesis) of landscape (lotic versus lentic) and transect position (distance from liquid water) on physicochemical properties of dry valley soils and sediments in hydrological margins*

	ln (water+1) (%g g^{-1})	pH	ln (conductivity+1) (μ Siemens cm^{-1})	$\ln(Cl^-+1)$ (mmol kg^{-1})	ln (SO_4^{2-}+1)	ln (Ca^{2+}+1)	ln (Mg^++1)	ln (Na^++1)	ln (K^++1)
Landscape type	5.20*	3.79	70.60****	58.09****	43.32****	39.83****	34.52****	73.62****	40.60****
	(0.04)	(0.02)	(0.27)	(0.23)	(0.21)	(0.18)	(0.16)	(0.28)	(0.19)
Transect position	278.60****	0.91	7.41****	7.70****	2.29	5.96***	3.84*	7.72****	4.16**
	(0.59)	(0.00)	(0.08)	(0.09)	(0.00)	(0.08)	(0.05)	(0.09)	(0.05)
Landscape* position	0.35	6.12***	1.53	3.83*	1.47	3.55*	2.39	3.15*	1.41
	(0.00)	(0.10)	(0.00)	(0.03)	(0.00)	(0.04)	(0.00)	(0.03)	(0.00)

Notes: *$P<0.05$, **$P<0.01$, ***$P<0.001$, ****$P<0.0001$.

Table 3.2. *ANOVA summaries (F statistic with P values, and partial R^2 in parenthesis) of landscape (lotic versus lentic) and transect position (distance from liquid water) of nutrients and organic matter content of dry valley soils and sediments in hydrological margins*

	$\ln(NH_4^+ + 1)$ (mg N kg^{-1})	$\ln(NO_3^- + 1)$ (mg N kg^{-1})	$\ln(PO_4^{3-} + 1)$ (mg N kg^{-1})	ln(SOC+1) (g C kg^{-1})	ln(TN+1) (g N kg^{-1})	C:N
Landscape type	14.66***	114.27****	1.43	13.45****	14.19****	6.99****
	(0.08)	(0.33)	(0.00)	(0.34)	(0.37)	(0.028)
Transect position	0.57	17.05****	2.62*	6.44***	4.74**	1.60
	(0.00)	(0.15)	(0.02)	(0.05)	(0.04)	(0.00)
Landscape position	2.75*	9.83****	0.70	3.27****	2.73****	1.54
	(0.04)	(0.05)	(0.00)	(0.26)	(0.22)	(0.00)

Notes: *$P<0.05$, **$P<0.01$, ***$P<0.001$, ****$P<0.0001$.

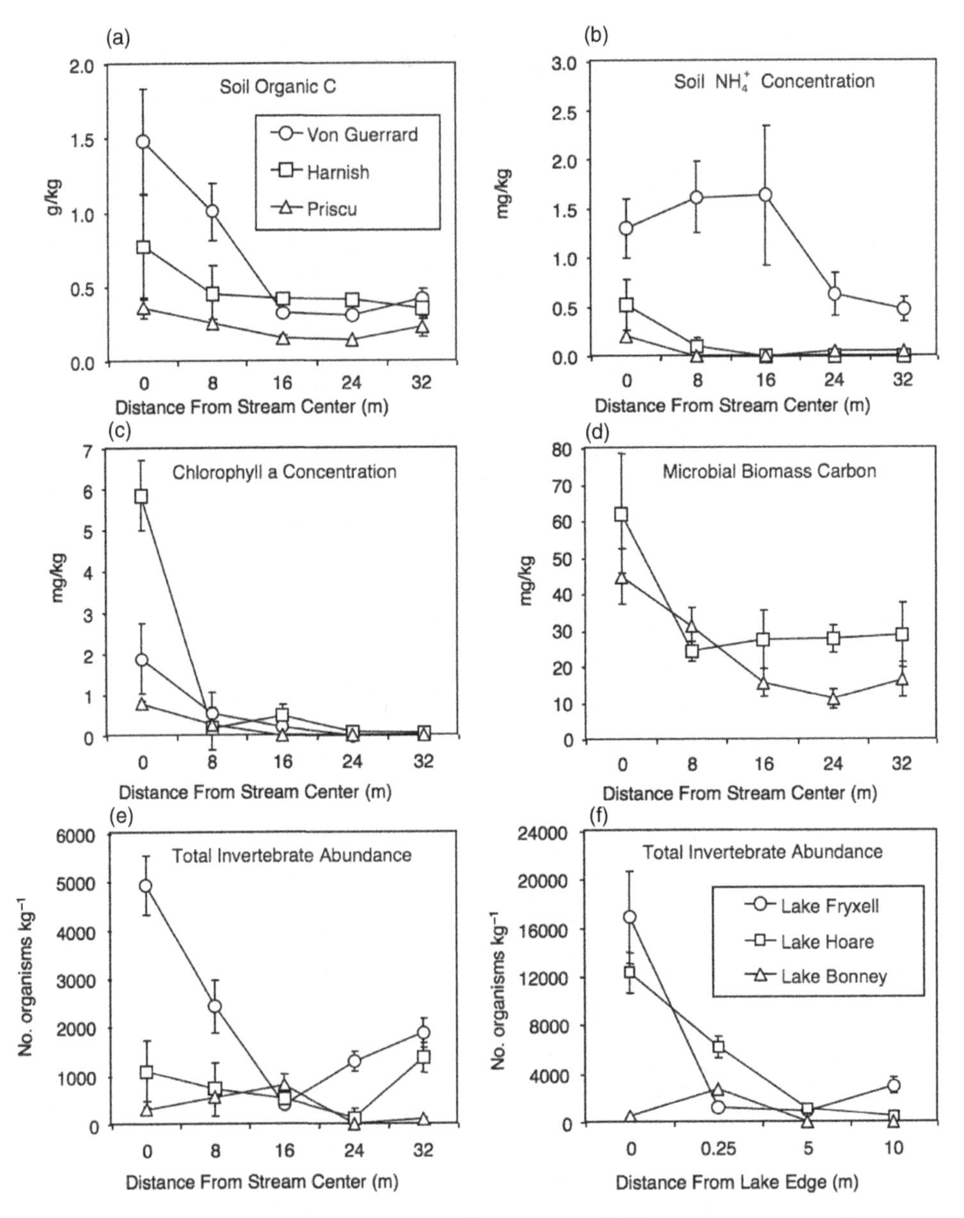

Fig. 3.4. Contemporary influences of stream and lake waters on nearshore biological and biochemical properties: (a) soil organic C content; (b) soil NH_4^+ concentration; (c) chlorophyll *a* concentration; (d) microbial biomass C; (e and f) invertebrate abundance in hydrological margins of (a–e) Von Guerrard, Harnish, and Priscu Stream and (f) Lakes Fryxell, Hoare, and Bonney in Taylor Valley, Antarctica. (Data replotted from Treonis et al., 1999; Ayres et al., 2007, and LTER unpublished.)

otherwise quite similar across dry valley hydrological margins. Electrical conductivity (a proxy for salinity) of sediment/soil extracts varied significantly between lake and stream environments, and most notably across the hydrological margins (Table 3.1), with the third transect positions consistently exhibiting the highest electrical conductivity (Fig. 3.3). Major ion (Cl^-, NO_3^-, SO_4^{2-}, K^+, Na^+, Ca^{2+}, Mg^{2+}) concentrations were also greater in lake relative to stream margin soil extracts (Table 3.1), with highest solute content typically occurring in the third transect position near the visual boundary of the wetted front (Fig. 3.3b–f). Substituting individual sites into the model in place of landscape type (stream versus lake) markedly improved the fit of the statistical models for these variables (e.g., $R^2 = 0.95$, $P < 0.0001$ for ANOVA of electrical conductivity by site), likely due to the influence of till composition and surface exposure age on solute concentration and composition (Bockheim, 1997; Barrett et al., 2007; Bate et al., 2008).

These trends in water content, electrical conductivity, and major ions are consistent with isotopic studies of nearshore environments which have shown that lake and stream margin pore waters generally exhibit an increasing enrichment of ^2H and ^{18}O with distance from the shore, and decreasing soil water content, suggesting evaporation of pore waters and concentration of solutes in these soils (Northcott et al., 2009). Together, these trends in physicochemical properties and isotopic composition of pore waters support the conclusion that salts are accumulating at the distal boundary of the wetted fronts in these hydrological margins due to a combination of water wicking and evapoconcentration of dissolved solutes. This effect is most pronounced in lake margins where electrical conductivity and soil solute content are higher on average than in lotic environments (Table 3.1, Fig. 3.3) suggesting that this effect is largely driven by the more stable boundary of the wetted front that exists in lake margins because the wetted margins in stream environments are largely determined by temporal variation in flow. Thus, paleolake shorelines are identifiable based upon surface geochemistry in cold desert environments.

Inorganic nitrogen concentrations of sediments and soils varied in and among hydrologic margins in the McMurdo Dry Valleys with distinct trends for NH_4^+ and NO_3^- (e.g., Barrett et al., 2002). Nitrate concentrations were greatest at the distal boundary of the wetted front (Fig. 3.3d) in both lake and stream environments and exhibited a significant correlation with electrical conductivity of sediment and soil extracts ($R^2 = 0.58$). Lake margins had significantly higher NO_3^- concentrations relative to stream sediments; a multiple regression model for NO_3^- including landscape type (stream versus lake) and conductivity accounted for 70% of the variation in NO_3^- content of hydrological margins suggesting that controls over NO_3^- transport are mainly physical. In contrast, NH_4^+ concentrations are higher in nearshore lake sediments than in lotic environments (Table 3.2) and typically highest within a few meters of open water (Fig. 3.4b). The differences between spatial trends in NO_3^- and NH_4^+ concentrations provide important constraints on source pools for nitrogen cycling in antarctic environments. Atmospheric deposition is the dominant source of NO_3^- in the McMurdo Dry Valleys (Campbell and Claridge, 1987; Michalski et al., 2005), whereas NH_4^+ may result from both physical (aerial deposition) and biological (decomposition of organic matter) processes (Barrett et al., 2002).

Soil organic carbon content is greatest in nearshore lotic environments and decreases with increasing distance from open water (Table 3.2, Fig. 3.4a), exhibiting similar trends to soil water content (Fig. 3.3a). Inorganic nutrients followed similar patterns to carbon (Fig. 3.4b). For example, $NH_4^+$and PO_4^{3-} had distinct spatial trends compared with other major ions (Cl^-, NO_3^-, SO_4^{2-}, K^+, Na^+, Ca^{2+}, Mg^{2+}), which appeared to be influenced primarily by physical conditions (i.e., wicking and evaporation) as well as by differences between stream and lake environments. Trends in chlorophyll *a* data (Fig. 3.4c) reflect the fact that denser colonization by microbial mats and less commonly mosses are associated with intermittently saturated sediments on the periphery of streams and lakes (Schwarz et al., 1993; Taton et al., 2003). Similarly, both stream channels where data are available exhibit similar lateral variation in microbial biomass, with the highest levels of microbial biomass carbon occurring in stream sediments and rapidly decreasing with increasing distance from the stream channel (Fig. 3.4d).

Differences in stream channel morphology, which influence the width of the wetted margin (Nortcott et al., in review), also influence the spatial distribution of biochemical properties (e.g., chlorophyll *a* concentration and microbial biomass). For example, in Fig. 3.4d the Harnish Creek data show that microbial biomass reaches a minimum within 8 m of the stream center, in contrast to the Priscu Stream data, which exhibit more gradual trends consistent with the lower grades of the Priscu Stream channel relative to Harnish Creek (Bate et al., 2008; Northcott et al., 2009).

These physical and biochemical gradients drive significant spatial variation in the distribution and diversity of metazoan invertebrate communities (Treonis et al., 1999; Ayres et al., 2007). Total invertebrate abundance is typically greatest in the saturated sediments on the periphery of streams and lakes (Fig. 3.4e, f), though variation in salinity among aquatic ecosystems also drives significant variance in soil communities (Treonis et al., 1999). Clearly, proximity to water sources facilitates higher microbial biomass, greater prevalence of invertebrates, and greater potential for biological activity in general (e.g., Zeglin et al., 2009), though this effect is most evident within one meter of open water; beyond that distance, physical processes associated with the wicking of water and evapoconcentration of salts appear to be the dominant geochemical processes.

Legacy effects of paleolakes on surface soil geochemistry

In the previous section we discussed contemporary processes in hydrological margins and how they generate distinct geochemical patterns in nearshore stream and lake environments. Here we extend these insights to consider how trends in surface salinity and organic matter distribution preserve a history of paleolake influences in contemporary dry valley soils. We present data from Taylor Valley collected during the austral summer of 2005–2006 focusing on spatial variation in soil salinity (measured as the electrical conductivity of a 1:5 soil:water extract) and soil organic carbon content along elevation transects spanning the range of contemporary and paleolake levels. To assess the spatial variation in soil geochemistry in Taylor Valley, we collected soil samples (0–10 cm) along transects on the south side of each of the three lake basins in Taylor Valley. Transect length ranged from ~800 m to >3500 m based upon valley morphology. Sample spacing ranged from 5–25 m and was determined individually for each basin depending on transect length and available sampling time.

Inundation and recession of proglacial lakes in the McMurdo Dry Valleys has been demonstrated by the presence of lacustrine organic matter preserved in perched deltas (Denton et al., 1989; Hall and Denton, 2000) and in low elevation soils (Burkins et al., 2000). For example, Fig. 3.5 illustrates elevational variation in Taylor Valley soil organic C content that is largely driven by proximity to contemporary lakes levels (indicated by arrow); similar trends have also been reported for upper Garwood Valley (Elberling et al., 2006). Isotopic evidence supports the conclusion that the proportion of soil organic matter derived from lakes increases at lower elevations (Burkins et al., 2000). Soil C concentrations vary over two orders of magnitude from the highest values at low elevations near contemporary lake shores to lower values at higher elevation, though the spatial resolution represented in this plot is insufficient to capture variability associated with landforms above 150 m in elevation. Fig. 3.6 plots soil organic C against linear distance from the contemporary shore of Lake Fryxell; similar to previous studies, these data show high levels of organic matter in low elevation soils, but also indicate variability in soil C content at higher elevations that is missed by lower spatial

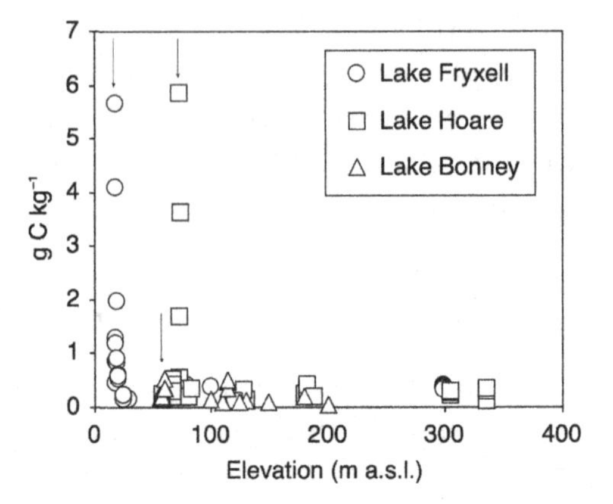

Fig. 3.5. Variation in soil organic matter (g organic C kg^{-1} soil) along elevation gradients above Lakes Fryxell, Hoare, and Bonney in Taylor Valley, Antarctica. Arrows indicate the contemporary elevations of shorelines of the lakes. (Data are replotted from Powers et al., 1998, Burkins et al., 2001, Barrett et al., 2005, and Ayres et al., 2007.)

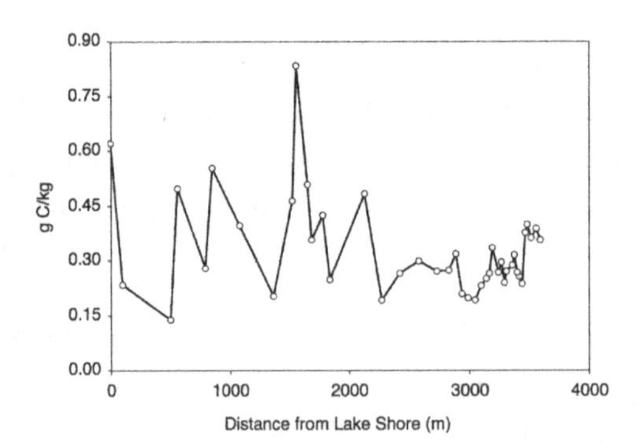

Fig. 3.6. Paleolake influences on soil organic matter on the south shore of Lake Fryxell in Taylor Valley, Antarctica.

resolution sampling. Thus, patterns in soil C are not strictly unimodal across elevation gradients, and may reflect complex geomorphic, sedimentologic, and hydrologic influences over surface soil chemistry in addition to the influence of paleolake levels.

Similar to soil organic matter, other geochemical parameters, particularly salinity, vary widely in Taylor Valley soils and are likely influenced by paleohydrologic activity including the rise and fall of paleolake levels. Fig. 3.7 shows soil salinity (electrical conductivity) plotted against linear distance from the modern lake shore for sampling transects in the Lake Fryxell, Lake Hoare, and Lake Bonney basins. Lake Fryxell conductivity

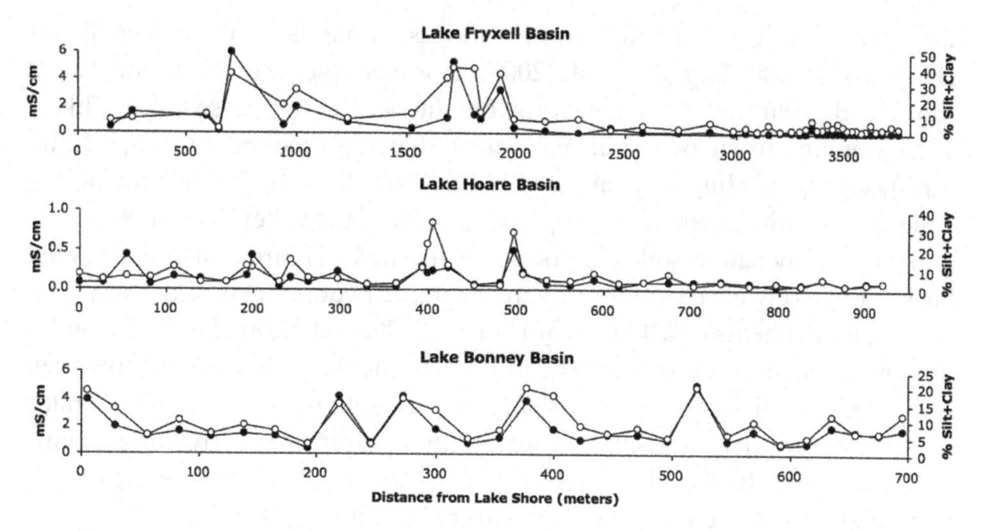

Fig. 3.7. Paleolake influences on surface soils across high resolution elevation transects in Taylor Valley, Antarctica. Electrical conductivity (filled circles) and soil particle size (open circles) versus linear distance from contemporary lake shores in each of Taylor Valley's three lake basins.

values exhibit more than two orders of magnitude variation, ranging from 25 to ~5000 $\mu S\,cm^{-1}$ and averaging ~500 $\mu S\,cm^{-1}$ when weighted for ground distance. Lake Hoare salinity values are also highly variable ranging from 20 to 900 $\mu S\,cm^{-1}$ with a distance weighted average of ~150 $\mu S\,cm^{-1}$. In both the Lake Fryxell and Lake Hoare basins, relatively low conductivity soils occur at higher elevations most distant from the modern shoreline transitioning to higher more variable conductivities at lower elevations. Soils in the Lake Bonney basin are equally variable but significantly more saline than those in the Lake Fryxell and Lake Hoare basins, ranging from 420 to >5000 $\mu S\,cm^{-1}$ and averaging >2000 $\mu S\,cm^{-1}$. Soils in the Lake Bonney basin show increases in average soil conductivity approaching the modern lake shoreline but are also characterized by highly variable conductivity values at higher elevations in the transect (Fig. 3.7c).

There is a strong covariation between soil salinity and soil particle-size distribution across the three lake basins (Fig. 3.7), suggesting a direct connection between soil geochemistry and the landscape processes (glacial, fluvial, lacustrine, aeolian) responsible for depositing and modifying (physically and chemically) surface sediments. The sample spacing of our transects was not sufficient to evaluate this relationship quantitatively; however, field observations show that soil types are distributed in association with dominant landforms, such as stream channels, ridges, mounds, and perched deltas (Hall et al., 2000; Higgins et al., 2000). For example, the consistently most saline soils near Lake Bonney form on silt- and clay-rich sediments likely corresponding to lacustrine or meltout till deposits (Sediment Groups 1 and 3 respectively of Higgins et al., 2000) and found forming hummocks on the valley floor, or hummocky ridges perpendicular to the valley axis. Similar fine-grained, highly saline soils also occur in the Lake Hoare and Lake Fryxell basins and likely correspond to the finer-grained component of the stratified lacustrine sediments (SGB facies of Hall et al., 2000) deposited during inundation by Glacial Lake Washburn. In addition to the strong relationship between soil salinity and texture, spatial variation in salinity also exhibits similar patterns to soil organic matter, with the two parameters correlating significantly within individual lake basins; the correlation between organic matter and salinity for Lake Fryxell soils illustrated in Figs. 3.6 and 3.7 is 0.81.

Whereas variation in soil organic matter is directly related to proximity to contemporary lake shores, mechanisms underlying the variation in soil salinity and its correlation with highly variable soil texture is less clear. Soil salinity integrates the effects of many surface processes including evapoconcentration along stream and lake margins, vertical movement within the soil column, present and historic stream activity, surface age and atmospheric deposition, glaciation, and lake inundation. Similarly soil texture integrates potential effects of lacustrine, glacial, stream, and aeolian deposition and erosion, as well as mechanical sorting during cryoturbation. At the broadest scale, variation in salinity and particle size among lake basins of Taylor Valley likely indicates the influence of till composition and deposition, surface age, and initial geochemical conditions prior to subsequent hydrological modification. For example, soils and sediments in the Lake Bonney basin are formed on Taylor II tills, considerably older than the Ross Sea and Taylor tills in eastern Taylor Valley (Bockheim, 2002). Thus, any paleohydrological influences on surface geochemistry are probably superimposed on existing variation in dry valley landscapes (Lyons et al., 2000).

Within individual lake basins, the finest-textured soils and coincident highest soil salinities first appear downslope at elevations of approximately 100–120 m at both Lake Fryxell and Lake Hoare and approximately 225–235 m elevation in the Lake Bonney basin. This suggests, albeit at different elevations, that similar mechanisms may be responsible for both the deposition and/or preservation of fine-textured sediments, as well as the selective introduction and/or removal of salts to or from those sediments in all three basins. For example, elevation gradients in soil properties may delineate the position of paleolake shores as lake levels declined over the past interglacial period

(Denton et al., 1989). Soils with higher levels of conductivity, organic matter, and fines (silt + clay fractions) may be indicative of stable shorelines formed successively from higher to lower elevations as lake levels episodically receded. Alternatively, salinity variation between soil textures may result from differential leaching rates associated with variation in soil permeability, or by selective retention of dissolved ions contributed by inundating lake waters. The difference in elevations where fine-textured and more saline sediments first appear in each basin may be attributable to the separation, at the present-day Defile (elevation ~200 m) in middle Taylor Valley, of a receding Lake Washburn into two lakes, one occupying the Lake Bonney basin, the other occupying the Lake Fryxell and Lake Hoare basins. In any scenario, it is likely that a historically active and complex surface hydrology is ultimately responsible for the extremely high variability in soluble ion concentrations across small distances in Taylor Valley soils.

Data from contemporary nearshore sediments and soils in the McMurdo Dry Valleys (Fig. 3.5; Gooseff et al., 2007; Northcott et al., 2009) support the hypothesis that soils with high levels of conductivity, organic matter, and fines are indicative of lake shores formed successively from higher to lower elevations as lake levels receded. Spikes in soil data (Figs. 3.6 and 3.7) may delineate the positions of lake shores of long duration, possibly reflecting periods of stable climate. Linking soil chemistry with specific landforms in Taylor Valley will be critical for understanding the effects of landscape history on contemporary geomorphology and soil ecological processes, e.g., controls over habitats suitable and the distribution and activity of biological communities (Poage et al., 2008; Zeglin et al., 2009; Takacs-Vesbach et al., this volume, Chapter 8). Such approaches may be useful for interpreting landscape history in the McMurdo Dry Valleys as well as in environments where liquid water once existed on the surface of Mars.

Considerations for future investigations of Mars

Nearshore variation in sediment and soil characteristics of the McMurdo Dry Valleys indicates an important influence of lakes and streams on contemporary surface biogeochemical dynamics dominated by physicochemical processes such as wicking and evaporation and to a lesser extent microbial influences on biochemical parameters such as organic matter, nutrient, and pigment concentrations. At broader scales, elevation gradients in soil properties of the dry valleys may delineate the position of paleolake shores as lake levels declined over the past interglacial period (e.g., Denton et al., 1989). Investigations of surface geochemistry and mineralogy have already provided extensive evidence of past aqueous environments on Mars (e.g., Squyres et al., 2004a) that may share similarities to the linked hydrology of soils and lakes in the McMurdo Dry Valleys. In future research, such information could be helpful in interpreting landscape history and the legacy of water on the Red Planet.

Understanding controls over the weathering and hydrologic processes that generate salts and control their spatial distribution in the McMurdo Dry Valleys could help focus efforts to identify and understand aqueous environments on Mars. Current (e.g., Mars Phoenix Lander) and future explorations of Mars targeted toward finding evidence for paleo or contemporary biology will use surface geochemical properties to identify legacy aqueous environments prior to explicit investigation for potential biology. Salinity is a useful diagnostic variable for elucidating landscape history for two reasons: (1) geochemical processes in the McMurdo Dry Valleys and on Mars are dominated by physical processes; and (2) electrical conductivity as a proxy variable for salinity is an easy and inexpensive measurement to make and thus greater numbers of samples can be collected and processed to provide greater spatial resolutions of sampling – useful considerations for designing future investigations of soils in both Antarctica and on Mars.

Acknowledgments

This work was funded by a National Science Foundation grant (#0338174) and the McMurdo Dry Valleys Long Term Ecological Research program. Logistical support for fieldwork was provided by Raytheon Polar Services and Petroleum Helicopters Inc.

References

Adams, B. J., Bardgett, R. D., Ayres, E., et al. (2006). Diversity and distribution of Victoria Land biota. *Soil Biology and Biochemistry*, **38**, 3003–3018.

Aislabie, J. M., Chhour, K. L., Saul, D. J., et al. (2006). Dominant bacteria in soils of Marble Point and Wright Valley, Victoria Land, Antarctica. *Soil Biology and Biochemistry*, **38**, 3041–3056.

Amundson, R. A., et al. (2008). On the in situ aqueous alteration of soil on Mars. *Geochimica et Cosmochimica Acta*, **72**, 3845–3864.

Ayers, E., Adams, B. J., Barrett, J. E., Virginia, R. A., and Wall, D. H. (2007). Soil and sediment biogeochemistry and faunal community structure across aquatic-terrestrial interfaces in a polar desert ecosystem. *Ecosystems*, doi: 10.1007/s10021–007–9035-x.

Baker, V. R. (2001). Water and the martian landscape. *Nature*, **412**, 228–235.

Baker, V. R., Strom, R. G., Gulick, V. C., et al. (1991). Ancient oceans, ice sheets and the hydrological cycle on Mars. *Nature*, **352**, 589–594.

Bandfield, J. L., Hamilton, V. E., and Christensen, P. R. (2000). A global view of Martian surface composition from MGS-TES. *Science*, **287**, 1626–30.

Barrett, J. E., Virginia, R. A., and Wall, D. H. (2002). Trends in resin and KCl-extractable soil nitrogen across landscape gradients in Taylor Valley, Antarctica. *Ecosystems*, **5**, 289–299.

Barrett, J. E., Wall, D. H., Virginia, R. A., et al. (2004). Biogeochemical parameters and constraints on the structure of soil biodiversity. *Ecology*, **85**, 3105–3118.

Barrett, J. E., Virginia, R. A., Parsons A. N., and Wall, D. H. (2005). Potential soil organic matter turnover in Taylor Valley, Antarctica. *Arctic Antarctic and Alpine Research*, **37**, 107–116.

Barrett, J. E., Virginia, R. A., Hopkins, D. W., et al. (2006). Terrestrial ecosystem processes of Victoria Land, Antarctica. *Soil Biology and Biochemistry*, **38**, 3019–3034.

Barrett, J. E., Virginia, R. A., Lyons, W. B., et al. (2007). Biogeochemical stoichiometry of Antarctic Dry Valley ecosystems. *Journal of Geophysical Research, Biogeosciences*, **112**, G01010.

Barrett, J. E., Virginia, R. A., Wall, D. H., et al. (2008). Persistent effects of a discrete climate event on a polar desert ecosystem. *Global Change Biology*, **14**, 2249–2261.

Bate, D. B., Barrett, J. E., Poage, M. A., and Virginia, R. A. (2008). Soil phosphorus cycling in an Antarctic Polar Desert. *Geoderma*, **144**, 21–31.

Bedard, J. H. J., Marsh, B. D., Hersum, T. G., Naslund, H. R., and Mukasa, S. B. (2007). Large-scale mechanical redistribution of orthopyroxene and plagioclase in the basement sill, Ferrar dolerites, McMurdo Dry Valleys, Antarctica. *Journal of Petrology*, **48**, 2289–2326.

Bell, J. F., McSeen, H. Y., Crisp, J. A., et al. (2000). Mineralogic and compositional properties of Martian soil and dust: results from Mars Pathfinder. *Journal of Geophysical Research, Planets*, **105**(E1), 1721–1755.

Benner, S. A., Devine, K. G., Matveeva, L. N., and Powell, D. H. (2000). The missing organic molecules on Mars. *Proceedings of the National Academy of Sciences of the United States of America*, **97**, 2425–2430.

Bockheim, J. G. (1997). Properties and classification of cold desert soils from Antarctica. *Soil Science Society of America*, **61**, 224–231.

Bockheim, J. G. (2002). Landform and soil development in the McMurdo Dry valleys, Antarctica: a regional synthesis. *Arctic Antarctic and Alpine Research*, **34**, 308–317.

Bockheim, J. G. (2007). Soil processes and development rates in the Quartermain Mountains, upper Taylor Glacier region, Antarctica. *Geografiska Annaler Series A, Physical Geography*, **89A**, 153–165.

Bockheim, J. G. and Hall, K. J. (2002). Permafrost, active-layer dynamics and periglacial environments of continental Antarctica. *South African Journal of Science*, **98**, 82–90.

Bockheim, J. G., Campbell, I. B., and McLeod, M. (2007). Permafrost distribution and active-layer depths in the McMurdo dry valleys, Antarctica. *Permafrost and Periglacial Processes*, **18**, 217–227.

Bomblies, A., McKnight, D. M., and Andrews, E. D. (2001). Retrospective simulation of lake-level rise in Lake Bonney based on recent 21-year record, indication of recent climate change in the McMurdo Dry Valleys, Antarctica. *Journal of Paleolimnology*, **25**, 477–492.

Burkins, M. B., Virginia, R. A., Chamberlain, C. P., and Wall, D. H. (2000). Origin and distribution of soil organic matter in Taylor Valley, Antarctica. *Ecology*, **81**, 2377–2391.

Burkins, M. B., Virginia, R. A., and Wall, D. H. (2001). Organic carbon cycling in Taylor Valley, Antarctica, quantifying soil reservoirs and soil respiration. *Global Change Biology*, **7**, 113–125.

Cabrol, N. A. and Grin, E. A. (1999). Distribution, classification, and ages of Martian impact crater lakes. *Icarus*, **142**, 160–172.

Campbell, I. B. (2003). Soil characteristics at a long-term ecological research site in Taylor Valley, Antarctica. *Australian Journal of Soil Research*, **41**, 351–364.

Campbell, I. B. and Claridge, G. G. C. (1987). *Antarctica: Soils, Weathering Processes and Environment*. Developments in Soil Science 16. Amsterdam, Netherlands: Elsevier.

Carr, M. H. (2006). *The Surface of Mars*. Cambridge, UK: Cambridge University Press.

Carr, M. H. and Head, J. W. (2003). Oceans on Mars: an assessment of the observational evidence and possible fate. *Journal of Geophysical Research, Planets*, **108**, 5042.

Chevrier, V. and Mathe, P. E. (2007). Mineralogy and evolution of the surface of Mars: a review. *Planetary and Space Science*, **55**, 289–314.

Chevrier, V., Mathe, P. -E., Rochette, P., and Gunnlaugsson, H. P. (2006). Magnetic study of an Antarctic weathering profile on basalt, implication for recent weathering on Mars. *Earth and Planetary Science Letters*, **244**, 501–514.

Chinn, T. H. (1993). Physical hydrology of the dry valley lakes. In *Physical and Biogeochemical Processes in Antarctic Lakes*, ed. W. J. Green and E. I. Freidmann. Antarctic Research Series 59. Washington, D.C.: American Geophysical Union, pp. 1–51.

Christensen, P. R., Bandfield, J. L., Hamilton, V. E., et al. (2001). Mars Global Surveyor thermal emission spectrometer experiment, investigation description and surface science results. *Journal of Geophysical Research, Planets*, **106**, 23 823–23 871.

Connell, L., Redman, R., Craig, S., and Rodriguez, R. (2006). Distribution and abundance of fungi in the soils of Taylor Valley, Antarctica. *Soil Biology and Biochemistry*, **38**, 3083–3094.

Convey, P., Block, W., and Peat, H. J. (2003). Soil arthropods as indicators of water stress in Antarctic terrestrial habitats? *Global Change Biology*, **9**, 1718–1730.

Cuneo, N. R., Taylor, E. L., Taylor, T. N., and Krings, M. (2003). In situ fossil forest from the upper Fremouw Formation (Triassic) of Antarctica: paleoenvironmental setting and paleoclimate analysis. *Palaeogegraphy Palaeoclimatology Palaeoecology*, **197**, 239–261.

Denton, G. H. and Hall, B. L. (2000). Glacial and paleoclimatic history of the Ross ice drainage system of Antarctica: preface. *Geografiska Annaler Series A, Physical Geography*, **82A**,139–141.

Denton, G. H., Bockheim, J. G., Wilson, S. C., and Stuiver, M. (1989). Late Wisconsin and Early Holocene glacial history, Inner Ross Embayment, Antarctica. *Quaternary Research*, **31**, 151–182.

Dickinson, W. W. and Rosen, M. R. (2003). Antarctic permafrost: an analogue for water and diagenetic minerals on Mars. *Geology*, **31**, 199–202.

Doran, P. T., McKay, C. P., Clow, G. D., et al. (2002). Valley floor climate observations from the McMurdo dry valleys, Antarctica 1986–2000. *Journal of Geophysical Research, Atmospheres*, **107**, Article 4772.

Doran, P. T., McKay, C. P., Fountain, A. G., et al. (2008). Hydrologic response to extreme warm and cold summers in the McMurdo Dry Valleys, East Antarctica. *Antarctic Science*, **20**, 499–509.

Elberling, B., Gregorich, E. G., Hopkins, D. W., et al. (2006). Distribution and dynamics of soil organic matter in an Antarctic dry valley. *Soil Biology and Biochemistry*, **38**, 3095–3106.

Foley, K. K., Lyons, W. B., Barrett, J. E., and Virginia, R. A. (2006). Pedogenic carbonate distribution within glacial till in Taylor Valley, Southern Victoria Land, Antarctica. In *Paleoenvironmental Records of Calcretes and Palustrine Carbonates*, ed. A. M. Alonso-Zara and L. H. Tanner. GSA Special Paper 416. Boulder, CO: Geological Society of America, pp. 89–103.

Foreman, C., Wolf, C. F., and Priscu, J. C. (2004). Impact of episodic warming events on the physical, chemical and biological relationships of lakes in the McMurdo Dry Valleys, Antarctica. *Aquatic Geochemistry*, **10**, 239–268.

Fountain, A. G., Lyons, W. B., Burkins, M. B., et al. (1999). Physical controls on the Taylor Valley ecosystem, Antarctica. *BioScience*, **49**, 961–971.

Gellert, R., Rieder, R., Anderson, R. C., et al. (2004). Chemistry of rocks and soils in Gusev crater from the alpha particle x-ray spectrometer. *Science*, **305**, 829–832.

Gooseff, M. N., Barrett, J. E., Doran, P. T., et al. (2003). Snow-patch influence on soil biogeochemical processes and invertebrate distribution in the McMurdo Dry Valleys, Antarctica. *Arctic Antarctic and Alpine Research*, **35**, 92–100.

Gooseff, M. N., Northcott, N. L., Barrett, J. E., et al. (2007). Controls on soil water dynamics in near-shore lake environments in an Antarctic polar desert. *Vadose Zone Journal*, **6**, 841–848.

Haberle, R. M., McKay, C. P., Schaeffer, J., et al. (2001). On the possibility of liquid water on present-day Mars. *Journal of Geophysical Research, Planets*, **106**, 23 317–23 326.

Hall, B. L. and Denton, G. H. (2000). Radiocarbon chronology of Ross Sea drift, eastern Taylor Valley, Antarctica: evidence for a grounded ice sheet in the Ross Sea at the last glacial maximum *Geografiska Annaler Series A, Physical Geography*, **82A**, 305–336.

Hall, B. L., Denton, G. H., and Hendy, C. H. (2000). Evidence from Taylor Valley for a grounded ice sheet in the Ross Sea, Antarctica. *Geografiska Annaler Series A, Physical Geography*, **82A**, 275–303.

Hagedorn, B., Sletten, R. S., and Hallet, B. (2007). Sublimation and ice condensation in hyperarid soils: modeling results using field data from Victoria Valley, Antarctica. *Journal of Geophysical Research*, **112**, F03017.

Harris, K. J., Carey, A. E., Lyons, W. B., Welch, K. A., and Fountain, A. G. (2007). Solution and isotope geochemistry of subsurface ice melt seeps in Taylor Valley, Antarctica. *Geological Society of America Bulletin*, **119**, 548–555.

Haskin, L. A., Wang, A., Jollif, B. L., et al. (2005). Water alteration of rocks and soils on Mars at the Spirit rover site in Gusev crater. *Nature*, **436**, 66–69.

Head, J. W., Hiesinger, H., Ivanov, M. A., et al. (1999). Possible ancient oceans on Mars: evidence from Mars Orbiter Laser Altimeter data. *Science*, **286**, 2134–2137.

Heldmann, J. L., Carlsson, E., Johansson, H., Mellon, M. T., and Toon, O. B. (2007). Observations of martian gullies and constraints on potential formation mechanisms. II. The northern hemisphere. *Icarus*, **188**, 324–344.

Hendy, C. H., Sadler, A. J., Denton, G. H., and Hall, B. L. (2000). Proglacial lake-ice conveyors: a new mechanism for deposition of drift in polar environments. *Geografiska Annaler Series A, Physical Geography*, **82A**, 249–270.

Higgins, S. M., Denton, G. H., and Hendy, C. H. (2000). Glacial geomorphology of Bonney drift, Taylor Valley, Antarctica. *Geografiska Annaler Series A, Physical Geography*, **82A**, 365–389.

Hvidberg, C. S. (2005). Polar caps. In *Water on Mars and Life*, ed. T. Tokano. Berlin: Springer, pp. 129–152.

Howard, A. D. (2000). The role of eolian processes in forming surface features of the Martian polar layered deposits. *Icarus*, **144**, 267–288.

Kasting, J. F. (1993). Earth's early atmosphere. *Science*, **259**, 920–926.

Kasting, J. F. and Siefert, J. L. (2002). Life and the evolution of Earth's atmosphere. *Science*, **296**, 1066–1068.

Kennedy, A. D. (1993). Water as a limiting factor in the Antarctic terrestrial environment: a biogeographical synthesis. *Arctic and Alpine Research*, **25**, 308–315.

Klein, H. P. (1978). The Viking biological experiments on Mars. *Icarus*, **34**, 666–674.

Klingelhöfer, G., Morris, R. V, Bernhardt, B., et al. (2004). Jarosite and hematite at Meridiani Planum from Opportunity's Mössbauer spectrometer. *Science*, **306**, 1740–1745.

Kuzmin, R. O. (2005). Ground ice in the Martian regolith. In *Water on Mars and Life*, ed. T. Tokano. Berlin: Springer, pp. 155–189.

Lancaster, N. (2002). Flux of eolian sediment in the McMurdo Dry Valleys, Antarctica: a preliminary assessment. *Arctic Antarctic and Alpine Research*, **34**, 318–323.

Lazcano, A. and Miller, S. L. (1994). How long did it take for life to begin and evolve to cyanobacteria? *Journal of Molecular Evolution*, **39**, 546–554.

Lazcano, A. and Miller, S. L. (1999). On the origin of metabolic pathways. *Journal of Molecular Evolution*, **49**, 424–431.

Leovy, C. (2001). Weather and climate on Mars. *Nature*, **412**, 245–249.

Lewis, K. J., Fountain, A. G., and Dana, G. L. (1998). Surface energy balance and meltwater production for a Dry Valley glacier, Taylor Valley, Antarctica. *Annals of Glaciology*, **27**, 603–609.

Lichtenberg, K. A., Arvidson, R. E., Poulet, F., et al. (2007). Coordinated analyses of orbital and Spirit Rover data to characterize surface materials on the cratered plains of Gusev Crater, Mars. *Journal of Geophysical Research, Planets*, **112**, E12S90.

Lundin, R., Barabash, S., Andersson, H., et al. (2004). Solar wind-induced atmospheric erosion at Mars: first results from ASPERA-3 on Mars Express. *Science*, **305**, 1933–1936.

Lyons, W. B., Welch, K. A., Nezat, C. A., et al. (1997). Chemical weathering rates and reactions in the Lake Fryxell Basin, Taylor Valley: comparison to temperate river basins. In *Ecosystem Processes in Antarctic Ice-free Landscapes*, ed. W. B. Lyons, C. Howard-Williams, and I. Hawes. Rotterdam, Netherlands: Balkema Press, pp. 147–154.

Lyons, W. B., Fountain, A. G., Doran, P. T., et al. (2000). Importance of landscape position and legacy, the evolution of the lakes in Taylor Valley, Antarctica. *Freshwater Biology*, **43**, 355–367.

Lyons, W. B., Welch, K. A., Carey, A. E., et al. (2005). Groundwater seeps in Taylor Valley, Antarctica: an example of a decadal subsurface melt event. *Annals of Glaciology*, **40**, 200–206.

Madronich, S., McKenzie, R. L., Bjorn, L. O., et al. (1998). Changes in biologically active ultraviolet radiation reaching the Earth's surface. *Journal of Photochemistry and Photobiology, Biology*, **46**, 5–19.

Mahaney, W. C., Dohm, J. C., and Baker, V. R. (2001). Morphogenesis of Antarctic paleosols, Martian analogue. *Icarus*, **154**, 113–130.

Malin, M. C. and Edgett, K. S. (2000). Evidence for recent groundwater seepage and surface runoff on Mars. *Science*, **288**, 2330–2335.
Mangold, N. (2005). High latitude patterned grounds on Mars: classification, distribution and climatic control. *Icarus*, **174**, 336–359.
Marchant, D. R. and Head, J. W. (2007). Antarctic dry valleys: microclimate zonation, variable geomorphic processes, and implications for assessing climate change on Mars. *Icarus*, **192**, 187–222.
Marchant, D. R., Lewis, A. R., Phillips, W. M., et al. (2002). Formation of patterned ground and sublimation till over Miocene glacier ice in Beacon Valley, southern Victoria Land, Antarctica. *Geological Society of America Bulletin*, **114**, 718–730.
Margulis, L. and Sagan, D. (1986). *Microcosmos: Four Billion Years of Microbial Evolution*. Berkeley, CA: University of California Press.
McKay, C. P. (1997). The search for life on Mars. *Origins Of Life And Evolution of the Biosphere*, **27**, 262–289.
McGinnis, L. D. (1981). *Dry Valley Drilling Project*. Antarctic Research Series 33. Washington, D.C.: American Geophysical Union, 465 pp.
McLennan, S. M., Bell, J. F., Calvin, W. M., et al. (2005). Provenance and diagenesis of the evaporite-bearing Burns formation, Meridani Planum, Mars. *Earth and Planetary Science Letters*, **240**, 95–121.
McSween, H. Y., Arvidson, R. E., Bell, J. F., et al. (2004). Basaltic rocks analyzed by the Spirit Rover in Gusev Crater. *Science*, **305**, 842–848.
Mellon, M. T. and Phillips, R. J. (2001). Recent gullies on Mars and the source of liquid water. *Journal of Geophysical Research, Planets*, **106,** 23 165–23 179.
Mellon, M. T., Feldman, W. C., and Prettyman, T. H. (2004). The presence and stability of ground ice in the southern hemisphere of Mars. *Icarus*, **169**, 324–340.
Michalski, G., Bockheim, J. G., Kendall, C., and Thiemens, M. (2005). Isotopic composition of Antarctic Dry Valley nitrate: implication for NO_y sources and cycling in Antarctica. *Geophysical Research Letters*, **32**, L13817.
Ming, D. W., Mittlefehldt, D. W., and Morris, R. V. (2006). Geochemical and mineralogical indicators for aquerous processes in the Columbia Hills of Gusev crater, Mars. *Journal of Geophysical Research, Planets*, **111**, E02S12.
Mitrofanov, I. G. (2005). Global distribution of subsurface water measured by Mars Odyssey. In *Water on Mars and Life*, ed. T. Tokano. Berlin: Springer, pp. 99–128.
Morris, R. V., Golden, D. C., Bell, J. F., et al. (2000). Mineralogy, composition, and alteration of Mars Pathfinder rocks and soils: evidence from multispectral, elemental, and magnetic data on terrestrial analogues, SNSC meteorite, and Pathfinder samples. *Journal of Geophysical Research, Planets*, **105**, 1757–1817.
Morris, R. V., Klingelhöfer, G., Bernhardt, B., et al. (2004). Mineralogy at Gusev Crater from the Mössbauer Spectrometer on the Spirit Rover. *Science*, **305**, 833–836.
Navarro-Gonzalez, R., Rainey, F. A, Molina, P., et al. (2003). Mars-like soils in the Atacama Desert, Chile, and the dry limit of microbial life. *Science*, **302**, 1018–1021.
Ng, F., Hallet, B., Sletten, R. S., and Stone, J. O. (2005). Fast-growing till over ancient ice in Beacon Valley, Antarctica. *Geology*, **33**, 121–124.
Nisbet, E. G. and Sleep, N. H. (2001). The habitat and nature of early life. *Science*, **409**, 1083–1091.
Northcott, M. L., Gooseff, M. N., Barrett, J. E., et al. (2009). Hydrologic characteristics of lake- and stream-side riparian margins in the McMurdo Dry Valleys, Antarctica. *Hydrological Processes*, **23**, 1255–1267.

Ori, G. G., Marinangeli, L., and Baliva, A. (2000). Terraces and Gilbert-type deltas in crater lakes in Ismenius Lacus and Memnonia (Mars). *Journal of Geophysical Research, Planets*, **105**, 17 629–17 641.

Parsons, A. N., Barrett, J. E., Wall, D. H., and Virginia, R. A. (2004). Soil carbon dioxide flux in Antarctic Dry Valley ecosystems. *Ecosystems*, **7**, 286–295.

Perron, J. T., Mitrovica, J. X., Manga, M., Matsuyama, I., and Richards, M. A. (2007). Evidence for an ancient martian ocean in the topography of deformed shorelines. *Nature*, **447**, 840–843.

Poage, M. A., Barrett, J. E., Virginia, R. A., and Wall, D. H. (2008). The influence of soil geochemistry on nematode distribution, McMurdo Dry Valleys, Antarctica. *Arctic Antarctic and Alpine Research*, **40**, 119–128.

Poreda, R. J., Hunt, A. G., Lyons, W. B., and Welch, K. A. (2004). The helium isotopic chemistry of Lake Bonney, Taylor Valley, Antarctica: timing of Late Holocene climate change in Antarctica. *Aquatic Geochemistry*, **10**, 353–371.

Powers, L. E., Freckman, D. W., and Virginia, R. A. (1995). Spatial distribution of nematodes in polar desert soils of Antarctica. *Polar Biology*, **15**, 325–333.

Powers, L. E., Ho, M. C., Freckman, D. W., and Virgina, R. A. (1998). Distribution, community structure, and microhabitats of soil invertebrates along an elevational gradient in Taylor Valley, Antarctica. *Arctic Antarctic and Alpine Research*, **30**, 133–141.

Quinn, R. C., Zent, A. P., Grunthaner, F. J., et al. (2005). Detection and characterization of oxidizing acids in the Atacama Desert using the Mars Oxidation Instrument. *Planetary and Space Science*, **53**, 1376–1388.

Rieder, R., Gellert, R., Anderson, R. C., et al. (2004). Chemistry of rocks and soils at Meridiani Planum from the alpha particle X-ray spectrometer. *Science*, **306**, 1746–1749.

Schwarz, A. M. J., Green, J. D., Green, T. G. A., and Seppelt, R. D. (1993). Invertebrates associated with moss communities at Canada Glacier, southern Victoria Land, Antarctica. *Polar Biology*, **13**, 157–162.

Seibert, N. M. and Kargel, J. S. (2001). Small-scale Martian polygonal terrain: implications for liquid surface water. *Geophysical Research Letters*, **28**, 899–902.

Sizemore, H. G. and Mellon, M. T. (2006). Effects of soil heterogeneity on martian ground-ice stability and orbital estimates of ice table depth. *Icarus*, **185**, 358–369.

Sletten, R. S., Hallet, B., and Fletcher, R. C. (2003). Resurfacing time of terrestrial surfaces by the formation and maturation of polygonal patterned ground. *Journal of Geophysical Research, Planets*, **108**, 8044.

Squyres, S. W. (1989). Water on Mars. *Icarus*, **79**, 229–288.

Squyres, S. W., Grotzinger, J. P., Arvidson, R. E., et al. (2004a). In situ evidence for an ancient aqueous environment at Meridiani Planum, Mars. *Science*, **306**, 1709–1714.

Squyres, S. W., Arvidson, R. E., Bell, III, J. F., et al. (2004b). The Opportunity Rover's Athena science investigation at Meridiani Planum, Mars. *Science*, **306**,1698–1703.

Sugden, D. E., Marchant, D. R., Potter, N., et al. (1995). Preservation of Miocene glacier ice in East Antarctica. *Nature*, **376**, 412–414.

Taton, A., Grubisic, S., Brambilla, E., De Wit, R., and Wilmotte, A. (2003). Cyanobacterial diversity in natural and artificial microbial mats of Lake Fryxell (McMurdo Dry Valleys, Antarctica): a morphological and molecular approach. *Applied and Environmental Microbiology*, **69**, 5157–5169.

Titus, T. N., Kieffer, H. H., and Christensen, P. R. (2003). Exposed water ice discovered near the South Pole of Mars. *Science*, **299**, 1048–1051.

Tosi, S., Onofri, S., Brusoni, M., Zucconi, L., and Vishniac, H. (2005). Response of Antarctic soil fungal assemblages to experimental warming and reduction of UV radiation. *Polar Biology*, **28**, 470–482.

Treonis, A. M., Wall, D. H., and Virginia, R. A. (1999). Invertebrate biodiversity in Antarctic dry valley soils and sediments. *Ecosystems*, **2**, 482–492.

Treonis, A. M., Wall, D. H., and Virginia, R. A. (2000). The use of anhydrobiosis by soil nematodes in the Antarctic Dry Valleys. *Functional Ecology*, **14**, 460–467.

Ugolini, F. C. and Bockheim, J. G. (2008). Antarctic soils and soil formation in a changing environment. *Geoderma*, **114**, 1–8.

Wentworth, S. J., Gibson, E. K., Velbel, M. A., and McKay, D. S. (2005). Antarctic Dry Valleys and indigenous weathering in Mars meterorites: implications for water and life on Mars. *Icarus*, **174**, 383–395.

Weber, K. A., Achenbach, L. A., and Coates, J. D. (2006). Microorganisms pumping iron: anaerobic microbial iron oxidation and reduction. *Nature Reviews Microbiology*, **4**, 752–764.

Westall, F. (2005). Early life on Earth and analogies to Mars. In *Water on Mars and Life*, ed. T. Tokano. Berlin: Springer, pp. 45–64.

Witherow, R. A., Lyons, W. B., Bertler, N. A. N., et al. (2006). The aeolian flux of calcium, chloride and nitrate to the McMurdo Dry Valleys landscape, evidence from snow pit analysis. *Antarctic Science*, **18**, 497–505.

Wood, S. A., Rueckert, A., Cowan, D. A., and Cary, S. C. (2008). Sources of edaphic cyanobacterial diversity in the Dry Valleys of Eastern Antarctica. *ISME Journal*, **2**, 308–320.

Wynn-Williams, D. D. and Edwards, H. G. M. (2000). Proximal analysis of regolith habitats and protective biomolecules in situ by laser Raman spectroscopy: overview of terrestrial Antarctic habitats and Mars analogs. *Icarus*, **144**, 486–503.

Yen, A. S., Gellert, R., Schroder, C., et al. (2005). An integrated view of the chemistry and mineralogy of Martian soils. *Nature*, **436**, 49–54.

Zeglin, L. H., Sinsabaugh, R. L., Barrett, J. E., Gooseff, M. N., and Takacs-Vesbach, C. D. (2009). Landscape distribution of microbial activity in the McMurdo Dry Valleys: linked biotic processes, hydrology and geochemistry in a cold desert ecosystem. *Ecosystems*, **12**, doi: 10.1007/s10021–009–9242–8.

Zuber, M. T. (2007). Mars at the tipping point. *Nature*, **447**, 785–786.

4

The antarctic cryptoendolithic microbial ecosystem

HENRY J. SUN, JAMES A. NIENOW,
AND CHRISTOPHER P. MCKAY

Introduction

The antarctic cryptoendolithic microbial ecosystem lives under sandstone surfaces in the dry valley region (Friedmann and Ocampo, 1976; Friedmann, 1977). It is relatively simple, consisting of cyanobacterial or algal primary producers, fungal consumers, and bacterial decomposers. It lacks animals and, possibly, also archaea. With rock temperatures rising above 0 °C only for a few weeks in the austral summer to allow photosynthetic productivity, this ecosystem is permanently poised on the edge of existence.

Before we talk about these specific rock-inhabiting organisms, it is useful to be familiar with all lithophytic life forms. *Epilithic* organisms live on rocks. *Endolithic* organisms grow inside rocks, with three subcategories that denote the mode of entry or the presence or absence of a protective surface crust (Golubic et al., 1981). *Euendolithic* algae and cyanobacteria actively bore into limestone in the intertidal zone and, occasionally, in deserts (Friedmann et al., 1993a; Garty, 1999). *Chasmoendolithic* organisms occupy weathering cracks and fissures in a variety of rocks. *Cryptoendolithic* organisms colonize pre-existing pore spaces in translucent rocks, most commonly sandstones (Friedmann and Ocampo-Friedmann, 1984; Bell, 1993, Nienow et al., 2002; Omelon et al., 2006). The colonized zone, in this case, is covered by a silicified surface crust. Other cryptoendolithic substrata include granite, gneiss, limestone, marble (see Nienow and Friedmann, 1993; van Thielen and Garbary, 1999; Cockell et al., 2002; Nienow et al., 2002 and the references therein), gypsum (Hughes and Lawley, 2003; Boison et al., 2004; Dong et al., 2007), halite and evaporite (Rothschild et al., 1994; Wierzchos et al., 2006) and sinters

Life in Antarctic Deserts and Other Cold Dry Environments: Astrobiological Analogs, ed. Peter T. Doran, W. Berry Lyons and Diane M. McKnight. Published by Cambridge University Press.

associated with geothermal environments (Walker et al., 2005; Phoenix et al., 2006). *Hypolithic* organisms colonize the rock–soil interface underneath desert pavements and elsewhere, usually under quartz or other translucent clasts (Schlesinger et al., 2003; Warren-Rhodes et al., 2006; Pointing et al., 2007). There are exceptions though: in the Canadian High Arctic and the Antarctic Peninsula, hypolithic cyanobacteria live under opaque dolomite, sustained presumably by diffusive light (Cockell and Stokes, 2004).

All major lithobiontic habitats occur in Antarctica (see Friedmann, 1982; Nienow and Friedmann, 1993; Broady, 1996; Nienow et al., 2002). In this chapter we focus on the most prevalent and best studied cryptoendolithic communities in the high mountainous regions of the McMurdo Dry Valleys (Ross Desert).

The Ross Desert

The high-altitude regions of the McMurdo Dry Valleys resemble a rocky desert, with exposed outcrops, barren soils, and stone pavements (Fig. 4.1). Unlike in most other deserts, however, the lifeless appearance in this polar desert is created more by low temperature than by the lack of precipitation. Here, the daily mean air temperature is −5 °C at the height of summer (mid-January) and fluctuates between −20 °C and −40 °C during the rest of the

Fig. 4.1. Landscape of Battleship Promontory, Dry Valleys, Antarctica.

year (Friedmann et al., 1987; McKay et al., 1993). Annual precipitation, all in the form of snow, is on the order of tens of centimeters (Nienow and Friedmann, 1993; Fountain et al., 1999). Most of the snow is not biologically available, however. It is either blown to the valleys or sublimates without melting (Friedmann, 1978; Friedmann et al., 1987; McKay et al., 1993). Year-round subzero temperatures, together with low water activity, result in a true desert that is not only devoid of plant life, but also, to a large extent, devoid of epilithic lichens. The soils are low in organic content and in microorganisms (Vishniac, 1993, 2002; Virginia and Wall, 1999). By and large, the interior of sandstone rocks is the only place that supports photosynthesis.

The rock substratum and biogenous weathering

The Beacon sandstone is by far the most widespread cryptoendolithic substratum in the McMurdo Dry Valleys region, although granite and a few other rock types are also present (Nienow and Friedmann, 1993). Colonized sandstone rocks display a distinct exfoliation pattern, where white, yellow, orange, and brown surface crusts, centimeters in size, are arranged in a patchwork-like, mosaic fashion (Fig. 4.2). This pattern is the result of interplay between a constructive aeolian weathering process that acts on the rock surface and a destructive weathering process that occurs within the rock.

Fig. 4.2. Surface of colonized sandstone rock on Battleship Promontory. Scale 10 cm.

The first process stains as well as stabilizes the rock surface, while the second process weakens sandstone cohesion and, on the estimated time scale of 10^4 years, causes the surface crust to exfoliate. The resultant mosaic pattern on the rock surface is a telltale sign of life below.

The colonized zone begins about one millimeter below the rock surface and extends to a depth of several, up to ten, millimeters. Multiple layers can develop within the microbial zone. The lichen-dominated community, which is most widespread, consists of a black, a white, and a green layer, followed by a reddish iron accumulation zone (Fig. 4.3a). The cyanobacterial community has a similar banding pattern, but lacks an accumulation zone (Fig. 4.3b). Communities dominated by eukaryotic algae do not possess multiple colored bands or an accumulation zone. In all three community types microorganisms

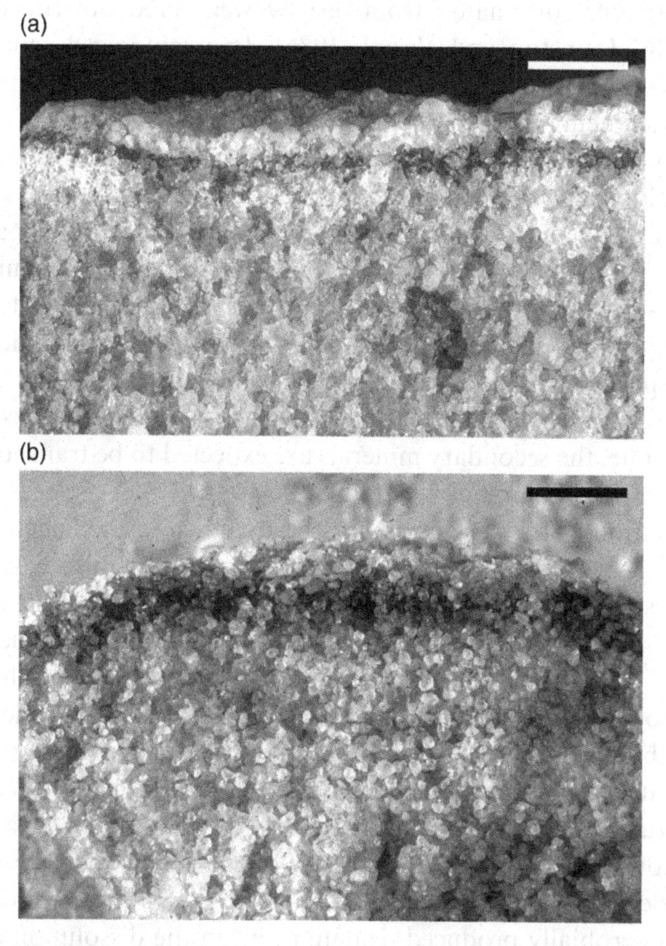

Fig. 4.3. (a) Cleaved sandstone to show the lichen-dominated community. (b) Cleaved rock to show the cyanobacterial *Hormathonema–Gloeocapsa* community. Scale 5 mm. (From Friedmann, 1982, by permission of the American Association for the Advancement of Science.)

grow adhered to sand grains. This physical relationship is shown in Fig. 4.4, using the lichen-dominated community as an example.

Colonized sandstone rocks have a well-sorted grain size between 0.15 and 0.8 mm and a porosity of 8 to 15% (Nienow et al., 1988a; Greenfield, 1988; Blackhurst et al., 2004). Feldspars and clays, including koalinite and illite, are present in the matrix (Blackhurst et al., 2004, 2005). Iron oxyhydroxides and iron oxides, including goethite and hematite, are present either as part of the matrix or as coating on sand grains (Weed and Ackert, 1986; Friedmann and Weed, 1987; Edwards et al., 2004). Minor elements present in the sandstone include aluminum, calcium, lead, magnesium, manganese, phosphorus, potassium, sodium, and titanium (Friedmann, 1982; Johnston and Vestal, 1989; Blackhurst et al., 2005). Some of these elements are probably derived from aeolian dust originated from nearby weathered dolerite (Weed and Ackert, 1986; Johnston and Vestal, 1986). In uncolonized sandstones the vertical distribution of constituent minerals is comparatively uniform, with a slight enrichment of clays and feldspars in the rock surface (Blackhurst et al., 2005). In contrast, in rocks containing the lichen-dominated community, the colonized zone is impoverished in metal elements relative to the surface crust above and the accumulation zone below (Johnston and Vestal, 1989). Blackhurst et al. (2004) reported that feldspars follow a similar profile. In a follow-up study, however, Blackhurst et al. (2005) reported that the colonized layer was enriched, instead of impoverished, in clay minerals. This particular specimen may represent the stage immediately after exfoliation, when the community is growing in what was previously the accumulation zone. With time, the secondary minerals are expected to be transported below the organisms (Blackhurst et al., 2005).

The formation of the surface crust is best studied in the lichen-dominated community. It begins with the deposition of wind-blown dust, which is rich in quartz, clays, and iron oxyhydroxides (Weed and Ackert, 1986; Friedmann and Weed, 1987; Weed and Norton, 1991). In time these materials are transformed into a thin, less than 100 μm thick, "veneer." The chemical and physical processes involved in this conversion are poorly understood but thought to be similar to those leading to the formation of desert varnish in arid and semiarid regions (Weed and Ackert, 1986).

The encrusted rock surface is resistant to wind-driven, grain-by-grain disintegration. Instead, colonized rocks undergo an exfoliative weathering process (Friedmann and Weed, 1987). The long-term presence of the organisms and microbially produced oxalate result in the dissolution of siliceous cementing substances between the sand grains (Johnston and Vestal, 1993). Freeze–thaw cycles and perhaps the expansion of the growing community may trigger the exfoliation of the surface crust. Most of the colonized zone is

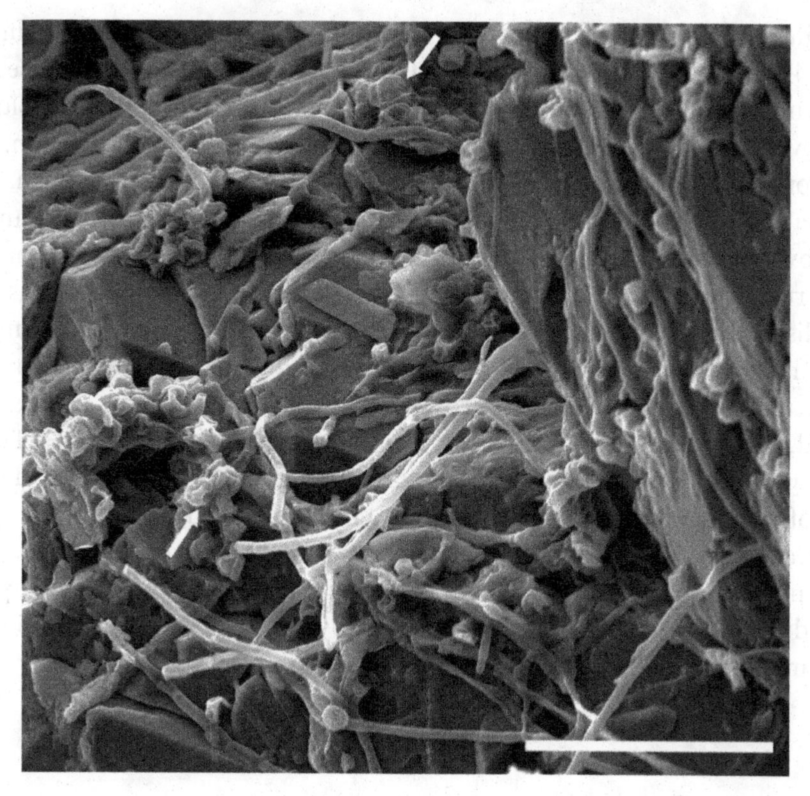

Fig. 4.4. Scanning electron micrograph of sandstone colonized by the lichen-dominated community. Arrows indicate lichen associations between algal cells and fungal hyphae. Scale 50 μm.

now lost to the environment. Subsequently, the remaining organisms grow deeper, while a fresh crust forms on the surface. The newly exposed surface is snow white, but in time it incorporates iron oxyhydroxides and hematite (Fe_2O_3) through aeolian weathering and darkens to yellow, orange, and brown. Some of the iron may be mobilized from the rock below by microbial activity (Edwards et al., 2004). The exfoliative weathering process is cyclic, alternating between long periods of crust formation and sudden exfoliation events, with each cycle estimated to last 10^4 years (Friedmann et al., 1993b). The resultant mosaic pattern on the rock surface, in essence a macroscopic manifestation of microbial activity below, can be preserved as trace fossil under certain conditions (Friedmann and Weed, 1987).

The surface color is, therefore, a relative indicator of community age (Fig. 4.5 top panel). White corresponds to the youngest community at the beginning of the exfoliation cycle, while dark brown corresponds to the oldest, prior to exfoliation. By quantifying biomass under crusts of different ages, Sun and Friedmann (1999) reconstructed the community growth curve.

Despite the ample time for growth, the community barely doubles in size (Fig. 4.5 bottom panel, solid part of the growth curve). At first glance, this confirms the long-held notion that the organisms grow extremely slowly. However, this result should not be taken as a measure of cell division. Cell division should be much faster if one considers what must happen after exfoliation. The colonized zone is "shaved off," and the remaining organisms are now too close to the lethal rock surface by about 5 mm. Yet, as soon as a new crust is formed (i.e., while the crust is still white), the community is fully established, complete with black-white-(green) zonation. The formation time of a white crust is, at most, two or three thousand years. For the micrometer-sized, nonmotile organisms to grow over a distance of 5 mm in this time period, they would have to divide at least once every two years. Therefore, a rapid growth phase must follow exfoliation (Fig. 4.5 bottom panel, dashed part of the curve). By the time the new crust is formed, the steady state is reached where growth is balanced by death and degradation. The small amount of biomass increase between the white and the brown crust is probably due to *de novo* pore spaces created by mineral dissolution (Fig. 4.5 bottom panel, solid part of the growth curve).

Microbiota

Early surveys described five different communities, named after the dominant phototrophs: (1) cyanobacterium *Chroococcidiopsis*, (2) lichens, (3) green alga *Hemichloris*, (4) red cyanobacterium *Gloeocapsa*, (5) cyanobacteria *Hormathonema–Gloeocapsa* (Friedmann et al., 1988). Each community supports numerous heterotrophic fungi and bacteria (Nienow and Friedmann, 1993; Broady, 1996; Siebert et al., 1996; Vishniac, 1996, 2002; de la Torre et al., 2003; Selbmann et al., 2005).

The Chroococcidiopsis *community*

This is the first cryptoendolithic community discovered in the dry valleys (Friedmann and Ocampo, 1976; referred to as *Gloeocapsa*). At first, this finding appeared to confirm the notion of a harsh environment in the dry valleys. In extreme hot deserts the single-celled, desiccation-resistant *Chroococcidiopsis* is often the dominant or sole photoautotroph (Friedmann and Galun, 1974, again, *Gloeocapsa*, instead of *Chroococcidiopsis*, was used; Friedmann and Ocampo-Friedmann, 1995; Billi et al., 2000). Later, however, as other, more widespread endolithic communities were discovered here, it was recognized that *Chroococcidiopsis* is rather rare and that, from a metabolic standpoint, the antarctic cold desert is not as harsh as some of the hot deserts.

The lichen-dominated community

The colonized zone consists of a black, a white, and a green layer (Fig. 4.3a). The black and white layers are populated by cryptoendolithic lichens. Characteristic lichen structures such as soredia, appressoria, and haustoria have been observed (Fig. 4.4; also Friedmann, 1982). Some of the fungi have been cultured and their lichenizing nature has been confirmed in reconstitution experiments (Ahmadjian and Jacobs, 1987; Koriem and Friedmann, 1987). The black layer is due to the presence of dark-pigmented parasymbionts, non-lichen-forming fungi that live in association with the lichens. The green layer below is formed by a free-living green alga, *Hemichloris antarctica*, and, occasionally, also *Chroococcidiopsis*. Sometimes, fungal hyphae extend into this zone, but they do not form lichen associations.

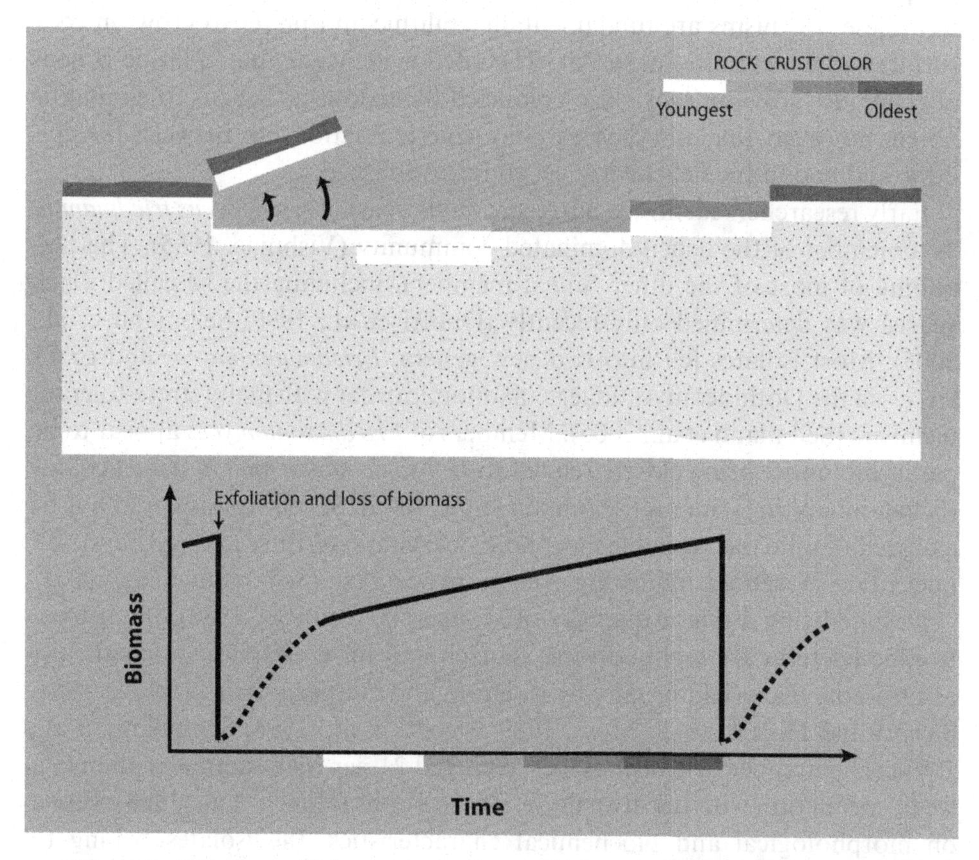

Fig. 4.5. Diagram of biogenous rock weathering and associated microbial growth. Following exfoliation and loss of most of the colonized zone, the remaining organisms grow into the sandstone under a new forming crust. Growth is presumably rapid initially, followed by an extended quasi-steady state. (Modified from Sun and Friedmann, 1999.)

The photobionts have been variously identified as *Pseudotrebouxia* and *Trebouxia*, single-celled green algae with a characteristic central plastid (Friedmann, 1982; Archibald et al., 1983). There is some confusion in the literature over the taxonomy of these two genera. The genus *Pseudotrebouxia* was originally erected by Archibald (1975) to accommodate *Trebouxia*-like species that reproduce by vegetative cell division in addition to autospore and zoospore formation. Gärtner (1985) questioned the existence of vegetative cell division and incorporated *Pseudotrebouxia* species into *Trebouxia*. Subsequent molecular work supports this decision (see Friedl and Büdel, 1996; Helms, 2003; DePriest, 2004, for discussions), but the synonym *Pseudotrebouxia* still persists in the literature.

The cryptoendolithic lichens are considered a morphogenetic adaptation of epilithic lichens (Friedmann et al., 1980a; Hale, 1987). The assumption is that these organisms are fundamentally epilithic and prefer to grow on rock surfaces where conditions permit. The cited evidence is that epilithic lichens occasionally grow on rocks also colonized by endolithic lichens. It should be noted, however, that the physical and genetic relationship between the epilithic and endolithic lichens has yet to be established.

Early researchers identified a psychrophilic yeast, *Cryptococcus friedmannii*, as a member of the lichen-dominated community (Vishniac, 1985). The taxonomy of most of the other fungal members, including the so-called black yeasts, was determined only recently (Onofri et al., 1999; Selbmann et al., 2005). Most isolates fell into two new genera, *Friedmanniomyces* and *Cryomyces*, in the Dothideomycetidae, a taxonomic order with many stress-tolerant members (Selbmann et al., 2005). Members of *Friedmanniomyces* appear to be parasymbiontic, being closely related to *Hobsonia santessonii*, a lichenicolous (lichen-inhabiting) fungus (Selbmann et al., 2005). Seven strains could not be assigned despite the available morphological characteristics and SSU and ITS nucleotide sequences, indicating even more new taxa (Selbmann et al., 2005).

Despite being a minor quantity of biomass (Greenfield, 1988), the heterotrophic bacteria are highly diverse. Hirsch and his colleagues systematically assessed the bacterial diversity by isolation and culturing (Hirsch et al., 1988; Siebert and Hirsch, 1988; Mevs, 1989; Siebert et al., 1996; Schumann et al., 1997). The number of morphotypes averaged 21 per rock sample, with only a weak correlation with the diversity of phototrophs (Hirsch et al., 1988). Based on morphological and biochemical characteristics, the isolates belong to *Deinococcus, Arthrobacter, Bifidobacterium, Brevibacterium, Corynebacterium, Geodermatophilous, Micrococcus, Micromonospora, Mycobacterium, Mycococcus, Nocardia*, and *Streptomyces* (Hirsch et al., 1988; Siebert and Hirsch, 1988). Except for *Deinococcus*, all the genera are in the class Actinobacteria.

Colorless flagellates were isolated from one sample from Linnaeus Terrace (Hirsch et al., 1988), despite the fact that protozoans and invertebrates are generally absent from the antarctic endolithic habitat. From a single boulder, Siebert et al. (1996) isolated 41 bacterial morphotypes, distinguished by Gram stain, cell morphology, carbon substrate utilization, organic acid secretion, antimicrobial compound production, and response to algal substances. A new actinomycete genus, *Friedmanniella*, was later described on the basis of biochemical characteristics and 16S rRNA sequences (Schumann et al., 1997). Hirsch et al. (1988) speculated that some of the bacteria may be chemolithotrophic. This idea was dismissed by Vestal (1988b), who found no evidence of chemolithotrophy utilizing ammonium, nitrite, sulfate, or thiosulfate.

Recently, the community was characterized by using the culture-independent approach (de la Torre et al., 2003). Of the 676 SSU rRNA sequences cloned, 25 are distinct. One dominant algal nuclear sequence is closely related to *Trebouxia jamesii*. *Trebouxia jamesii*, originally described as *Pseudotrebouxia jamesii* and a common phycobiont in European lichens (Ettl and Gärtner, 1995), was also found in antarctic lichens *Umbilicaria* (Romeike et al., 2002). A second common phylotype belongs to a plastid, probably of *T. jamesii*. The most common fungal sequence was almost identical (97%) to that of the lichen-forming fungus *Texosporium sancti-jacobi*. Interestingly, some of the epilithic lichens from Linnaeus Terrace were identified as *Buellia* (Friedmann et al., 1980a; Hale, 1987), also a close relative of *Texosporium* (Helms, 2003).

The remaining phylotypes are distributed among fungi, mitochondria, Actinobacteria, α-Proteobacteria, γ-Proteobacteria, Cytophagales, Planctomycetales, and unknown bacteria (de la Torre et al., 2003). The total number of bacterial phylotypes is similar to the number of bacterial morphotypes identified by Hirsch et al. (1988). Interestingly, no archean sequences were detected.

The Hemichloris *community*

The green alga *Hemichloris*, which forms the bottom green layer in the lichen-dominated community, can form a community of its own under the lower surface of overhanging sandstone ledges. Despite being shaded and freed from the lichens, the organism does not grow closer to the rock surface but remains at a depth of one centimeter, suggesting that it is narrowly adapted to low light (Nienow et al., 1988b). Unlike the endolithic lichens, *Hemichloris* does not leach iron. Two *Hemichloris* species have been described, *H. antarctica* (Tschermak-Woess and Friedmann, 1984) and *H. polyspora* (Tschermak-Woess et al., 2006), both of which are considered endemic to Antarctica.

Another green algal member of this community is *Stichococcus bacillaris*, which is characterized as rod shaped and possessing a single parietal plastid (Nienow and Friedmann, 1993; Broady, 1996; Hughes, 2006). Members of the genus *Stichococcus* are commonly found in subaerial habitats (Ettl and Gärtner, 1995). But the occurrence of these organisms in endolithic habitats is new. Other photosynthetic organisms found in the *Hemichloris* community include the xanthophycean *Heterococcus endolithicus* (Darling et al., 1987) as well as cyanobacteria *Gloeocapsa* sp. and *Chroococcidiopsis* sp. The cyanobacteria reside below the green algae and sometimes form a separate band. Other unidentified cyanobacteria were isolated by Hirsch et al. (1988). Nonpigmented fungi are present, but they do not appear to form lichen associations with the algae or the cyanobacteria.

The red Gloeocapsa *community*

The dominant *Gloeocapsa* species possesses a reddish, lamellated sheath (Friedmann et al., 1988). It is often accompanied by a smaller, yellowish *Gloeocapsa*. Other cyanobacteria, including *Gloeocapsa*, *Eucapsis*, and *Microchaete*, are encountered occasionally. The ecological preference of this community is similar to that of the lichen-dominated community. Both colonize the upper, drier parts of the sandstone boulders and produce a leached zone.

The Hormathonema–Gloeocapsa *community*

This community is found only on Battleship Promontory. In contrast to the lichen-dominated community, which occupies drier parts of rocks and boulders, this community is restricted to rocks that are close to the ground and permanently moistened. The colonized zone consists of an upper, dark-brown band and a lower greenish band (Fig. 4.3b). The upper layer is dominated by *Hormathonema*, *Gloeocapsa* and, sometimes, *Anabaena* (Friedmann et al., 1988). The lower band is dominated by *Aphanocapsa*, occasionally accompanied by *Gloeocapsa* and *Anabaena*, and *Lyngbya*. Enrichment cultures show an even greater diversity. Hirsch et al. (1988) isolated 17 cyanobacteria and six eukaryotic algae from one specimen. There appears to be a correlation between cyanobacterial diversity and rock pH, with alkaline conditions supporting more species (Hirsch et al., 1988).

Hirsch et al. (1988) also found a great diversity of heterotrophs. Three samples contained flagellated protozoans. Nearly all samples contained filamentous fungi and/or yeasts. As many as 39 morphologically distinct heterotrophic bacteria were present in a single sample. Two samples from Battleship Promontory contained no cyanobacteria or eukaryotic algae, but they contained a small number of bacteria. These bacteria are either chemolithotrophic or survive on relic organic material (Hirsch et al., 1988).

The culture-independent assessment by de la Torre et al. (2003) revealed 26 unique phylotypes. Only four belong to cyanobacteria, considerably fewer than the number of cyanobacterial morphotypes described by Friedmann et al. (1988) and Hirsch et al. (1988). All four sequences cluster with the *Lyngbya–Phormidium–Plectonema* (LPP) group. Filamentous cyanobacteria do exist in the community, but they are rare compared to the unicellular forms such as *Gloeocapsa* or *Hormathonema* (Friedmann et al., 1988). There are three possible explanations for the discrepancy between the classical approach and the DNA-based method. First, the unicellular cyanobacteria could be resistant to cell lysis and DNA extraction and, consequently, might have escaped detection. Second, morphology may not be a stable character in this group of organisms. It is known that some cyanobacteria can switch between the unicellular, *Gloeocapsa*-like morphology and the filamentous, *Scytonema*-like growth form (Marton and Galun, 1976). Third, species composition varies from one part of the specimen to another. The different studies may have sampled different communities (de la Torre et al., 2003).

Of the other phylotypes recovered by de la Torre et al. (2003), four belong to α-Proteobacteria, one to the *Thermus–Deinococcus* group, 13 to Actinobacteria, one to Acidobacteria, two to green nonsulfur bacteria, and one to the Cytophagales–Flavobacteria–Bacterioides group. One phylotype has no known relatives. The number of Actinobacteria is consistent with the work of Hirsch et al. (1988). One of the α-Proteobacteria, which represented 31% of the 480 sequence clones, showed 94% sequence identity with *Blastomonas ursincola* (de la Torre et al., 2003). *Blastomonas ursincola* (=*Erythromonas ursincola*) is a bacteriochlorophyll α-containing member of the Sphingomonadales (Sly and Hugenholtz, 2005). Hughes and Lawley (2003) isolated a yellow-pigmented *Sphingomonas* from gypsum crusts, suggesting this group may be common in Antarctica. The purported photosynthetic activity in these organisms is interesting but requires substantiation (de la Torre et al., 2003). Species of *Blastomonas* need reduced carbon for growth and are generally considered to be photoheterotrophic. In addition, they make only a small amount of bacteriochlorophyll and lack ribulose bis-phosphate carboxylase, a key enzyme in CO_2 fixation (Sly and Hugenholtz, 2005). De la Torre et al. (2003) also speculated that a member of the *Thermus–Deinococcus* group may be "engaged in primary productivity," because of its high abundance in their clone library (26% of 480).The Actinobacteria is another group that is possibly capable of some degree of autotrophy (Hirsch et al., 1988). No eukaryotic or archean sequences were detected (de la Torre et al., 2003).

Ecology

The physicochemical environment in the cryptoendolithic habitat

The physical environment, both in and outside the sandstone, has been the subject of direct measurement by automatic meteorology stations (Friedmann et al., 1987; McKay et al., 1993) and of mathematical models (Nienow et al., 1998a, 1998b). The data and findings from these efforts were detailed elsewhere (Friedmann et al., 1987; McKay et al., 1993; Nienow and Friedmann, 1993). Here we restrict our discussion to a few key points.

Temperature

Temperatures high enough to support metabolic activity occur mostly between mid November and mid February. During this period, the maximum daily air temperature approaches 0 °C, and the rocks are several degrees warmer due to insolation (Friedmann et al., 1987; McKay et al., 1993). Not all rocks are equally warmed, however, due to differences in surface orientation and albedo. North-facing surfaces can be as much as 13 °C warmer than south-facing surfaces (Friedmann et al., 1987; Nienow et al., 1988a; McKay et al., 1993). This difference is thought to be the key reason that colonization is more prevalent in northern exposures than in southern exposures (Kappen et al., 1981; Friedmann, 1982; McKay and Friedmann, 1985). Similarly, insufficient insolation may be the reason why some light-colored rocks on Linnaeus Terrace are not colonized. It should be kept in mind, however, that none of these restrictions are absolute. On Battleship Promontory, a relatively warmer location, most surfaces are colonized regardless of orientation and color. At Two Step Cliffs on Alexander Island, east-facing surfaces, instead of north-facing ones, are the most extensively colonized (Hughes and Lawley, 2003). Finally, although there is a steep temperature gradient between a rock surface and the overlying air, there are no sharp temperature transitions within a rock that can explain the sometimes sharp boundaries in colonization (Nienow et al., 1988a).

Light

Less than 1% of the incident light penetrates the colonized sandstone (Nienow et al., 1988b). This has two ecological implications. First, it means that photosynthesis can take place only in summertime when photon flux is sufficient to overcome the extinction loss. The period of most vigorous primary productivity should be from December to January, when the average daily maximum of photosynthetic photon flux (PPF) varies between 1100 and 1250 $\mu mol\ m^{-2}\ s^{-1}$ (Friedmann et al., 1987; McKay et al., 1993). Second, as we will discuss below, small differences in access to sunlight can determine whether a rock is or is not colonized.

Within the colonized zone there is a steep light gradient with depth. Measured gross extinction coefficients in the visible light are 1.2–3.0 mm^{-1}, about 70 to 95% loss per millimeter (Nienow et al., 1988b). With depth, there also appears to be a spectral shift towards the red. The quantitative and qualitative change may be the cause for the zonation or banding pattern observed in some of the communities. The degree of light penetration apparently defines the depth of colonization. The importance of light in structuring endolithic communities has been confirmed by Hughes and Lawley (2003) and Matthes et al. (2001) in gypsum and limestone systems respectively.

Ultraviolet light is blocked by the iron oxide in the sandstone surface (Nienow et al., 1988b). The organisms below are, therefore, protected from UV radiation. This doesn't necessarily mean, however, that the endolithic lifestyle is an ecological response of UV avoidance. The fact that epilithic lichens can exist here in full exposure suggests that survival does not require the benefit of being inside the rock. Being inside sandstone does not always carry the benefit of UV protection either, as is the case with the *Hormathonema–Gloeocapsa* community due to the absence of an oxidized crust. These organisms protect themselves by secreting the UV screen compound scytonemin (Wynn-Williams et al., 1999).

Moisture

Snow is the only form of precipitation in the antarctic desert. However, in a topographically dynamic landscape with strong wind, not all surfaces have equal access to snow. Differences in access to snow appear to control the distribution of colonization under sandstone surfaces in a way that is more important than temperature (H. J. Sun, unpublished data). Snow does not accumulate on cliffs. These surfaces, therefore, are too dry to support life regardless of orientation. Sloped surfaces are well positioned to catch snow and are usually always colonized. The situation of a horizontal surface depends on its topographic context. If the surface is situated in a low-lying area or leeward of a large boulder such that it is prone to burial by drift snow and to consequent deprivation of sunlight, it tends to preclude life. In contrast, a flat surface that is well elevated and without the risk of burial is suitable for cryptoendolithic colonization.

A thin layer of snow can melt on rocks in the austral summer (Kappen et al., 1981; Friedmann et al., 1987; McKay et al., 1993). The melt is quickly absorbed by the porous sandstone and retained for days or even weeks. Several factors contribute to water retention. First, evaporation is very slow, 1 to 2 mm $m^{-2} d^{-1}$, due to low temperature (Kappen et al., 1981; Nienow et al., 1988a). Second, the surface crust may retard evaporation (Kappen and Friedmann, 1983), although this effect is not fully supported by field data (Kappen et al., 1981). Third, thin films of water may be retained on internal surfaces (Palmer and Friedmann, 1990). This matric effect may also depress the freezing point of water, thereby extending the time during which liquid water is available (Meyer et al., 1988; Hughes and Lawley, 2003).

Nutrients

There is no evidence of nitrogen limitation in the antarctic cryptoendolithic communities. To the contrary, the organisms are awash in nitrate and ammonium ions of atmospheric origin (Friedmann and Kibler, 1980; Vestal, 1988c; Johnston and Vestal, 1991). As a result, although nitrogen-fixing cyanobacteria exist in the communities, biological nitrogen fixation is rare (Friedmann and Kibler, 1980). The addition of nitrogen nutrient does not stimulate carbon fixation in colonized rocks (Johnston and Vestal, 1991).

No evidence exists for phosphorus limitation either. Phosphorus is abundant in the sandstone in mineral-bound forms and as part of the rock matrix (Johnston and Vestal, 1989; Banerjee et al., 2000). Biologically available forms are scarce though (Johnston and Vestal, 1989; Banerjee et al., 2000), probably due to sequestration by metal oxides in the system (Johnston and Vestal, 1989). As with nitrogen, the addition of phosphate does not stimulate carbon fixation (Johnston and Vestal, 1991).

One possible explanation for the lack of nutrient limitation is that the organisms, being naturally slow growing, do not have a great demand for nutrients (Friedmann and Kibler, 1980). In the case of phosphorus, the need is, at least in part, satisfied by recycling. Thus, Banerjee et al. (2000) found high rates of phosphatase activity in all communities examined and estimated that it would take only 30 hours of activity at 5 °C to extract enough phosphorus to double the community mass (Banerjee et al., 2000).

Adaptations

The antarctic cryptoendolithic ecosystem clearly owes its existence to the protective rock substratum. Protection can only go so far, however, and survival in this extreme habitat still requires active adaptation on the part of the organisms.

Temperature

Contrary to expectation, the antarctic endolithic organisms are not psychrophiles, but psychrotolerant mesophiles. The majority of the bacteria isolated by Hirsch and his colleagues grow best between 20 °C and 30 °C, although they are able to grow at lower temperatures, between 4 °C and 9 °C (Hirsch et al., 1988; Siebert and Hirsch, 1988; Mevs, 1989, Siebert et al., 1996). Similar temperatures are also required by the growth of the black fungi (Selbmann et al., 2005), yeast (Vishniac, 1985, 1996), and eukaryotic algae (Tschermak-Woess and Friedmann, 1984; Darling et al., 1987; Ocampo-Friedmann et al., 1988; Tschermak-Woess et al., 2006).

The fact that the antarctic organisms are, as a rule, mesophiles, with optimum growth temperatures between 20 °C and 30 °C, is puzzling. The most frequently occurring rock temperatures in the austral summer, when the organisms are metabolically active, are much lower, between −10 °C and 5 °C (Friedmann et al., 1993). These values do not even overlap with the optimal range of the organisms, so it would appear that the organisms are ill adapted to their environment. However, a different picture emerges if one looks at the temperature relationship of the community. Maximal photosynthetic gain in colonized rock occurs, depending on light level, between −2 °C and 5 °C (Friedmann et al., 1993). Unlike its members, the community is clearly well adapted.

How could the community be optimal while the member species are suboptimal? Friedmann and Sun (2005) hypothesized that microbial communities can adapt to a wide range of thermal regimes, sometimes outside the optimal range of the member species, by regulating their Rp/c, the ratio of algal primary producers and fungal consumers. In each thermal regime, there is an optimum Rp/c that maximizes the fitness of the community. In warmer habitats, however, respiration increases more steeply than does photosynthesis, resulting in a net carbon loss. To compensate, the community must increase its Rp/c. Sun and Friedmann (2005) validated this prediction in the cosmopolitan lichen *Cladina rangiferina* and found that Rp/c is indeed correlated with habitat temperature. The community adaption theory also explains why lichens, in general, are well adapted to their thermal regime. The high degree of adaptability in lichens is attributed to community adaptation and, only to a limited degree, to genetic adaptation or photosynthetic acclimation, two mechanisms that are available to individual photosynthetic organisms (Sun and Friedmann, 2005).

Another adaptation key to survival in the antarctic desert is the ability to withstand freeze–thaw cycles. In the laboratory, cultures of *Hemichloris antarctica* and *Trebouxia* showed no signs of cellular damage or reduction in photosynthetic carbon fixation after exposure to a rapid succession of freeze–thaw cycles (Meyer et al., 1988). A strain of *Stichococcus* isolated from a relatively milder antarctic habitat also showed resistance to freeze–thaw stress, although to a lesser degree (Meyer et al., 1988).

Light

The low flux of photons in the endolithic habitat is an advantage, not a disadvantage. Without the sheltering by the rock, exposure to full sunlight would cause photoinhibition and cellular damage (Weber et al., 1996; Hughes, 2006). However, while the visible light is reduced, the transmission of UV may still be significant in some communities (Jorge Villar et al., 2005a). Many microorganisms can produce UV screen compounds in response to exposure (Potts, 1994; Wynn-Williams et al., 2002). A number of such compounds have been detected in colonized rocks and in isolated organisms. These include melanin in dark-colored fungi (Selbmann et al., 2005), carotenoids in cyanobacteria and/or algae (Jorge Villar et al., 2005a, 2005b; Tschermak-Woess et al., 2006), and scytonemin in cyanobacteria (Wynn-Williams et al., 1999; Jorge Villar et al., 2005a, 2005b). Somewhat anomalously, Jorge Villar et al. (2005b) found evidence for scytonemin in "the black area (fungal hyphae)" of a pale colonized stone, presumably from Battleship Promontory, but no evidence of melanin or mycosporine, two UV-protective pigments commonly associated with terrestrial lithophytic fungi (Gorbushina et al., 2003; Selbmann et al., 2005). Erokhina et al. (2004) found an increase in the ratio of carotenoids to chlorophyll in a strain of endolithic *Trebouxia*. This is either an adaptation to prevent photo-oxidative damage, as the authors suggested, or an adaptation to increase light harvesting efficiency under low light conditions. The same group also found evidence of chromatic adaptation in some of the *Gloeocapsa* species (Erokhina et al., 2002).

Moisture

Little is known about how the antarctic cryptoendoliths cope with desiccation. The rock matrix and lichen structures may condense water under conditions of high humidity (Palmer and Friedmann, 1990). Many of the organisms, including the black fungi (Selbmann et al., 2005), *Hemichloris antarctica* (Tschermak-Woess and Friedmann, 1984), and all of the cyanobacteria (Friedmann et al., 1988), possess thick polysaccharide sheaths. Such sheaths are known to protect cells from desiccation damage (Potts, 1994) and, possibly, from freeze–thaw cycles (Tamaru et al., 2005). Hershkovitz et al. (1991) found that a strain of *Chroococcidiopsis* from the Negev Desert accumulated trehalose and sucrose under conditions of osmotic stress. The production of small, metabolically compatible solutes is a known desiccation-protection mechanism used by many microorganisms (Potts, 1994) and possibly by those in the antarctic desert.

Turnover time and productivity

Multiple lines of evidence suggest that the endolithic communities are extremely long lived. The radiocarbon age of the lichen-dominated community is around 1000 years (Bonani et al., 1988). In a living system, this value represents the mean residence time of carbon, not true age (McKay et al., 1986). By dividing standing lipid biomass with laboratory-determined lipid synthesis rates, Johnston and Vestal (1991) arrived at turnover times of 17 000–19 000 years. The rock surface exposure age was estimated, based on the thickness of varnish coatings, to be several thousand years (Friedmann and Weed, 1987). Taking all evidence into consideration, Friedmann et al. (1993) estimated that the community age – the duration of exfoliation of individual crusts – is around 10 000 years.

Biomass estimates varied widely, from 0.01 to 655 $g\,C\,m^{-2}$ (Kappen and Friedmann, 1983; Touvila and LaRock, 1987; Vestal, 1988a). Dividing the average value, 33.3 $g\,C\,m^{-2}$, by the community age yields a carbon incorporation rate of 3 $mg\,C\,m^{-2}\,y^{-1}$ (Nienow and Friedmann, 1993). The rate of carboxylation (gross primary productivity) in the community, calculated from laboratory determined CO_2 fixation rates and long-term climate record, is much higher, about 1200 $mg\,C\,m^{-2}\,y^{-1}$ (Friedmann et al., 1993). Subtracting dark respiration, the remainder is the rate of carbon gain, 600 $mg\,C\,m^{-2}\,y^{-1}$. The annual increase in biomass thus represents only 0.5% of the estimated annual net productivity. What happens to the rest? One suggested sink is the leaching of low molecular weight organic compounds to deeper rock layers and ultimately to soils and lakes in the valleys (Friedmann et al., 1993). However, the concentration of organic nitrogen and amino acids is extremely low below the colonized zone (H. J. Sun, unpublished data). This suggests that leached compounds represent a limited quantity and that the short-term gross primary production was overestimated.

At the end of the 10 000 year growth cycle, much of the accumulated biomass is exfoliated to the soil. In the higher elevation parts of Taylor Valley this process is significant enough to impart an endolithic signature on soil organic matter (Friedmann et al., 1980; Burkins et al., 2000). In the lower elevations, soils contain no material of endolithic origin, and much of the soil organic matter appears to be relic from past productivity (Burkins et al., 2000).

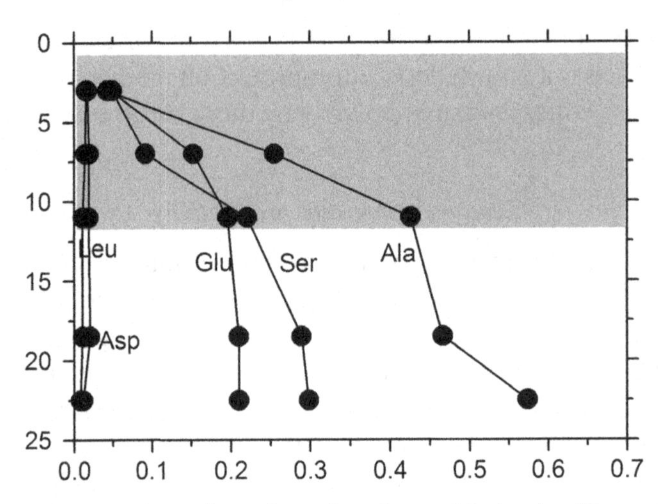

Fig. 4.6. D/L enantiomeric ratios of amino acids in the *Hormathonema–Gloeocapsa* community. The D/L ratios for alanine, glutamic acid, and serine are due to accumulated peptidoglycan remnants, not to age-related racemization. Dark band represents the colonized zone. (H. J. Sun, unpublished data.)

Amino acid racemization and pseudoracemization

Recent studies show that both the *Hormathonema–Gloeocapsa* and the lichen-dominated communities contain large amounts of D-amino acids (Fig. 4.6) (H. J. Sun, unpublished data). This finding is unusual because amino acids in proteins are present exclusively in the L-enantiomeric configuration. Amino acids do racemize: in time L-enantiomers can reversibly convert to corresponding D-enantiomers until the two forms reach equal proportions (thermodynamic equilibrium). But racemization is an extremely slow process, especially at low temperature (racemization rate is an exponential function of temperature). Calculations based on the estimated community age and the long-term temperature record show that racemization is negligible and could not explain the high amino acid D/L ratios of the communities.

A new theory, termed pseudoracemization, was proposed to explain the D-enrichment. In pseudoracemization, the D-amino acids originate from bacterial and cyanobacterial cell walls (peptidoglycans). Peptidoglycans are macromolecules consisting of long glycan strands crossed linked by short peptides. The linking peptides are composed of D- as well as L-enantiomers of certain amino acids, notably alanine and glutamic acid. Peptide bonds formed between D–D, D–L, and L–D enantiomers are resistant to common proteases (Rogers, 1974). Compared with proteins, peptidoglycans are harder to degrade (Nagata et al., 2003; Veuger et al., 2006). After death, peptidoglycans accumulate while the easily degraded proteins are recycled. Thus, the high D/L ratios in the antarctic communities are the consequence of chiral-selective degradation of proteins over peptidoglycans and other D-enantiomer-containing peptides. The implications of pseudoracemization are far reaching. It can also explain why certain D-amino acids accumulate in oceans and soils (Lee and Bada, 1977; Bada and Hoopes, 1979; Griffin and Kimber, 1988; Kimber et al., 1990; Amelung and Zhang, 2001; Dittmar et al., 2001; Pedersen et al., 2001; Amelung and Brodowski, 2002; Brinton et al., 2002; Grutters et al., 2002; Harada et al., 2002; Amelung, 2003; Lomstein et al., 2006).

Implications for possible life on Mars

Mars today is a cold dry desert with a thin atmosphere. There is evidence that it had liquid water flowing on its surface in the past, but even then Mars seems to have been a cold world (McKay and Stoker, 1989). Thus, Antarctica, particularly the cold dry desert regions, provides the best terrestrial analog for astrobiological studies of Mars (McKay, 1993). The utility of such terrestrial analogs lies in the fact that biological phenomena do not lend themselves to the simple quantification and generalizations possible in planetary physics. Studies of microbial ecosystems in analog environments are the only way to provide a realistic basis for speculations of extraterrestrial biota.

For all life forms on Earth, the essential environmental requirement is liquid water (McKay and Davis, 1991). McKay and Davis (1991) suggested a conceptual history of liquid water on Mars. Over 3.8 billion years (Ga) ago, the planet had flowing liquid water and mean temperatures above freezing. This is the period when life could have originated on Mars. The extent of the carbon dioxide atmosphere at that time is unknown. However, no matter whether it was thick or thin, the atmosphere would have been diminished very rapidly due to carbonate formation in the presence of liquid water and loss to space (e.g., Manning et al., 2006). As the atmosphere thinned and the temperature dropped, a second epoch began where the average temperatures were below freezing, but peak temperatures were above freezing. These conditions would have allowed ice-covered lakes similar to those now seen on the floors of the dry valleys to exist on Mars (McKay and Davis, 1991). Eventually temperatures would have fallen so low that air temperatures never rose above freezing (McKay and Davis, 1991). Endolithic microbial communities similar to those found in Antarctica could have continued to survive (McKay et al., 1992). Proceeding further toward the present epoch, the amount of atmospheric carbon dioxide would have fallen so low that the total pressure would be near the triple point of water (Manning et al., 2006). Water is not stable as a liquid below this pressure, and life on the surface would have become extinct.

The current foci of astrobiological interest in Mars are on the possible separate origin of life there, on the reconstruction of possible past ecosystems, and on search strategies for remnants of past life. Even though the Viking mission and further spacecraft exploration may indicate that there is little chance of life extant on Mars today, the fact that over two thirds of the martian surface dates back to 3.8 Ga may suggest that Mars – not Earth – may hold the best record of the origin of life on Earth-like planets.

Antarctic ecosystems are relevant to Mars astrobiology in two distinct ways. First, current ecosystems provide models for possible past martian ecosystems (Friedmann, 1986; Friedmann and Koriem, 1989). Second, studies of the remains of extinct ecosystems – both biological and chemical – can be used to aid the development of methods to locate and identify analogous signs of life on Mars, if they exist. For example, in Antarctica, the fossil record corresponds to ecosystems still extant and hence conclusions derived from studies of fossil material can be verified (Friedmann and Weed, 1987). On Mars, this type of methodological check may not be available. Thus, antarctic studies may provide valuable training for work on Mars.

The cryptoendolithic microbial ecosystem provides an example of what may have been the final habitat on the surface of Mars (Freidmann, 1982; Friedmann et al., 1987). It suggests that life is more robust than previously realized and that on Mars life would not necessarily have ceased as average temperatures fell to below zero. The cryptoendolithic microbial community provides an analog of a possible ecosystem that may have existed on Mars after the ice-covered lake habitats vanished (McKay, 1993).

Dedication

The authors wish to dedicate this chapter to our mentor and the discoverer of the antarctic cryptoendolithic microbial ecosystem, Dr. Imre Friedmann, who died on June 11, 2007, at the age of 86.

Acknowledgments

HJS thanks the Exobiology Program, the NASA Astrobiology Institute, and the NASA EPSCoR office for support and Lisa Wable for graphics assistance.

References

Ahmadjian, V. and Jacobs, J. B. (1987). Studies on the development of synthetic lichens. *Bibliotheca Lichenologica*, **25**, 47–58.

Amelung, W. (2003). Nitrogen biomarkers and their fate in soil. *Journal of Plant Nutrition and Soil Science*, **166**, 677–686.

Amelung, W. and Brodowski, S. (2002). *In vitro* quantification of hydrolysis-induced racemization of amino acid enantiomers in environmental samples using deuterium labeling and electron-impact ionization mass spectrometry. *Analytical Chemistry*, **74**, 3239–3246.

Amelung, W. and Zhang, X. (2001). Determination of amino acid enantiomers in soils. *Soil Biology and Biochemistry*, **33**, 553–562.

Archibald, P. A. (1975). *Trebouxia* de Pulmaly (Chlorophyceae, Chlorococcales) and *Pseudotrebouxia* gen. nov. (Chlorophyceae, Chlorosarcinales). *Phycologia*, **14**, 125–137.

Archibald, P. A., Friedmann, E. I., and Ocampo-Friedmann, R. (1983). Representatives of the cryptoendolithic flora of Antarctica. *Journal of Phycology*, **19**, 7.

Bada, J. L. and Hoopes, E. A. (1979). Alanine enantiomeric ratio in the combined amino acid fraction in seawater. *Nature*, **282**, 822–823.

Banerjee, M., Whitton, B. A., and Wynn-Williams, D. D. (2000). Phosphatase activities of endolithic communities in rocks of the Antarctic dry valleys. *Microbial Ecology*, **39**, 80–91.

Bell, R. A. (1993). Cryptoendolithic algae of hot semiarid lands and deserts. *Journal of Phycolology*, **29**, 133–139.

Billi, D., Friedmann, E. I., Hofer, K. G., et al. (2000). Ionizing-radiation resistance in the desiccation-tolerant cyanobacterium *Chroococcidiopsis*. *Applied and Environmental Microbiology*, **66**, 1489–1492.

Blackhurst, R. L., Jarvis, K., and Grady, M. M. (2004). Biologically-induced elemental variations in Antarctic sandstones: a potential test for martian micro-organisms. *International Journal of Astrobiology*, **3**, 97–106.

Blackhurst, R. L., Genge, M. J., Kearsley, A. T., and Grady, M. M. (2005). Cryptoendolithic alteration of Antarctic sandstones: pioneers or opportunists? *Journal of Geophysical Research*, **110**, E12S24, doi: 10.1029/2005JE002463.

Boison, G., Mergel, A., Jolkver, H., and Bothe, H. (2004). Bacterial life and dinitrogen fixation at a gypsum rock. *Applied and Environmental Microbiology*, **70**, 7070–7077.

Bonani, G., Friedmann, E. I., Ocampo-Friedmann, R., McKay, C. P., and Wölfli, W. (1988). Preliminary report on radiocarbon dating of cryptoendolithic microorganisms. *Polarforschung*, **58**, 199–200.

Brinton, K. L. F., Tsapin, A. I., Gilichinsky, D., and McDonald, G. D. (2002). Aspartic acid racemization and age-depth relationships for organic carbon in Siberian permafrost. *Astrobiology*, **2**, 77–82.

Broady, P. A. (1996). Diversity, distribution, and dispersal of Antarctic terrestrial algae. *Biodiversity and Conservation*, **5**, 1307–1335.

Burkins, M. B., Virginia, R. A., Chamberlain, C. P., and Wall, D. H. (2000). Origin and distribution of soil organic matter in Taylor Valley, Antarctica. *Ecology*, **81**, 2377–2391.

Cockell, C. S. and Stokes, M. D. (2004). Widespread colonization by polar hypoliths. *Nature*, **431**, 414.

Cockell, C. S., Lee, P., Osinski, B., Horneck, G., and Broady, P. (2002). Impact-induced microbial endolithic habitats. *Meteoritics and Planetary Science*, **37**, 1287–1298.

Darling, R. B., Friedmann, E. I., and Broady, P. A. (1987). *Heterococcus endolithicus* sp. nov. (Xanthophyceae) and other terrestrial *Heterococcus* species from Antarctica: morphological changes during life history and response to temperature. *Journal of Phycology*, **23**, 598–607.

de la Torre, J. R., Goebel, B. M., Friedmann, E. I., and Pace, N. R. (2003). Microbial diversity of cryptoendolithic communities from McMurdo Dry Valleys, Antarctica. *Applied and Environmental Microbiology*, **69**, 3858–3867.

DePriest, P. T. (2004). Early molecular investigations of lichen-forming symbionts: 1986–2001. *Annual Review of Microbiology*, **58**, 273–301.

Dittmar, T., Fitznar, H. P., and Kattner, G. (2001). Origin and biogeochemical cycling of organic nitrogen in the eastern Arctic Ocean as evident from D- and L-amino acids. *Geochimica et Cosmochimica Acta*, **65**, 4103–4114.

Dong, H., Rech, J. A., Jiang, H., Sun, H., and Buck, B. J. (2007). Endolithic cyanobacteria in soil gypsum: occurrences in Atacama (Chile), Mojave (USA), and Al-Jafr Basin (Jordan) deserts. *Journal of Geophysical Research, Biogeosciences*, **112**, G02030, doi: 10.1029/2006JG000385.

Edwards, H. G. M., Wynne-Williams, D. D., and Jorge Villar, S. E. (2004). Biological modification of haematite in Antarctic cryptendolithic communites. *Journal of Raman Spectroscopy*, **35**, 470–474.

Erokhina, L. G., Shatilovich, A. V., Kaminskaya, O. P., and Gilinchinskii, D. A. (2002). The absorption and fluorescence spectra of the cyanobacterial phycobionts of cryptoendolithic lichens in the high-polar regions of Antarctica. *Microbiology*, **71**, 601–607.

Erokhina, L. G., Shatilovich, A. V., Kaminskaya, O. P., and Gilinchinskii, D. A. (2004). Spectral properties of the green alga *Trebouxia*, a phycobiont of cryptoendolithic lichens in the Antarctic dry valleys. *Microbiology*, **73**, 420–424.

Ettl, H. and Gärtner, G. (1995). *Syllabus der Boden-, Luft-, und Flechtenalgen*. Jena, Germany: Gustav Fischer, 721 pp.

Fountain, A. G., Lyons, W. B., Burkins, M. B., et al. (1999). Physical controls on the Taylor Valley ecosystem, Antarctica. *BioScience*, **49**, 961–971.

Friedl, T. and Büdel, B. (1996). Photobionts. In *Lichen Biology*, ed. T. H. Nash, III. Cambridge, UK: Cambridge University Press, pp. 8–24.

Friedmann, E. I. (1977). Microorganisms in Antarctic desert rocks from dry valleys and Dufek Massif. *Antarctic Journal of the United States*, **12**, 26–29.

Friedmann, E. I. (1978). Melting snow in the dry valleys is the source of water for endolithic microorganisms. *Antarctic Journal of the United States*, **13**, 162–163.

Friedmann, E. I. (1982). Endolithic microorganisms in the Antarctic cold desert. *Science*, **215**, 1045–1053.

Friedmann, E. I. (1986). The Antarctic cold desert and the search for traces of life on Mars. *Advances in Space Research*, **6**, 167–172.

Friedmann, E. I. and Galun, M. (1974). Desert algae, lichens, and fungi. In *Desert Biology*, Vol. 2, ed. G. W. Brown, Jr. New York: Academic Press, pp. 165–212.

Friedmann, E. I. and Kibler, A. P. (1980). Nitrogen economy of endolithic microbial communities in hot and cold deserts. *Microbial Ecology*, **6**, 95–108.

Friedmann, E. I. and Koriem, A. (1989). Life on Mars: how it disappeared (if it ever was there). *Advances in Space Research*, **9**, 167–172.

Friedmann, E. I. and Ocampo, R. (1976). Endolithic blue-green algae in the Dry Valleys: primary producers in the Antarctic desert ecosystem. *Science*, **193**, 1247–1249.

Friedmann, E. I. and Ocampo-Friedmann, R. (1984). Endolithic microorganisms in extreme dry environments: an analysis of a lithobiontic microbial habitat. In *Current Perspectives in Microbial Ecology*, ed. M. J. Klug and C. A. Reddy. Washington, D.C.: American Society for Microbiology, pp. 177–185.

Friedmann, E. I. and Ocampo-Friedmann, R. (1995). A primitive cyanobacterium as pioneer microorganism for terraforming Mars. *Advances in Space Research*, **15**, 243–246.

Friedmann, E. I. and Sun, H. J. (2005). Communities adjust their temperature optima by shifting producer-to-consumer ratio, shown in lichens as models. I. Hypothesis. *Microbial Ecology*, **49**, 523–527.

Friedmann, E. I. and Weed, R. (1987). Microbial trace-fossil formation, biogenous, and abiotic weathering in the Antarctic cold desert. *Science*, **236**, 703–705.

Friedmann, E. I., Garty, J., and Kappen, L. (1980a). Fertile stages of cryptoendolithic lichens in the dry valleys of southern Victoria Land. *Antarctic Journal of the United States*, **12**, 6–30.

Friedmann, E. I., LaRock, P. A., and Brunson, J. P. (1980b). Adenosine triphosphate (ATP), chlorophyll, and organic nitrogen in endolithic microbial communities and in adjacent soils in the dry valleys of Southern Victoria Land. *Antarctic Journal of the United States*, **15**, 164–166.

Friedmann, E. I., McKay, C. P., and Nienow, J. A. (1987). The cryptoendolithic microbial environment in the Ross Desert of Antarctica: nanoclimate data, 1984 to 1986. *Polar Biology*, **7**, 273–287.

Friedmann, E. I., Hua, M., and Ocampo-Friedmann, R. (1988). Cryptoendolithic lichen and cyanobacterial communities of the Ross Desert, Antarctica. *Polarforshung*, **58**, 251–259.

Friedmann, E. I., Hua, M., and Ocampo-Friedmann, R. (1993a). Terraforming Mars: dissolution of carbonate rocks by cyanobacteria. *Journal of the British Interplanetary Society*, **46**, 291–292.

Friedmann, E. I., Kappen, L., Meyer, M. A., and Nienow, J. A. (1993b). Long-term productivity in the cryptoendolithic microbial community of the Ross Desert, Antarctica. *Microbial Ecology*, **25**, 51–69.

Gärtner, G. (1985). Taxonomische Probleme bei den Flechtenalgengattungen *Trebouxia* and *Pseudotrebouxia* (Chlorophyceae, Chlorellales). *Phyton (Austria)*, **25**, 101–111.

Garty, J. (1999). Lithobionts in the eastern Mediterranean. In *Enigmatic Microorganisms and Life in Extreme Environments*, ed. J. Seckbach. The Hague: Kluwer Academic Publishers, pp. 255–276.

Golubic, S., Friedmann, E. I., and Schneider, J. (1981). The lithobiontic ecological niche, with special reference to microorganisms. *Journal of Sedimentary Petrology*, **51**, 475–478.

Gorbushina, A. A., Whitehead, K., Dornieden, T., et al. (2003). Black fungal colonies as units of survival: hyphal mycosporines synthesized by rock-dwelling microcolonial fungi. *Canadian Journal of Botany*, **81**, 131–138.

Greenfield, L. G. (1988). Forms of nitrogen in Beacon sandstone rocks containing endolithic microbial communities in Southern Victoria Land, Antarctica. *Polarforschung*, **58**, 211–218.

Griffin, C. V. and Kimber, R. W. L. (1988). Racemization of amino-acids in agricultural soils – an age effect. *Australian Journal of Soil Research*, **26**, 531–536.

Grutters, M., van Raaphorst, W., Epping, E., et al. (2002). Preservation of amino acids from in situ-produced bacterial cell wall peptidoglycans in northeastern Atlantic continental margin sediments. *Limnology and Oceanography*, **47**, 1521–1524.

Hale, M. E. (1987). Epilithic lichens in the Beacon sandstone formation Victoria Land, Antarctica. *Lichenologist*, **19**, 269–287.

Harada, N., Kondo, T., Fukuma, K., et al. (2002). Is amino acid chronology applicable to the estimation of the geological age of siliceous sediments? *Earth and Planetary Science Letters*, **198**, 257–266.

Helms, G. W. F. (2003). Taxonomy and symbiosis in associations of Physciaceae and Trebouxia. Doctoral Dissertation, Georg-August Universität, Göttingen, Germany, 158 pp.

Hershkovitz, N., Oren, A., and Cohen, Y. (1991). Accumulation of trehalose and sucrose in cyanobacteria exposed to matric water stress. *Applied and Environmental Microbiology*, **57**, 645–648.

Hirsch, P., Hoffmann, B., Gallikowski, C. A., Mevs, U., Siebert, J., and Sittig, M. (1988). Diversity and identification of heterotrophs from Antarctic rocks of the McMurdo dry valleys (Ross Desert). *Polarforschung*, **58**, 261–270.

Hughes, K. A. (2006). Solar UV-B radiation, associated with ozone depletion, inhibits the Antarctic terrestrial microalga, *Stichococcus bacillaris*. *Polar Biology*, **29**, 327–336.

Hughes, K. A. and Lawley, B. (2003). A novel Antarctic microbial endolithic community within gypsum crusts. *Environmental Microbiology*, **5**, 555–565.

Johnston, C. G. and Vestal, J. R. (1986). Does iron inhibit cryptoendolithic microbial communities? *Antarctic Journal of the United States*, **21**, 225–226.

Johnston, C. G. and Vestal, J. R. (1989). Distribution of inorganic species in two Antarctic cryptoendolithic microbial communities. *Journal of Geomicrobiology*, **7**, 137–153.

Johnston, C. G. and Vestal, J. R. (1991). Photosynthetic carbon incorporation and turnover in Antarctic cryptoendolithic communities: are they the slowest growing communities on Earth? *Applied and Environmental Microbiology*, **57**, 2308–2311.

Johnston, C. G. and Vestal, J. R. (1993). Biochemistry of oxalate in the Antarctic cryptoendolithic lichen-dominated community. *Microbial Ecology*, **25**, 305–319.

Jorge Villar, S. E., Edwards, H. G. M., and Cockell, C. S. (2005a). Raman spectroscopy of endoliths from Antarctic cold desert environments. *The Analyst*, **130**, 156–162.

Jorge Villar, S. E., Edwards, H. G. M., and Worland, M. R. (2005b). Comparative evaluation of Raman spectroscopy at different wavelengths for extremophile exemplars. *Origins of Life and Evolution of Biospheres*, **35**, 489–506.

Kappen, L. and Friedmann, E. I. (1983). Ecophysiology of lichens in the dry valleys of Southern Victoria Land, Antarctica. II. CO_2 gas exchange in cryptoendolithic lichens. *Polar Biology*, **1**, 227–232.

Kappen, L., Friedmann, E. I., and Garty, J. (1981). Ecophysiology of lichens in the dry valleys of Southern Victoria Land, Antarctica. I. Microclimate of the cryptoendolithic lichen habitat. *Flora*, **171**, 216–235.

Kimber, R. W. L., Nannipieri, P., and Ceccanti, B. (1990). The degree of racemization of amino-acids released by hydrolysis of humic protein complexes – implications for age assessment. *Soil Biology and Biochemistry*, **22**, 181–185.

Koriem, A. and Friedmann, E. I. (1987). Resynthesis of lichens from Antarctic cryptoendolithic isolates (Abstract). Paper presented at International Symposium on Modern Approaches in the Biology of Terrestrial Microorganisms and Plants in the Antarctic. Kiel, Germany, September 7–11.

Lee, C. and Bada, J. L. (1977). Dissolved amino acids in equatorial Pacific, Sargasso Sea, and Biscayne Bay. *Limnology and Oceanography*, **22**, 502–510.

Lomstein, B. A., Jorgensen, B. B., Schubert, C. J., and Niggemann, J. (2006). Amino acid biogeo- and stereochemistry in coastal Chilean sediments. *Geochimica et Cosmochimica Acta*, **70**, 2970–2989.

Manning, C. V., McKay, C. P., and Zahnle, K. J. (2006). Thick and thin models of the evolution of carbon dioxide on Mars. *Icarus*, **180**, 38–59.

Marton, K. and Galun, M. (1976). In vitro dissociation and reassociation of the symbionts of the lichen *Heppia echinulata*. *Protoplasma*, **87**, 135–143.

Matthes, U., Turner, S. J., and Larson, D. W. (2001). Light attenuation by limestone rock and its constraint on the depth distribution of endolithic algae and cyanobacteria. *International Journal of Plant Science*, **162**, 263–270.

McKay, C. P. (1993). Relevance of Antarctic microbial ecosystems to exobiology. In *Antarctic Microbiology*, ed. E. I. Friedmann. New York: Wiley-Liss, pp. 593–601.

McKay, C. P. and Davis, W. L. (1991). The duration of liquid water habitats on early Mars. *Icarus*, **90**, 214–221.

McKay, C. P. and Friedmann, E. I. (1985). The cryptoendolithic microbial environment in the Antarctic cold desert: temperature variations in nature. *Polar Biology*, **4**, 19–25.

McKay, C. P. and Stoker, C. R (1989). The early environment and its evolution on Mars: implications for life. *Reviews of Geophysics*, **27**, 189–214.

McKay, C. P., Long, A., and Friedmann, E. I. (1986). Radiocarbon dating of open systems with bomb effect. *Journal of Geophysical Research*, **91**(B3), 3836–3840.

McKay, C. P., Friedmann, E. I., Wharton, R. A., and Davis, W. L. (1992). History of water on Mars: a biological perspective. *Advances in Space Research*, **12**, 231–238.

McKay, C. P., Nienow, J. A., Meyer, M. A., and Friedmann, E. I. (1993). Continuous nanoclimate of the Ross Desert cryptoendolithic environment. *Antarctic Research Series*, **61**, 201–207.

Mevs, V. (1989). Taxonomie und ökophysiologische Eigenschaften ausgewählter Aktinomyceten aus der kontinentalen Antarktis. Doctoral dissertation, University of Kiel, Germany.

Meyer, M. A., Huang, G. -H., Morris, G. J., and Friedmann, E. I. (1988). The effect of low temperatures on Antarctic endolithic green algae. *Polarforschung*, **58**, 113–119.

Nagata, T., Meon, B., and Kirchman, D. L. (2003). Microbial degradation of peptidoglycan in seawater. *Limnology and Oceanography*, **48**, 745–754.

Nienow, J. A. and Friedmann, E. I. (1993). Terrestrial lithophytic communities. In *Antarctic Microbiology*, ed. E. I. Friedmann. New York: Wiley-Liss, pp. 343–412.

Nienow, J. A., McKay, C. P., and Friedmann, E. I. (1988a). The cryptoendolithic microbial environment in the Ross Desert of Antarctica: mathematical models of the thermal regime. *Microbial Ecology*, **16**, 253–270.

Nienow, J. A., McKay, C. P., and Friedmann, E. I. (1988b). The cryptoendolithic microbial environment in the Ross Desert of Antarctica: light in photosynthetically active region. *Microbial Ecology*, **16**, 271–289.

Nienow, J. A., Friedmann, E. I., and Ocampo-Friedmann, R. (2002). Endolithic microorganisms in arid regions. In *Encyclopedia of Environmental Microbiology*, Vol. 2, ed. G. Bitton. New York: John Wiley, pp. 1100–1112.

Ocampo-Friedmann, R., Meyer, M. A., Chen, M., and Friedmann, E. I. (1988). Temperature response of Antarctic cryptoendolithic photosynthetic microorganisms. *Polarforschung*, **58**, 121–124.

Omelon, C. R., Pollard, W. H., and Ferris, F. G. (2006). Environmental controls on microbial colonization of high Arctic cryptoendolithic habitats. *Polar Biology*, **30**, 19–29.

Onofri, S., Pagano, S., Zucconi, L., and Tosi, S. (1999). *Friedmanniomyces endolithicus* (Fungi, Hyphomycetes), anam.-gen. and sp. nov. from continental Antarctica. *Nova Hedwigia*, **68**, 175–181.

Palmer, Jr., R. J. and Friedmann, E. I. (1990). Water relations and photosynthesis in the cryptoendolithic microbial habitat of hot and cold deserts. *Microbial Ecology*, **19**, 111–118.

Pedersen, A. G. U., Thomsen, T. R., Lomstein, B. A., and Jorgensen, N. O. G. (2001). Bacterial influence on amino acid enantiomerization in a coastal marine sediment. *Limnology and Oceanography*, **46**, 1358–1369.

Phoenix, V. R., Bennett, P. C., Engel, A. S., Tyler, S. W., and Ferris, F. G. (2006). Chilean high-altitude hot-spring sinters: a model system for UV screening mechanisms by early Precambrian cyanobacteria. *Geobiology*, **4**, 15–28.

Pointing, S. B., Warren-Rhodes, K. A., Lacap, D. C., Rhodes, K. L., and McKay, C. P. (2007). Hypolithic community shifts occur as a result of liquid water availability along environmental gradients in China's hot and cold hyperarid deserts. *Environmental Microbiology*, **9**, 414–424.

Potts, M. (1994). Desiccation tolerance of prokaryotes. *Microbiological Reviews*, **58**, 755–805.

Rogers, H. J. (1974). Peptidoglycans (mucopeptides): structure, function, and variations. *Annals of the New York Academy of Science*, **235**, 29–51.

Romeike, J., Friedl, T., Helms, G., and Ott, S. (2002). Genetic diversity of algal and fungal partners in four species of *Umbilicaria* (lichenized ascomycetes) along a transect of the Antarctic peninsula. *Molecular Biology and Evolution*, **19**, 1209–1217.

Rothschild, L. J., Giver, L. J., White, M. R., and Mancinelli, R. L. (1994). Metabolic activity of microorganisms in evaporites. *Journal of Phycology*, **30**, 431–438.

Schlesinger, W. H., Pippen, J. S., Wallenstein, M. D., et al. (2003). Community composition and photosynthesis by photoautotrophs under quartz pebbles, southern Mojave Desert. *Ecology*, **84**, 3222–3231.

Schumann, P., Prauser, H., Rainey, F. A., Stackebrandt, E., and Hirsch, P. (1997). *Friedmanniella antarctica* gen. nov. et sp. nov., an LL-diaminopimelic acid-containing actinomycete from Antarctic sandstone. *International Journal of Systematic Bacteriology*, **47**, 278–283.

Selbmann, L., de Hoog, G. S., Mazzaglia, A., Friedmann, E. I., and Onofri, S. (2005). Fungi at the edge of life: cryptoendolithic black fungi from Antarctic desert. *Studies in Mycology*, **51**, 1–32.

Siebert, J. and Hirsch, P. (1988). Characterization of 15 selected coccal bacteria isolated from Antarctic rock and soil samples from the McMurdo dry valleys (South-Victoria Land). *Polar Biology*, **9**, 37–44.

Siebert, J., Hirsch, P., Hoffmann, B., Gliesche, C. G., Peissl, K., and Jendrach, M. (1996). Cryptoendolithic microorganisms from Antarctic sandstone of Linnaeus Terrace (Asgard Range): diversity properties and interactions. *Biodiversity and Conservation*, **5**, 1337–1363.

Sly, L. I. and Hugenholtz, P. (2005). Blastomonas. In *Bergey's Manual of Systematic Bacteriology*, 2nd edition, Vol. 2: The Proteobacteria, Part C: *The Alpha-, Beta-, Delta-*, and Epsilonproteobacteria. New York: Springer, pp. 258–263.

Sun, H. J. and Friedmann, E. I. (1999). Growth on geological time scales in the Antarctic cryptoendolithic microbial community. *Geomicrobiology Journal*, **16**, 193–202.

Sun, H. J. and Friedmann, E. I. (2005). Communities adjust their temperature optima by shifting producer-to-consumer ratio, shown in lichens as models. II. Experimental verification. *Microbial Ecology*, **49**, 528–535.

Tamaru, Y., Takani, Y., Yoshida, T., and Sakamoto, T. (2005). Crucial role of extracellular polysaccharides in desiccation and freezing tolerance in the terrestrial cyanobacterium *Nostoc commune*. *Applied and Environmental Microbiology*, **71**, 7327–7333.

Tschermak-Woess, E. and Friedmann, E. I. (1984). *Hemichloris antarctica*, gen. et sp. nov. (Chlorococcales, Chlorophyta), a cryptoendolithic alga from Antarctica. *Phycologia*, **23**, 443–454.

Tschermak-Woess, E., Hua, M., Gärtner, G., and Hesse, M. (2006). Observations in *Hemichloris antarctica* Tschermak-Woess and Friedmann (Chlorophyceae) and the occurrence of a second *Hemichloris* species, *Hemichloris polyspora* n. sp. *Plant Systematics and Evolution*, **258**, 27–37.

Tuovila, J. and LaRock, P. A. (1987). Occurrence and preservation of ATP in Antarctic rocks and its implications in biomass determinations. *Geomicrobiology Journal*, **5**, 105–118.

van Thielen, N. and Garbary, D. J. (1999). Life in the rocks: endolithic algae. In *Enigmatic Microorganisms and Life in Extreme Environments*, ed. J. Seckbach. Dordrecht, Netherlands: Kluwer Academic Publishers, pp. 245–253.

Vestal, J. R. (1988a). Biomass of the cryptoendolithic microbiota from the Antarctic desert. *Applied and Environmental Microbiology*, **54**, 957–959.

Vestal, J. R. (1988b). Carbon metabolism of the cryptoendolithic microbiota from the Antarctic desert. *Applied and Environmental Microbiology*, **54**, 960–965.

Vestal, J. R. (1988c). Primary production of the cryptoendolithic microbiota from the Antarctic desert. *Polarforschung*, **58**, 193–198.

Veuger, B., van Oevelen, D., Boxchker, H. T. S., and Middelburg, J. J. (2006). Fate of peptidoglycan in an intertidal sediment: an *in situ* C^{13}-labeling study. *Limnology and Oceanography*, **51**, 1572–1580.

Virginia, R. A. and Wall, D. H. (1999). How soils structure communities in the Antarctic dry valleys. *BioScience*, **49**, 973–983.

Vishniac, H. S. (1985). *Crytococcus friedmannii*, a new species of yeast from the Antarctic. *Applied and Environmental Microbiology*, **54**, 960–965.

Vishniac, H. S. (1993). Soil microbiology. In *Antarctic Microbiology*, ed. E. I. Friedmann. New York: Wiley-Liss, pp. 297–341.

Vishniac, H. S. (1996). Biodiversity of yeasts and filamentous microfungi in terrestrial Antarctic ecosystems. *Biodiversity and Conservation*, **5**, 1365–1378.

Vishniac, H. S. (2002). Desert environments: soil microbial communities in cold deserts. In *Encyclopedia of Environmental Microbiology*, Vol. 2, ed. G. Bitton. New York: John Wiley, pp. 1023–1029.

Walker, J. J., Spear, J. R., and Pace, N. R. (2005). Geobiology of a microbial endolithic community in the Yellowstone geothermal environment. *Nature*, **434**, 1011–1014.

Warren-Rhodes, K. A., Rhodes, K. L., Pointing, S. B., et al. (2006). Hypolithic cyanobacteria, dry limit of photosynthesis, and microbial ecology in the hyperarid Atacama Desert. *Microbial Ecology*, **52**, 389–398.

Weber, B., Wessels, D. C. J., and Büdel, B. (1996). Biology and ecology of cryptoendolithic cyanobacteria of a sandstone outcrop in the Northern Province, South Africa. *Algological Studies*, **83**, 565–579.
Weed, R. and Ackert, Jr., R. P. (1986). Chemical weathering of Beacon supergroup sandstones and implications for Antarctic glacial chronology. *South African Journal of Science*, **82**, 513–516.
Weed, R. and Norton, S. A. (1991). Siliceous crusts, quartz rinds and biotic weathering of sandstones in the cold desert of Antarctica. In *Diversity of Environmental Biogeochemistry*, ed. J. Berthelin. Amsterdam, Netherlands: Elsevier, pp. 327–339.
Wierzchos, J., Ascaso, C., and McKay, C. P. (2006). Endolithic cyanobacteria in halite rocks from the hyperarid core of the Atacama Desert. *Astrobiology*, **6**, 415–422.
Wynn-Williams, D. D., Edwards, H. G. M., and Garcia-Pichel, F. (1999). Functional biomolecules of Antarctic stromatolitic and endolithic cyanobacterial communities. *European Journal of Phycology*, **34**, 381–391.
Wynn-Williams, D. D., Edwards, H. G. M., Newton, E. M., and Holder, H. M. (2002). Pigmentation as a survival strategy for ancient and modern photosynthetic microbes under high ultraviolet stress on planetary surfaces. *International Journal of Astrobiology*, **1**, 39–49.

5

Antarctic McMurdo Dry Valley stream ecosystems as analog to fluvial systems on Mars

MICHAEL N. GOOSEFF, DIANE M. MCKNIGHT, MICHAEL H. CARR, AND JENNY BAESEMAN

Introduction

The stream systems of the McMurdo Dry Valleys of Antarctica represent a relatively simple end member of terrestrial hydrologic systems. Many Dry Valley streams are prominent landscape features, especially in summer when they carry glacial meltwater from the alpine and outlet glaciers to the perennially ice-covered lakes on the valley floors (Fig. 5.1). Observations beginning in 1968 indicate that these channels carry water for 8–12 weeks each year, though some are only wetted in warm, high flow years, and others have been deactivated because of changes to flow routing. In addition to obvious channels incised in the landscape, smaller, less frequent fluvial features may become active in the Dry Valleys, such as small rivulets (shallow, broad gullies that are not wetted annually) carrying snowmelt or meltwater from buried ice down steep valley walls in particularly warm summers. Although these fluvial systems are relatively unique on Earth, the surface of Mars holds evidence of ancient fluvial features that are similar to snowmelt rivulets observed in the Dry Valleys.

In this chapter, we compare the contemporary status and function of streams of the Dry Valleys with those that may have existed on ancient Mars. Our current understanding of martian fluvial processes is limited to what can be inferred by the "leftover" drainages that are readily observed, some of which are quite large. The Dry Valley streams are a good analog for those on Mars because of the similarity in surface features – mostly expanses of open, bare ground, and because, common to both locations, streamflow is, or was in the case of Mars, intermittent. Certainly, in the current epoch, streamflow

Life in Antarctic Deserts and Other Cold Dry Environments: Astrobiological Analogs, ed. Peter T. Doran, W. Berry Lyons and Diane M. McKnight. Published by Cambridge University Press.

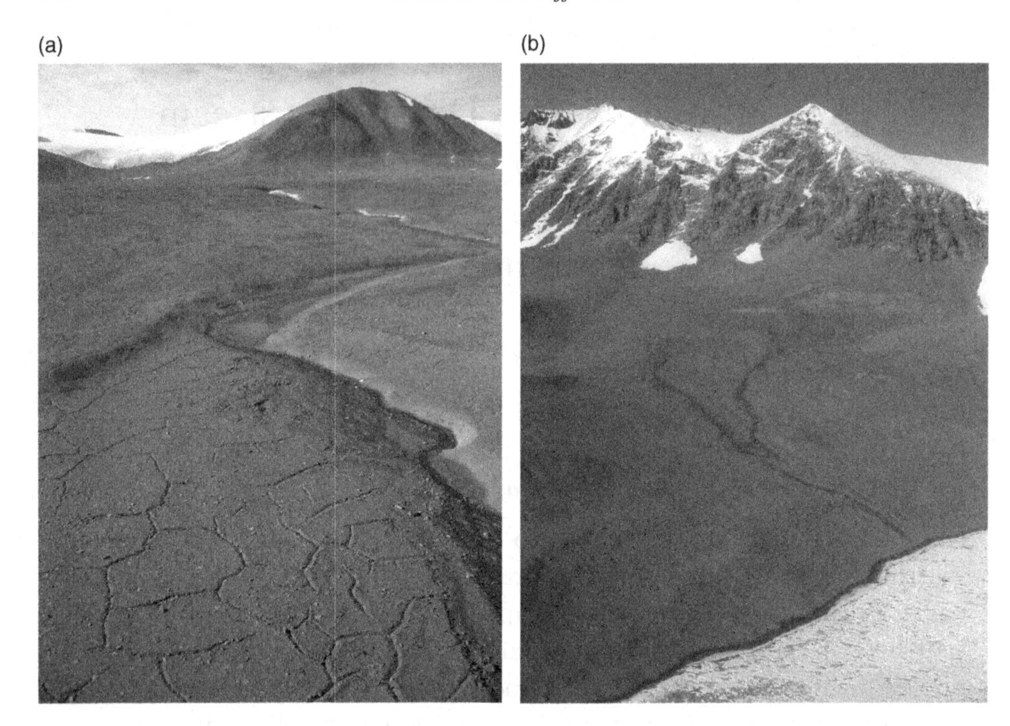

Fig. 5.1. (a) An upstream looking view of a channel in Taylor Valley, Antarctica, on the south side of Lake Fryxell at low flow. The stream originates from the glaciers of the Kukri Hills. In the middle ground, the incised stream channel can be seen with some collected snow, as it serves as a topographic lee. In the foreground, the wetted channel and lateral wetted shores are evident, as are ice polygons in the soils more distal from the channel. (b) Melt of subsurface ice creates seeps and rivulets above the south shore of the West Lobe of Lake Bonney. This site is the same as "Wormherder Creek" in Harris et al., 2007.

is more predictable in the Dry Valleys than on Mars, where most of the drainages are very ancient. However, there most likely have been periods in the past 12 million years when streamflow was a rare event in the Dry Valleys.

Research on Dry Valley streamflow began on the Onyx River in 1968. In addition to quantifying the inputs of water, solutes, and sediments to lakes, much research on stream ecosystems of the Dry Valleys has focused on the ecological importance of processes occurring in the streambed and in underlying hyporheic zones (areas of adjacent sediment through which stream water exchanges, Fig. 5.2). Dry Valley streams, which are underlain by loose, unconsolidated substrate, develop fairly extensive hyporheic zones. One motivation for studying hyporheic processes in Dry Valley streams has been that understanding gained from these simple streams can inform studies

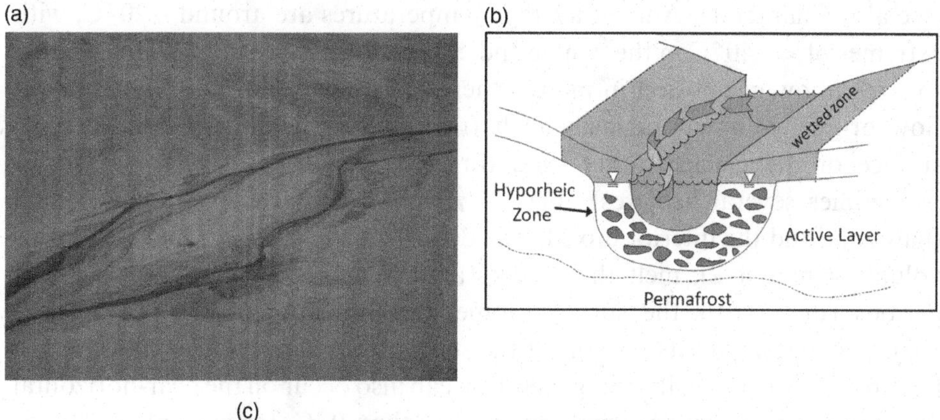

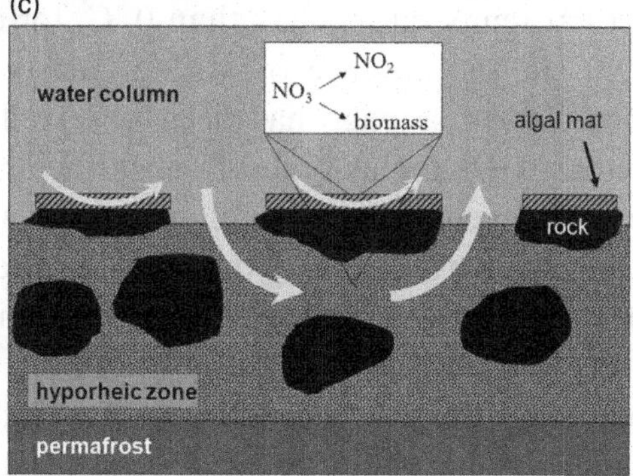

Fig. 5.2. (a) Onyx River channel in the middle of Wright Valley, Antarctica, with extensive wetted zones around the channel, the water for which comes from the river channel. Channel length is approximately 400 m in this image. (b) Conceptual model of the coupled stream–hyporheic systems of the Dry Valleys in cross section and (c) indicating active exchange of water between the flowing water column and, firstly, benthic algal mats growing primarily on the larger substrate of the streambeds and, secondly, the subchannel thawed active layer, which creates a hyporheic zone.

of temperate streams, where hyporheic sediments host extensive microbial communities that are responsible for potentially a large fraction of the in-stream biogeochemical cycling (Fischer et al., 2005; Sliva and Williams, 2005). The landscape similarities between these systems suggest that martian stream ecosystems, to the extent they may have existed, may also be greatly dependent on shallow subsurface processes.

Fluvial processes in the Dry Valleys

Due to very little precipitation (<10 cm as snow annually), streamflow in the Dry Valleys is dominated by the melting of ice during the austral summer (see also Chapter 1). Annual average temperatures are around −20 °C, with extremes of <−40 °C in the winter and >5 °C during the summer. In contrast to streamflow generation in many other glacial environments, most streamflow originates as supraglacial melt (i.e., generated and flowing over the surface of the glacier rather than through and under it) from extensive (sometimes several hectares) ablation zones of the glaciers. In the coastal thaw zone and the interior mixed zone described by Marchant and Head (this volume, Chapter 2), melt that occurs at the surface during warm summer periods runs across the surface, sometimes flowing through supraglacial channels, and makes its way off of the steep glacier termini in small waterfall features. However, meltwater generation can also occur on the near-horizontal ablation zone surfaces at temperatures less than 0 °C, for example as low as −15 °C (Chinn, 1993). In the absence of visible surface meltwater, some shallow (<10 cm) percolating seepage has been observed along a surficial channel (Fountain et al., 1998). This type of subsurface melt in low albedo ice can be attributed to a greenhouse effect, in which radiative heat flux into the ice surpasses conductive heat loss to the surface (Brandt and Warren, 1993). Another important aspect of meltwater generation in the Dry Valleys is that melting on the vertical faces of the glaciers can be the dominant source of streamflow when low summer temperatures limit meltwater generation on the much larger glacier surfaces (Fountain et al., 1998; Conovitz et al., 1998; Lewis et al., 1999; Ebnet et al., 2005).

Characteristics of Dry Valley streams are largely driven by glacier and landscape locations, rather than fluvial geomorphic processes, as observed in temperate stream networks on Earth. Drainage density (number of streams per area) of the Dry Valleys is very low, and largely a function of glacier location, as glaciers generate streamflow. The length and slope of Dry Valley stream channels are primarily determined by landscape position and valley morphology rather than by erosive fluvial processes. For example, streams in the Lake Fryxell basin range from <1 km to >11 km in length, as dictated by the distance of the source glacier from the lake in the valley floor. Furthermore, the Onyx River, the longest stream in the Dry Valleys, and in all of Antarctica, has a length of 32 km with ~12 intermittent tributaries. The Onyx River is one of the few second-order stream systems in the Dry Valleys. Most of the streams remain as first-order streams before reaching the outlet to a lake or the ocean (Fig. 5.1a). Stream slope development is limited by the low peak flows (typically $Q_{peak} \approx 1\,m^3\,s^{-1}$) that provide insufficient power to cause substantial erosion to the channel beds, except in the relatively steeper reaches. Below the steeper reaches, the channels are poorly defined and typically divided into multichannel networks perched on small deltas near the inlet to the lake. The greatest abundance of perennial microbial mats occurs in reaches of moderate gradient where the rocks of the streambed form a stone pavement. This pavement provides a stable substrate, minimizing the movement of fine sediment and loss of mat due to abrasion (Alger et al., 1997; McKnight et al., 1998). Stone pavements are formed by repeated freezing and thawing of the saturated hyporheic alluvium that eventually forces the rocks upward with the largest flat side exposed (French, 1996). Because the quantity of sand and gravel transported by these streams is quite limited and occurs infrequently, a given configuration of a channel may persist for decades, until high flows during a warm summer reconfigure the channel (McKnight et al., 2007).

Some portion of the meltwater that enters the stream channel at the glacier face does not reach the lake, but either is lost to the atmosphere by evaporation or soaks into the hyporheic zone (Cozzetto et al., 2006; Marchant and Head, this volume, Chapter 2). During the summer, the daily peak in solar radiation warms the stream water to temperatures as high as 15–20 °C and the extremely dry air and strong winds enhance evaporation. Once meltwater generation begins, the hyporheic zone becomes saturated as flow advances downstream. The hyporheic zone becomes visible as an extensive wetted band of alluvium adjacent to the stream. This stored hyporheic water represents a reservoir of moisture for stream ecosystems during cold summers with low, intermittent streamflow. The wetted boundary of the surface expression of the hyporheic zone is typically marked by a salt crust due to evaporative losses from these locations (Gooseff et al., 2006; Marchant and Head, this volume, Chapter 2).

Within the hyporheic zone, water moves along heterogeneous flow paths that depart from the channel, moves through the sediments, and then returns back to the channel (Fig. 5.2b). This process is called hyporheic exchange (Fig. 5.2c). Depth of the hyporheic zone below the stream is regulated by the relatively impermeable permafrost at depths of less than 1 m (Gooseff et al., 2003; Conovitz et al., 2006). The warming of the land surface in summer progressively increases the storage of water in the hyporheic zone by causing the expansion of the thaw depth.

The water that moves through the hyporheic zone is in continuous contact with mineral surfaces, which enhances weathering rates and brings both major ions and required nutrients to the stream surface (Gooseff et al., 2002). Hyporheic exchange can also mobilize salts that are deposited from the atmosphere and are present in the salt crusts at the edges of the saturated zone. The chemistry of hyporheic water is generally enriched in weathering products (e.g., Na, K, Cl, and Ca) compared with stream water (Gooseff et al., 2003). The role of weathering was confirmed by the direct observation of weathered mineral surfaces after a month-long incubation of pristine mineral surfaces in the hyporheic zone of a flowing Dry Valley stream (Maurice et al., 2002). Further, in Dry Valley streams weathering rates are rapid in comparison with rates determined in temperate environments and account for the downstream increase in solute concentrations. These weathering processes are ultimately an important source of solutes to the lakes on the valley floors.

Melting subsurface ice or high elevation (near valley ridge lines) snowpacks also provide liquid water to rivulets that progress down either paleochannels or hollows toward valley floors (Fig. 5.1b). Many of these discharge little water but provide extensive wetted zones with small wetted channels (Marchant and Head, this volume, Chapter 2). These snowmelt rivulets are perhaps not greatly important to the hydrologic cycle and balance of the Dry Valleys, but given our understanding of subsurface ecology in other streams of the Dry Valleys, these may be important locations of biogeochemical transformations.

The paleohydrologic history of the Dry Valleys greatly impacts the modern landscape by creating ephemeral surface water environments disconnected from the fluvial networks. Approximately 5000 years BP, Taylor Valley was inundated by Glacial Lake Washburn (Doran et al., 1994). Neighboring Wright Valley is thought to have been inundated by a ~480 m water column of Glacial Lake Wright approximately 20 000 years ago (Hall et al., 2001). A complex series of bedrock canyons known as the Labyrinth exists at the western end of Wright Valley. It has been recently documented that the Labyrinth was most likely formed by draining of a subglacial lake from the East Antarctic Ice Sheet approximately 14.4–12.4 Ma ago (Lewis et al., 2006). The Labyrinth contains many small meltwater ponds, most of them fed by melt from Wright Upper Glacier at the head of the Labyrinth (Mikucki et al., this volume, Chapter 6).

There are no extensive groundwater aquifers that connect directly to the Dry Valley streams. Permafrost, furthermore, is >100 m deep underneath the Dry Valleys (Stuiver et al., 1981), and its continuous presence, as well as the cold surface conditions, prevent the development of extensive groundwater aquifers. It has been proposed that deep, very briny groundwater may be connected to the depths of Lake Vanda, in Wright Valley (Cartwright and Harris, 1981), and perhaps to Don Juan Pond, near Lake Vanda. However, the predominant form of subsurface water is found in the hyporheic zones adjacent to streams and rivulets.

Fluvial history of Mars

Conditions on the surface of Mars today are very inhospitable for life, as we understand it, mainly because of the scarcity of liquid water. The atmosphere is thin, being composed largely of CO_2 with an average surface pressure of 5.6 mbar, and containing only 10 precipitable micrometers of H_2O. Mean annual temperatures are close to 215 K (−58 °C) at the equator and 160 K (−113 °C) at the poles. The ground is frozen to form a cryosphere that is on average several kilometers thick. Temperatures right at the surface fluctuate widely during the day, commonly exceeding 273 K (0 °C) close to noon, but the fluctuations damp out rapidly at depth so that above freezing temperatures are unlikely at depths greater than 1–2 cm below the surface. The cryosphere may be thinner locally in areas of anomalously high heat flow, but no such areas have been identified. Under these conditions liquid water can exist only transiently near the surface in anomalous situations. Yet, despite the restrictive conditions, Mars has a rich array of fluvial and lacustrine features, which suggests that conditions on Mars may have been different at times in the past.

The history of Mars has been divided into three different eras: the Noachian (4.1 to 3.7 Ga ago), the Hesperian (3.7 to 3.0 Ga ago) and the Amazonian (3.0 Ga ago to the present). The period before the Noachian has not been named. The ages are only approximate and based on crater counts. Three types of fluvial features are generally recognized: (1) valleys 1–5 km across that branch upstream to form hierarchical networks similar to terrestrial river systems (Fig. 5.3), (2) outflow channels tens to thousands of kilometers across that start full size and have few, if any, tributaries, and (3) gullies, tens of meters across, that occur on steep slopes (Carr, 2006). Branching valley networks are best developed in Noachian terrain, the largest outflow channels are upper Hesperian in age, and most gullies are Amazonian in age, but all three eras have a mix of fluvial features.

The Noachian terrains provide the best evidence for warm surface conditions and an active hydrologic cycle. Noachian terrain forms most of the high-standing, heavily cratered terrain of the southern highlands, and most of the terrain is dissected by valley networks. The terrain is poorly graded with many local lows into which the valleys drain, but where there are long regional slopes, valleys may be over 2000 km in length. The valleys typically have rectangular to U-shaped cross sections, 50–200 m depths, and maintain their 1–5 km widths over large distances. Drainage densities (i.e., stream length per unit area) vary greatly with values up to 10^{-1} km^{-1}, close to the low end of the terrestrial range. Lakes probably were common when the

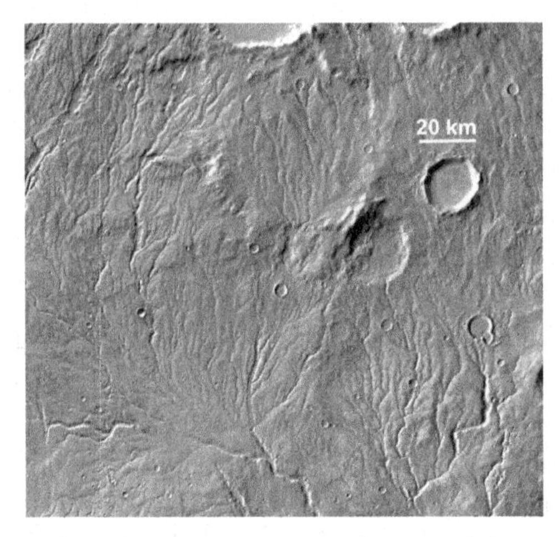

Fig. 5.3. Valley networks in Noachian terrain at 42S, 267E. The dense drainage pattern suggests precipitation and surface runoff. (Image credit: THEMIS.)

valleys formed. Impact craters commonly have breached walls both where valleys enter and leave the crater, indicating possible fluvial causes. As better data are acquired, more delta-like deposits are being found where streams enter local lows (Malin and Edgett, 2003) (Fig. 5.4). Despite the pervasive dissection of Noachian terrains, the drainage system appears immature. Large drainage systems, thousands of kilometers across in all directions, like the Amazon and Mississippi did not develop. Drainage basins have low concavities, and the tendency for higher-elevation slices through a basin to be more circular, as is typical of terrestrial basins, is less apparent in martian basins.

The origin of the valley networks is controversial. There is almost universal agreement that the valleys were cut by water, but there has been reluctance to ascribe the valleys to precipitation and surface runoff because of the difficulty of warming Mars sufficiently to allow rainfall. This difficulty is particularly challenging for early in the planet's history when the solar constant was significantly lower than it is today (Haberle, 1998). In addition, many of the more prominent, fresh-appearing valleys have open drainage systems with stubby tributaries, typical of valleys formed mainly by groundwater sapping. It has been suggested, therefore, that the valleys could have formed under cold conditions by ice-covered streams fed by groundwater (Squyres and Kasting, 1994). However, this scenario now seems much less likely, at least for the Noachian era (Craddock and Howard, 2002). The better Orbiter imaging shows that dense, area-filling networks are common (Fig. 5.3).

Fig. 5.4. Delta in Holden Crater at 26.6 S, 325.6 E. A valley has cut through the south rim of the crater, just visible at the bottom of the picture, and deposited its sediment load to form a fan within the crater. (Image credit: THEMIS.)

The rover results from Meridiani indicate subaqueous deposition of sediments and groundwater oscillations (Grotzinger et al., 2005). Further, thick salt deposits within the canyons, at Meridiani and elsewhere (Gendrin et al., 2005) may indicate large bodies of water at the surface. Detection of phyllosilicate weathering products in Noachian terrains (Bibring et al., 2006) indicates warm conditions, and high erosion rates in the Noachian (Golombek and Bridges, 2000) are also consistent with precipitation and surface runoff. The case for at least episodes when Noachian Mars was warm with an active hydrologic cycle appears strong. If Mars was warm, then given the likely large inventory of water, there must have been large bodies of water in low areas such as Hellas and the northern plains that acted as both sources and sinks for liquid water. Although, while shorelines have been tentatively identified around both these lows (Clifford and Parker, 2001; Moore and Wilhelms, 2001), observational evidence for such former seas or oceans is weak.

The cause of the warm conditions postulated for the Noachian remains puzzling. Modeling suggests that raising surface temperatures with a CO_2–H_2O greenhouse falls far short of those required to stabilize liquid water at the surface without the participation of other greenhouse gases such as SO_2 or CH_4, which have very short lifetimes in the atmosphere (Haberle, 1998). Moreover, large deposits of carbonate, which would be expected if there were ever a thick CO_2 atmosphere, have not been detected despite intense searches from orbit. Another possibility is that warm conditions occurred episodically for short periods of time as a result of large impacts (Segura et al., 2002) or large volcanic eruptions. Such short-lived episodes would be consistent with the seemingly immature nature of the drainage system.

At the end of the Noachian, 3.7 Ga ago, the rate of valley formation declined dramatically, leaving most Hesperian surfaces undissected by valley networks. The rate did not, however, decline to zero. Some of the most prominent valleys, such as Nirgal Vallis and Nanedi Vallis are Hesperian in age, and some Hesperian surfaces on volcanoes and adjacent to the vast equatorial canyons, Valles Marineris, are heavily dissected. Nevertheless, the most typical fluvial feature of the Hesperian era is the outflow channel. Outflow channels are very different from valley networks. They start full size, have striated floors, sculpted walls and contain teardrop-shaped islands (Fig. 5.5). Most are enormous in size ranging up to hundreds of kilometers across. Instead of branching upstream, many start abruptly at faults or rubble-filled depressions. Others emerge from the eastern end of Valles Marineris.

Outflow channels are thought to have been formed by large floods. They start full-size and have numerous features in common with large terrestrial floods. Estimates of their peak discharges range up to 10^8 m^3 s^{-1}. Floods may

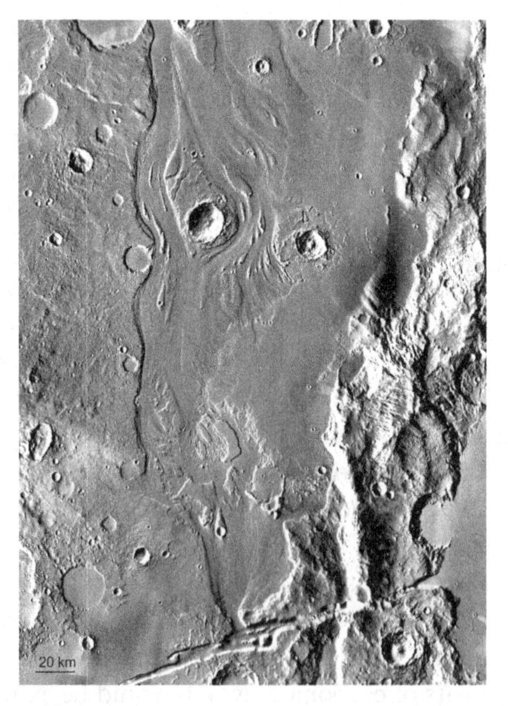

Fig. 5.5. The source of the outflow channel Mangala Vallis. It starts a 7-km-wide gap in a graben wall. The fault may have disrupted the cryosphere, thereby allowing groundwater trapped under pressure below ground ice to access the surface. (Image credit: THEMIS.)

have had different causes. The common observation that channels start at rubble-filled depressions suggests massive eruptions of groundwater confined under pressure beneath a thick cryosphere. Cold conditions during the Hesperian and a thick cryosphere are consistent with low weathering (Haskins et al., 2005) and erosion rates. The large hydrostatic pressures needed to generate the apparent large discharges could have been induced in a variety of ways such as by large impacts, volcanic eruptions, and tectonic forces. The large outflow channels that emerge from the east end of the canyons may have been caused by rapid draining of lakes within the canyons, which could also have formed by groundwater eruptions, but here the water was temporarily contained within the large rift valleys. Whatever their cause, the floods must have left behind large lakes or seas in the lows at their ends, which are mostly in the low-lying northern plains. The lakes would have frozen and ultimately sublimated away. The former presence of terminal lakes or seas is supported by burial of ridges and craters by sediments (Kreslavsky and Head, 2002), the presence of possible subglacial features in these areas (Kargel et al., 1995) and possible shorelines (Clifford and Parker, 2001).

The dramatic change in the rates of erosion, weathering and dissection at the end of the Noachian, around 3.7 Ga ago, has been ascribed to a change from warm surface conditions favoring precipitation in the Noachian to cold conditions with groundwater trapped under pressure below a thick cryosphere that favor formation of outflow channels. This scenario cannot be the whole story, however, because of the occasional young valley networks (Mangold et al., 2004), and the heavily dissected surfaces on some volcanoes. These younger valleys have been variously attributed to short warm episodes triggered by floods or large impacts, hydrothermal circulation caused by volcanic activity, melting of ice deposits formed during the floods, or during periods of high obliquity (see below).

The last 3 Ga of Mars' history is called the Amazonian era. Valley networks are even rarer than in the Hesperian, but are present, as for example, on Alba Patera, and outflow channels continued to form occasionally, as for example Athabasca Vallis (Burr et al., 2002). However, these younger outflow channels are generally smaller than the older channels and generally start at faults. The so-called gully is the most common water-worn feature of the Amazonian (Fig. 5.6). Gullies are typically meters to tens of meters wide and hundreds of meters long and found on slopes such as crater walls and central peaks (Malin and Edgett, 2000). They appear to form preferentially on pole-facing slopes. When first discovered, gullies were attributed to recent groundwater seepage, but another possibility is that they form by melting of snow deposited during periods of high obliquity (Costard et al., 2002). The

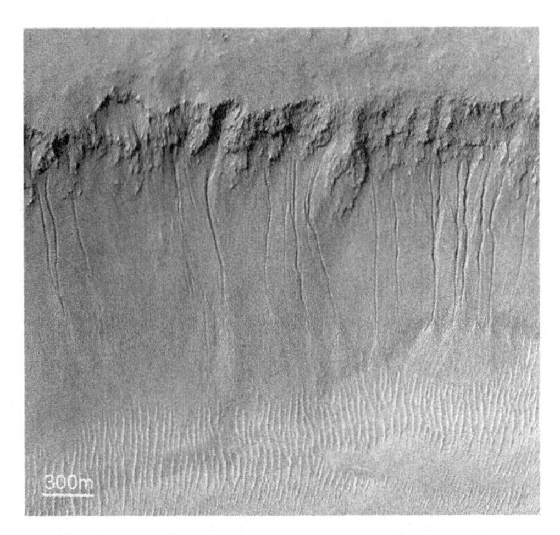

Fig. 5.6. Gullies on the wall of Nirgal Vallis at 29.7 S, 321.1 E. (Image credit: MOC M0322990.)

obliquity of Mars (the tilt of the spin axis toward the Sun) undergoes large oscillations (Laskar et al., 2004). The average obliquity is close to 40°, but there is a 63% probability of having reached 60° in the last 1 Ga. At large obliquities, water is driven off the poles to accumulate at lower latitudes. Melting of the snow during the long warm summers at high to mid latitudes may have then caused the gullies. High obliquities may also have caused extensive glaciation at low to mid latitudes (Head et al., 2005).

Thus, Mars has had a long and diverse history of water erosion. The most compelling evidence for warmer climates is from the Noachian, where warm conditions are supported not only by widespread dissection, but also by mineralogic evidence from orbit and from ground observations by the Mars rovers. The evidence for warm conditions after the end of the Noachian is more conflicting. Outflow channels which formed mainly 3.0–3.7 Ga ago, but also up to the geologically recent past, appear to require cold conditions and a thick cryosphere. But rare young valley networks and highly dissected surfaces on volcanoes suggest that there may have been occasional warm episodes.

Stream ecosystems in the Dry Valleys

The current stream ecosystems in the Dry Valleys provide analogs to guide exploration for past life in fluvial systems on Mars. In the Dry Valleys, there are some stream reaches where the streambed is covered by benthic microbial mats the entire width of the stream, and into the wetted margins. The visible abundant microbial life in these benthic mats is comprised primarily of cyanobacteria (formerly classified as blue-green algae), diatoms, and other microbes. These benthic microbial mats persist in a freeze-dried state during the austral winter. Photosynthesis resumes shortly after rewetting, when streamflow is renewed. Black-colored mats commonly occur at the stream edge and are composed almost entirely of one species of filamentous cyanobacteria of the genus *Nostoc*, whereas orange-colored mats occur in the main flowing part of the channel and are composed of about 15 different species of the genera *Phormidium* and *Oscillatoria.* Diatoms are found in both black- and orange-colored mats and have a greater diversity of species compared with the cyanobacteria. The 40 species in the Dry Valley stream diatom flora (Alger et al., 1997; Spaulding, 2005) include 24 endemic species that increase in relative abundance under conditions of low flow during cold summers. These benthic microbial mats also contain occasional invertebrates, such as nematodes or tardigrades, but the numbers are low and grazing is not a major factor controlling the abundance of the mats. Rather, the presence of abundant perennial mats is controlled by the geomorphological features of the streams that allow for the formation of a stable stone pavement arrangement of the rocks in the streambed. In these reaches, large flat rock surfaces are exposed and the streambed is relatively flat and even, so that at low flow the water spreads out across the channel, maximizing the area of the streambed covered by shallow water. Monitoring of the abundance of mats in these stream reaches indicates that the limitation on mat coverage is not due to low flow in cold summers, but rather due to infrequent scour or burial by deposited sediment associated with extreme flows.

In general, the presence of phototrophic microbial life is much less apparent on the streambed surfaces which are not armored by a stone pavement. In steep gradient reaches, large rocks appear jumbled and are wedged together in the stream center, as if the alluvium had been carried away by previous high flows. Here, isolated patches of green filamentous algae of the genus *Prasiola* are found on the shaded, but not buried, underside of the large cobble, where these mats receive light when the sun is at a low angle and are also protected from scour by the sediment. When flow has ceased, these dry mats appear as a green gauze glued onto the underside of the rock. In these steeper channels, another habitat for benthic mats is the seeps that drain the stream banks and hyporheic zone, where orange and black mats can be found.

In comparison with the steep reaches, there are even fewer obvious benthic mats in the areas where the stream gradient is low and the streambed is composed of sandy, unstable alluvium. In these streams, orange mats can develop in the center of the stream if steady, low flow conditions persist. These mats are then washed away or buried during periods of high flow, or can be covered with fine wind-blown sediment and form an endolithic crust if no flow occurs for an entire summer.

One important connection between the Dry Valley streams and fluvial environments on Mars is the potential for persistence of microbial communities through extended periods without surface moisture. In the Dry Valleys, these periods would correspond to periods with sufficiently cold summers, without meltwater generation from the alpine glaciers, even in the warmer coastal zone of the Dry Valleys. To address this question experimentally, an abandoned channel in the Fryxell Basin was "reactivated" by redirecting the flow from an upper channel (McKnight et al., 2007). Although this channel had not received substantial flow for approximately two decades, cyanobacterial mats became abundant in the reactivated channel within a week, indicating that the mats had been preserved in a cryptobiotic state in the channel. These mats soon achieved rates of primary production and nitrogen fixation that exceeded those of other benthic mats in the established streams. Experiments in which mats from the reactivated channel and another stream were incubated in water from both of the streams indicated that the greater solute concentrations in the reactivated channel stimulated net primary productivity of mats from both streams. These results indicate that the observed rapid rates of microbial growth were sustained by the high solute concentrations from mobilization of accumulated weathering products, which lasted for several years after the diversion. These stream-scale experimental results together with experimental laboratory mat rejuvenation indicate that the cryptobiotic preservation of cyanobacterial mats in abandoned channels in the Dry Valleys allows for rapid response of these stream ecosystems to climatic and geomorphological change, similar to other arid zone stream ecosystems. Zones on Mars where river networks of liquid water were present would then represent an area of interest to look for life such as these desiccated mats. Hyporheic zones also are important analog habitats in the Dry Valleys.

The Dry Valleys also provide opportunities to explore how long these encrusted buried mats can remain in an inactive dehydrated state on the scales of thousands of years. Recently several freeze-dried desiccated algal mats from Taylor, Wright, and Victoria Valleys that have been ^{14}C-dated to range from 8600 to 26 500 years before present (Hall et al., 2002) have been reactivated. Bacterial counts based on DAPI (4′,6-diamidino-2-phenylindole – a fluorescent DNA-binding dye) and live/dead staining indicated that between 1% and 20% of cells were viable (D. Antibus, L. Leff, C. Blackwood, and J. Baeseman, unpublished data) (Fig. 5.7). As expected, total cell number and viability percentage declined with increased age. Both photosynthetic and heterotrophic bacteria were isolated from the mats. Microbes in the Dry Valleys,

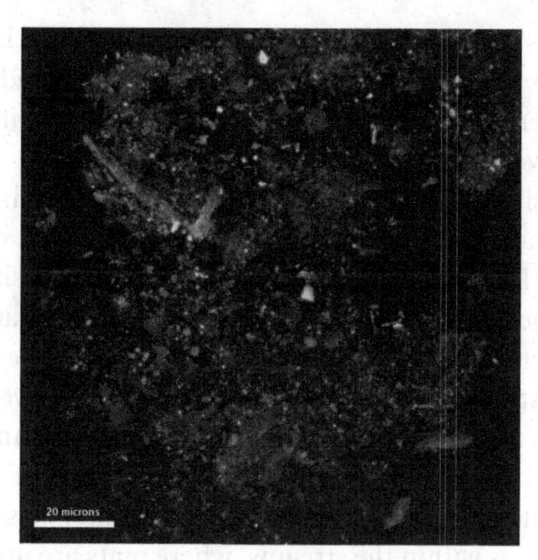

Fig. 5.7. This is an image of a 8643 ± 72-year-old microbial mat retrieved from the Dry Valleys after growth in liquid culture medium for 70 days. The confocal laser scanning microscope (CLSM) used was an Olympus Fluoview 500. A small piece of the mat was broken off the edge of a larger piece in liquid medium, so the entire depth of the sample is represented. The entire sample is approximately 100 μm thick. This image is a combination of approximately 12 images taken from the surface of the mat to a depth of 20 μm at 2 μm intervals. Beyond this depth, it was not possible to image the interior of the mat. The mat was stained with a live-dead kit (Invitrogen, Molecular Probes) using Propidium iodide and Syto 9. The dye was excited with a laser at 488 nm and fluorescence was collected separately in red and green channels. Dead cells (compromised membranes) appear red in the original. The color version of this figure can be found at http://www.cambridge.org/uk/catalogue/catalogue.asp?isbn=9780521889193.

as with many desert regions, have had to overcome many stresses and adapt to an extreme environment to survive. Many organisms in the Dry Valleys have evolved to process a strategy of desiccation or anhydrobiosis to sustain during periods of no water. The drying of cells, however, causes the oxidation of proteins, DNA, and membrane structures, as well as the breakdown of many essential sugars and amines (Potts, 1999). The evolutionary mechanism for the ancient mats that were revived in the study mentioned above is currently being investigated, but it is clear that these organisms have found a way to assure the preservation of their basic organic structure which will be important when thinking of possible past and present life on Mars.

It is important that the rehydration of these mats that have been dehydrated/freeze-dried for thousands of years be taken into consideration for life on Mars. If life is present or was present years ago, it is possible that similar survival mechanisms could be employed by these organisms. Areas in the Dry Valleys, as well as those on Mars, where mats are likely to be found, such as the stone pavement sections or the areas of sediment deposition from high-flow periods, warrant more examination for life, be it currently metabolically active, or in a desiccated state.

While the diversity of the phototrophic microorganisms in the benthic mats of Dry Valley streams has been documented, albeit based on morphological traits, much less is known about the distribution and diversity of microorganisms present on the sand grains of the hyporheic zone. Maurice et al. (2002) found diverse microorganisms, including denitrifying bacteria, growing on experimental substrates of freshly cleaved mica that were placed in the hyporheic zone of a stream with abundant benthic mats and examined after about 40 days of flow. Microbial cell counts on sediments from the hyporheic zone of streams with visible mats are equal to or sometimes higher than those of streams, or patches within the streams, where mats are not visible. Ranging from 10^5 to 10^8 cells per gram, cell counts are within the range of many temperate stream hyporheic zones. The microbial species composition of these communities is not yet fully known, but preliminary work suggests nitrogen form (i.e., as nitrate, ammonium, nitrite, etc.) and concentration are significantly more important to increased diversity than dissolved organic carbon (J. Baeseman and B. Ward, unpublished data).

Dry Valley stream ecosystems regulate nutrient flux to the lakes. Where there are abundant microbial mats, direct nutrient uptake causes the concentrations of both nitrate and phosphate to be significantly lower than in other stream reaches (McKnight et al., 2004). These microbial mats are also zones of nitrate reduction (Gooseff et al., 2004). However, in addition to microbial activity in the benthic mats, nitrogen transformations are driven by microbial communities in the hyporheic zone. Maurice et al. (2002) found that the surface of the feldspar incubated in the stream was colonized by nitrate reducers. A recent nitrogen injection to Huey Creek revealed the consumption of nitrate along a stream reach void of benthic mats, and a production of nitrous oxide, an important intermediate in the denitrification process. This indicates that there may be certain reaches in these shallow streams that can become anoxic, providing a wider range of habitats than just photosynthetic and/or oxygenic metabolisms.

These connections between the surface and subsurface environments are not unique to the Dry Valleys, but are perhaps more important to ecosystem function than in their temperate counterparts (Treonis et al., 1999; Maurice et al., 2002; Ayers et al., 2007). These connections are also important in considering habitat conditions in past fluvial systems on Mars.

Comparisons and contrasts between Mars and the Dry Valleys

Streams in the McMurdo Dry Valleys and those of Mars share several characteristics – they exist in cold, dry climates, they are composed of similar substrates, discharge is intermittent, and they are surrounded by barren landscapes whose geomorphic evolution is affected by extensive near-surface ice. Given these commonalities, the probable ecosystems of martian streams were likely to be quite similar to those of the Dry Valleys. As with Dry Valley stream ecosystems, the subsurface microbial communities could have been important to cycling of nutrients or other solutes in martian streamflow. There is no current evidence for microbial mats on streambed surfaces of the ancient martian channels, though those too may have existed, anticipating the next occurrence of flowing water in a cryptobiotic state. Both ecosystems likely evolved to cope with the stress of extended periods of dryness, and severe cold in between periods of flow, further emphasizing the likelihood of extensive subsurface communities of microbes that are likely to be crucial to ecosystem function.

Despite some of the dissimilarities between streams of Mars and those of the McMurdo Dry Valleys, for example, it is not clear what patterns of nutrients and other biogeochemicals would be found in martian streams; if there are extensive fluvial environments on Mars, they are likely to be similar to those of the McMurdo Dry Valleys. In both environments, the physical factors greatly control the biodiversity and habitat. The species composition of both ecosystems may be very different, but the general patterns of distribution and function are likely to be very similar.

To date, our understanding of martian streams and stream processes is made possible by assessing landscape-scale features from satellite data, whereas our understanding of Dry Valley stream processes comes from fieldwork and our assessment of a range of features, including sediment grain-scale weathering. This mismatch in scales of investigation, of course, hinders a complete comparison of these systems. Despite these differences, it is likely that the similar forcing functions of climate, geology, and sparse organic material in the surrounding landscapes of both environments will result in similar stream ecosystem function.

References

Alger, A. S., McKnight, D. M. Spaulding, S. A., et al. (1997). *Ecological Processes in a Cold Desert Ecosystem: the Abundance and Species Distribution of Algal Mats in Glacial Meltwater Streams in Taylor Valley*. Institute of Arctic and Alpine Research, Occasional Paper 51. Boulder, CO: University of Colorado, 108 pp.

Ayers, E. B., Adams, B. J., Barrett, J. E., Virginia, R. A., and Wall, D. H. (2007). Unique similarity of faunal communities across aquatic–terrestrial interfaces in a polar desert ecosystem. *Ecosystems*, **10**(4), 523–535.

Bibring, J., Langevin, Y., Mustard, J. F., et al. (2006). Global mineralogical and aqueous history derived from OMEGA/Mars Express data. *Science*, **312**, 400–404.

Brandt, R. E. and Warren, S. G. (1993). Solar-heating rates and temperature profiles in Antarctic snow and ice. *Journal of Glaciology*, **39**, 99–110.

Burr, D. M., Grier, J. A., McEwen, A. S., and Keszthelyi, L. P. (2002). Repeated aqueous flooding from the Cerberus Fossae: evidence for very recently extant, deep groundwater on Mars. *Icarus*, **159**, 53–73.

Carr, M. H. (2006). *The Surface of Mars*. Cambridge, UK: Cambridge University Press, 308 pp.

Cartwright, K. and Harris, H. J. H. (1981). Hydrogeology of the Dry Valleys region, Antarctica. In *Dry Valley Drilling Project*, ed. L. D. McGinnis. Antarctic Research Series. Washington, D.C.: American Geophysical Union, pp. 193–214.

Chinn, T. J. (1993). Physical hydrology of the dry valley lakes. In *Physical and Biogeochemical Processes in Antarctic Lakes*, ed. W. J. Green and E. I. Freidmann. Antarctic Research Series. Washington, D.C.: American Geophysical Union, pp. 1–51.

Clifford, S. M. and Parker, T. J. (2001). The evolution of the Martian hydrosphere: implications for the fate of a primordial ocean and the current state of the northern plains. *Icarus*, **154**, 40–79.

Conovitz, P. A., McKnight, D. M., MacDonald, L. H., Fountain, A. G., and House, H. R. (1998). Hydrological processes influencing streamflow variation in Fryxell Basin, Antarctica. In *Ecosystem Dynamics in a Polar Desert: The McMurdo Dry Valleys, Antarctica*, ed. J. C. Priscu. Antarctic Research Series 72. Washington, D.C.: American Geophysical Union, pp. 93–108.

Conovitz, P. A., Macdonald, L. H., and McKnight, D. M. (2006). Spatial and temporal active layer dynamics along three glacial meltwater streams in the McMurdo Dry Valleys, Antarctica. *Arctic Antarctic and Alpine Research*, **38**, 42–53.

Costard, F., Forget, F., Mangold, N., and Peulvast, J. P. (2002). Formation of recent Martian debris flows by melting of near-surface ground ice at high obliquity. *Science*, **295**, 110–113.

Cozzetto, K., McKnight, D. M., Nylen, T., and Fountain, A. G. (2006). Experimental investigations into processes controlling stream and hyporheic temperatures, Fryxell Basin, Antarctica. *Advances in Water Resources*, **29**, 130–153.

Craddock, R. A. and Howard, A. D. (2002). The case for rainfall on a warm, wet early Mars. *Journal of Geophysical Research*, **107**(E11), doi: 10.1029/2001JE001505.

Doran, P. T., Wharton, Jr., R. A., and Lyons, W. B. (1994). Paleolimnology of the McMurdo Dry Valleys, Antarctica. *Journal of Paleolimnology*, **10**(2), 85–114.

Ebnet, A. F., Fountain, A. G., Nylen, T. H., McKnight, D. M., and Jaros, C. I. (2005). A temperature-index model of stream flow at below freezing temperatures in Taylor Valley Antarctica. *Annals of Glaciology*, **40**, 76–82.

Fischer, H., Kloep, F., Wilzcek, S., and Pusch, M. T. (2005). A river's liver: microbial processes within the hyporheic zone of a large lowland river. *Biogeochemistry*, **76**(2), 349–371.

Fountain, A. G., Dana, G. L., Lewis, K. J., et al. (1998). Glaciers of the McMurdo Dry Valleys, southern Victoria Land, Antarctica. In *Ecosystem Dynamics in a Polar Desert: The McMurdo Dry Valleys, Antarctica*, ed. J. C. Priscu. Antarctic Research Series 72. Washington, D.C.: American Geophysical Union, pp. 65–75.

French, H. M. (1996). *The Periglacial Environment*, 2nd edition. Harlow, UK: Addison Wesley Longman.

Golombek, M. P. and Bridges, N. T. (2000). Erosion rates on Mars and implications for climate change: constraints from the Pathfinder landing site. *Journal of Geophysical Research*, **105**, 1841–1853.

Gooseff, M. N., McKnight, D. M., Lyons, W. B., and Blum, A. E. (2002). Weathering reactions and hyporheic exchange controls on stream water chemistry in a glacial meltwater stream in the McMurdo Dry Valleys. *Water Resources Research*, **38**(12), 1279, doi: 10.1029/2001WR000834.

Gooseff, M. N., McKnight, D. M., Runkel, R. L., and Vaughn, B. H. (2003). Determining long time-scale hydrologic flow paths in Antarctic streams. *Hydrological Processes*, **17**(9), 1691–1710.

Gooseff, M. N., McKnight, D. M., Runkel, R. L., and Duff, J. H. (2004). Denitrification and hydrologic transient storage in a glacial meltwater stream, McMurdo Dry Valleys, Antarctica. *Limnology and Oceanography*, **49**(5), 1884–1895.

Gooseff, M. N., Lyons, W. B., McKnight, D. M., et al. (2006). A stable isotopic investigation of a polar desert hydrologic system, McMurdo Dry Valleys, Antarctica. *Arctic Antarctic and Alpine Research*, **38**, 60–71.

Gendrin, A., Mangold, N., Bibring, J., et al. (2005). Sulfates in martian layered terrains: the OMEGA/Mars Express view. *Science*, **302**, 1587–1591.

Grotzinger, J. P., Arvidson, R. E., Bell, J. F., et al. (2005). Stratigraphy, sedimentology and depositional environment of the Burns Formation, Meridiani Planum, Mars. *Earth Planetary Science Letters*, **240**, 11–72.

Haberle, R. M. (1998). Early climate models. *Journal of Geophysical Research*, **103**, 28 467–28 479.

Hall, B. L., Denton, G. H., and Overturf, B. (2001). Glacial Lake Wright, a high-level Antarctic lake during the LGM and early Holocene. *Antarctic Science*, **31**(1), 53–60.

Hall, B. L., Denton, G. H., Overturf, B., and Hendy, C. H. (2002). Glacial Lake Victoria, a high-level Antarctic lake inferred from lacustrine deposits in Victoria Valley. *Journal of Quaternary Science*, **17**, 697–706.

Harris, K., Carey, A. E., Welch, K. A., Lyons, W. B., and Fountain, A. G. (2007). Solute and isotope geochemistry of near-surface ice melt flows in Taylor Valley, Antarctica. *Geological Society of America Bulletin*, **119**, 548–555.

Haskins, L. A., et al. (34 authors) (2005). Water alteration of rocks and soils on Mars and the Spirit rover site in Gusev crater. *Nature*, **436**, 66–69.

Head, J. W., Neukum, G., Jaumann, R., et al. (2005). Tropical to mid-latitude snow and ice accumulation, flow and glaciation on Mars. *Nature*, **434**, 346–351.

Kargel, J. S., Baker, V. R. Beget, J. E., et al. (1995). Evidence for ancient continental glaciation in the Martian northern plains. *Journal of Geophysical Research*, **100**, 5351–5368.

Kreslavsky, M. A. and Head, J. W. (2002). Fate of outflow channel effluents in the northern lowlands of Mars: the Vastitas Borealis Formation as a sublimation residue from frozen, ponded bodies of water. *Journal of Geophysical Research*, **107**, doi: 10.1029/2001JE001831.

Laskar, J., et al. (2004). Long term evolution and chaotic diffusion of the insolation quantities of Mars. *Icarus*, **170**, 343–364.

Lewis, K. J., Fountain, A. G., and Langevin, P. L. (1999). The importance of terminus cliff melt on stream flow, Taylor Valley, McMurdo Dry Valleys. *Global and Planetary Change*, **22**, 105–115.

Lewis, A. R., Marchant, D. R., Kowalewski, D. E., Baldwin, S. L., and Webb, L. E. (2006). The age and origin of the Labyrinth, western Dry Valleys, Antarctica: evidence for extensive middle Miocene subglacial floods and freshwater discharge to the Southern Ocean. *Geological Society of America Bulletin*, **34**(7), 513–516.

Malin, M. C. and Edgett, K. S. (2000). Evidence for recent groundwater seepage and surface runoff on Mars. *Science*, **288**, 2330–2335.

Malin, M. C. and Edgett, K. S. (2003). Evidence for persistent flow and aqueous sedimentation on early Mars. *Science*, **302**, 1931–1934.

Mangold, N., Quantin, C., Anson, V., Delacourt, C., and Allemand, P. (2004). Evidence for precipitation on Mars from dendritic valleys in the Valles Marineris area. *Science*, **305**, 78–81.

Maurice, P. A., McKnight, D. M., Leff, L., Fulghum, J. E., and Gooseff, M. (2002). Direct observations of aluminosilicate weathering in the hyporheic zone of an Antarctic Dry Valley stream. *Geochimica et Cosmochimica Acta*, **66**(8), 1335–1347.

McKnight, D. M., Alger, A., and Tate, C. M. (1998). Longitudinal patterns in algal abundance and species distribution in meltwater streams in Taylor Valley, Southern Victoria Land, Antarctica. In *Ecosystem Dynamics in a Polar Desert: The McMurdo Dry Valleys, Antarctica*, ed. J. C. Priscu. Antarctic Research Series 72. Washington, D.C.: American Geophysical Union, pp. 109–127.

McKnight, D. M., Runkel, R. L., Tate, C. M., Duff, J. H., and Moorhead, D. L. (2004). Inorganic N and P dynamics of Antarctic glacial meltwater streams as controlled by hyporheic exchange and benthic autotrophic communities. *Journal of the North American Benthological Society*, **23**, 171–188.

McKnight, D. M., Tate, C. M. Andrews, E. D., et al. (2007). Reactivation of a cryptobiotic stream ecosystem in the McMurdo Dry Valleys, Antarctica: a long-term geomorphological experiment. *Geomorphology*, **89**, 186–204.

Moore, J. M. and Wilhelms, D. E. (2001). Hellas as a possible site of ancient ice-covered lakes on Mars. *Icarus*, **154**, 258–276.

Potts, M. (1999). Mechanisms of desiccation tolerance in cyanobacteria. *European Journal of Phycology*, **34**, 319–328.

Segura, T. L., Toon, O. B., Colaprete, A., and Zahnle, K. (2002). Environmental effects of large impacts. *Science*, **298**, 1977–1980.

Sliva, L. and Williams, D. D. (2005). Exploration of riffle-scale interactions between abiotic variables and microbial assemblages in the hyporheic zone. *Canadian Journal of Fisheries and Aquatic Sciences*, **62**(2), 276–290.

Spaulding, S., Esposito, R., Lubinski, D., et al. (2005). Antarctic Freshwater Diatoms web site, McMurdo Dry Valleys LTER, visited 2 Oct 2009 at http://huey.colorado.edu/diatoms/.

Squyres, S. W. and Kasting, J. F. (1994). Early Mars: how warm and how wet? *Science*, **265**, 744–748.

Stuiver, M., Yang, I. C., Denton, G. H., and Kellogg, T. B. (1981). Oxygen isotope ratios of Antarctic permafrost and glacier ice. In *Dry Valley Drilling Project*, ed. L. D. McGinnis. Antarctic Research Series 33. Washington, D.C.: American Geophysical Union, pp. 131–140.

Treonis, A. M., Wall, D. H., and Virginia, R. A. (1999). Invertebrate biodiversity in Antarctic Dry Valley soils and sediments. *Ecosystems*, **2**(6), 482–492.

6

Saline lakes and ponds in the McMurdo Dry Valleys: ecological analogs to martian paleolake environments

JILL MIKUCKI, W. BERRY LYONS, IAN HAWES,
BRIAN D. LANOIL, AND PETER T. DORAN

Introduction

On the basis of the prevalence of cold environments in our solar system, the search for extraterrestrial life is focused largely on icy habitats. The McMurdo Dry Valleys (MDV) area is a polar desert with a mean annual temperature below freezing and extremely low humidity (Wharton et al., 1995) and thus offers a suitable earthly analog to our nearest exobiological candidate, Mars. Water is thought to have been abundant on Mars early in its geological history (Solomon et al., 2005; Squyres et al., 2006; Head and Marchant, this volume, Chapter 2) and perhaps may even have flowed across the surface more recently (Hauber et al., 2005; Head et al., 2005; Malin et al., 2006). Life as we know it requires the presence of liquid water to mediate biochemical reactions for energy as well as a reasonably stable environment in which to grow; therefore the search for extraterrestrial life has been largely a search for environments where liquid water can be maintained for some duration (e.g., Carr, 1983).

There is significant geomorphological evidence, and mounting physical evidence supporting the presence of paleolakes on ancient Mars (Squyres et al., 2006). This intrigues exobiologists because paleolakes would provide a suitable habitat for early martian life forms (Carr, 1983; Wharton et al., 1995; Doran et al., 2004). Lakes on the martian surface would have become progressively colder over geological time, developing seasonal and eventually perennial ice covers (Carr, 1983). This process of freezing would have necessarily concentrated any solutes present in the water (Wharton et al., 1995) resulting in stratified lakes similar to those we see in the MDV today.

Life in Antarctic Deserts and Other Cold Dry Environments: Astrobiological Analogs, ed. Peter T. Doran, W. Berry Lyons and Diane M. McKnight. Published by Cambridge University Press.

We know from the study of the MDV lakes that microbial life can persist and thrive under a permanent ice cover even when temperatures are routinely below freezing, light is sharply attenuated or absent, and chemical gradients become extreme (e.g., Spiegel and Priscu, 1998; Lyons et al., 1998b; Fountain et al., 1999). In this chapter we provide a brief introduction to the numerous paleolake features detected on Mars. We then describe the MDV lake and pond ecosystems to highlight the utility of the MDV as an analog to putative martian ecologies.

Paleolakes on Mars

Numerous lake basins have been recognized on Mars since the return of images from the Viking spacecraft (Goldspiel and Squyres, 1991; Cabrol and Grin, 1999, 2001). These lakes were identified by the presence of valley networks and channels that converge towards closed basin depressions. Orbital imagery also shows signs that water once pooled within natural basins and sediment deposits exist within these putative lake bottoms (Goldspiel and Squyres, 1991). Such findings are of considerable interest for the planning of Mars surface exploration missions because a wealth of information on climate, geochemistry, and biology are routinely retrieved from lake sediments on Earth.

The case for past (and perhaps modern) hydrology on Mars is based largely on the interpretation of distinct surface morphology. Using orbital image analysis, Cabrol and Grin (1999) described three different martian lake basins including closed, open, and chain system lakes. These authors identified 179 distinct paleolake structures from Viking Orbiter data (Cabrol and Grin, 1999), all of which were located between 47.5° N and 64.7° S latitudes (Fig. 6.1). The highest concentration of lakes was discovered between 20° N and 20° S latitude. A significant portion of identified lakes are impact craters centered in the highly cratered uplands; these features are associated with abundant fluvial valley networks. Another peak in lake abundance is located in the northern hemisphere and related to large outflow channels. These outflow channels are thought to have flooded many of the craters at one time (Cabrol and Grin, 1999). Regional clusters of lake features have been interpreted as dynamic water systems, and labeled as watersheds, paleochannel courses, and extensive basins (Cabrol and Grin, 2005; Fasset and Head, 2008). The presence of dynamic water systems on Mars is further supported by new higher resolution data supplied by the Mars Global Surveyor (MGS), Mars Orbiter Camera (MOC), and Mars Orbiter Laser Altimeter (MOLA) (Cabrol and Grin, 2005). The presence of diverse lake types suggests that any associated biogeochemistry would be similarly diverse and reflect lake

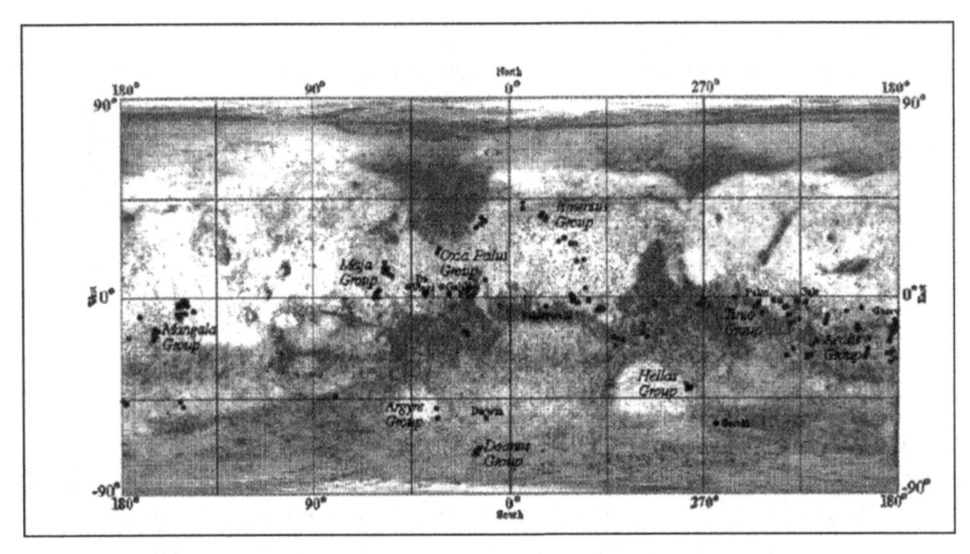

Fig. 6.1. Distribution map of impact martian crater lakes. (Reprinted from Cabrol and Grin, 2001, with permission from Elsevier.)

position and formation analogous to the dramatic variation among the MDV lakes exists today.

Speculation as to possible water sources that fed these surface paleolakes include an ancient planetary sea (Baker et al., 1991) or permafrost melted by geothermal heat or an impactor (Scott et al., 1991; Newsome et al., 1996). Volume calculations based on fluvial features demonstrate that surface discharge of water was abundant enough to maintain perennial lakes (Baker et al.,1991; Goldspiel and Squyres, 1991; Scott et al., 1991); however, these estimates cannot accurately predict whether there was enough water to fill ancient lakes or sufficient only to allow for shallow playas. The source and flow rates of water to these lakes can define the subsequent lake environment. For example, basins associated with high flows, such as may occur with an outburst flood from rapidly melted permafrost, would only retain water for a limited time within their basins. This phenomenon is particularly true for open basin systems (Cabrol and Grin, 2001). One result of this short residence time could be the presence of ephemeral, localized ponds. By contrast, basins not directly in the path of outburst flow, such as a closed basin system, would be supplied by slower moving water through meandering channels. The reduced velocity of water flow would allow for standing lakes over extended periods of time. Evidence for such standing lakes has been identified in association with chaotic terrain features (Cabrol and Grin, 2001). As environmental conditions changed on Mars and periods of active surface hydrology began to wane, flowing water would have become intermittent or completely absent (Cabrol and Grin, 2001). Evaporation would then have increased during these periods of extended dry and cold conditions. The result would have been a stage in which lake solutes became highly concentrated and precipitates may have been deposited on the lake bottom. This theory of lake evolution is supported by the detection in several crater basins with high albedo features, which are suggestive of evaporites (Cabrol and Grin, 2001). Other large evaporite deposits have been identified more recently with the Mars Express OMEGA imaging spectrometer. These deposits are associated with dune features in the north pole region of Mars (Fishbaugh et al., 2007). Their mode of deposition has yet to be determined but could be associated with subglacial release events. Finally, the story of water on Mars may not be exclusive to the ancient past. Evidence for recent water activity through gullies and glaciers opens the possibility for modern short-term ponds as well (Cabrol and Grin, 2005).

The McMurdo Dry Valleys ice-covered lakes and ponds

The closest earthly analog to Mars based on climate (dry desert) and temperature (mean annual temperatures well below freezing) are the dry valleys of Antarctica. The dry valleys represent the 2% of the antarctic continent that is ice free (Moorhead et al., 1999). Here we focus our discussion on a well-studied portion of this dry valley region, the McMurdo Dry Valleys (MDV). The geochemistry and physical constraints of this polar desert reflect regional climatic history and the metabolic processes of the microbial inhabitants now and over millennial timescales. The MDV provide an appropriate analog to extrapolate putative martian ecologies that may have existed in paleolakes systems or perhaps extant subglacial niches. There has been extensive research on the MDV lake systems (e.g., Vincent, 1988; Spiegel and Priscu, 1998; Priscu et al., 1999) and in this chapter we broadly discuss the chemistry and biology of five major lakes and the ponds with the goal of providing a depiction of the potential geochemical diversity that may have existed in martian paleolakes; such discussions can help inform future exobiological missions to Mars. The presence of an ice cover and a complex paleohistory result in the unusual stratification and geochemical profiles observed in the MDV today which in turn affect the activity, function, and diversity of life in the lakes.

Here we describe the paleohistory, geochemistry, and microbiology of Lakes Fryxell, Hoare, Bonney, Vanda, and Vida. Lake Vida is frozen solid down to a brine lens at the lake bottom and may be analogous to the last stages of a freezing martian limnology. We include a discussion on the ecology of the permanent ice covers of the MDV lakes and the upland ponds. Then, we describe important characteristics of the microbial mats found in the MDV lakes and ponds, which are of particular exobiological interest because their extensive growth creates structures that may be preserved as fossils, similar to earthly stromatolites. Finally, we focus on two brine systems in the MDV that provide insight into the potential controls on microbial life on Mars exerted by highly saline conditions. One of these is Blood Falls, a subglacial iron-rich brine that emanates from beneath the Taylor Glacier. The other is Don Juan Pond, one of the coldest, saltiest water bodies on Earth. Thus, the following discussion broadly illustrates how life forms can persist and even thrive in the MDV under comparable martian conditions.

History/evolution of the Taylor Valley lakes

The lakes of the McMurdo Dry Valleys are ancient systems. Lakes in Taylor Valley (TV) have existed for at least 300 000 years and are perhaps much older; evidence suggests the present lake stands are from the Pleistocene and earlier (Barrett and Hambrey, 1992; Hendy, 2000). Throughout the MDV history the size and chemistry of the lakes have fluctuated with climate (Hendy, 2000). Small lakes have given way to large proglacial lakes. These large lakes were positioned adjacent to the Ross Ice Sheet during glacial times and were accompanied by sea level lowering. Still larger proglacial lakes of the East Antarctic Ice Sheet existed during interglacial periods (Hendy, 2000). This ~100-ka cycle has been documented through the dating and characterization of TV lacustrine sediment (Hendy, 2000).

Today all of the MDV lakes are perennially ice covered. Currently TV has three major lake basins: Bonney to the west, Hoare in the mid portion of the valley, and Fryxell in the east, closest to McMurdo Sound and the Ross Sea (Fig. 6.2). The Hoare Basin stands at ~90 m above sea level and is separated from Fryxell Basin by the Canada Glacier. Geophysical data suggests that Lake Hoare would drain into the Fryxell Basin if the Canada Glacier were to retreat. Increasing lake levels would also allow Hoare waters to flow around Canada Glacier into Lake Fryxell (Hendy, 2000). The Bonney Basin is the largest drainage basin in TV; Lake Bonney is divided into west and east lobes by the Bonney Riegel.

The distinct lake basins we see today were not always the case. A large lake, Glacial Lake Washburn (GLW), filled the TV by 23 800 years ago and may have existed up until 8750 years ago (these dates are based on ^{14}C analyses from organic matter in former deltas; Hall and Denton, 2000).

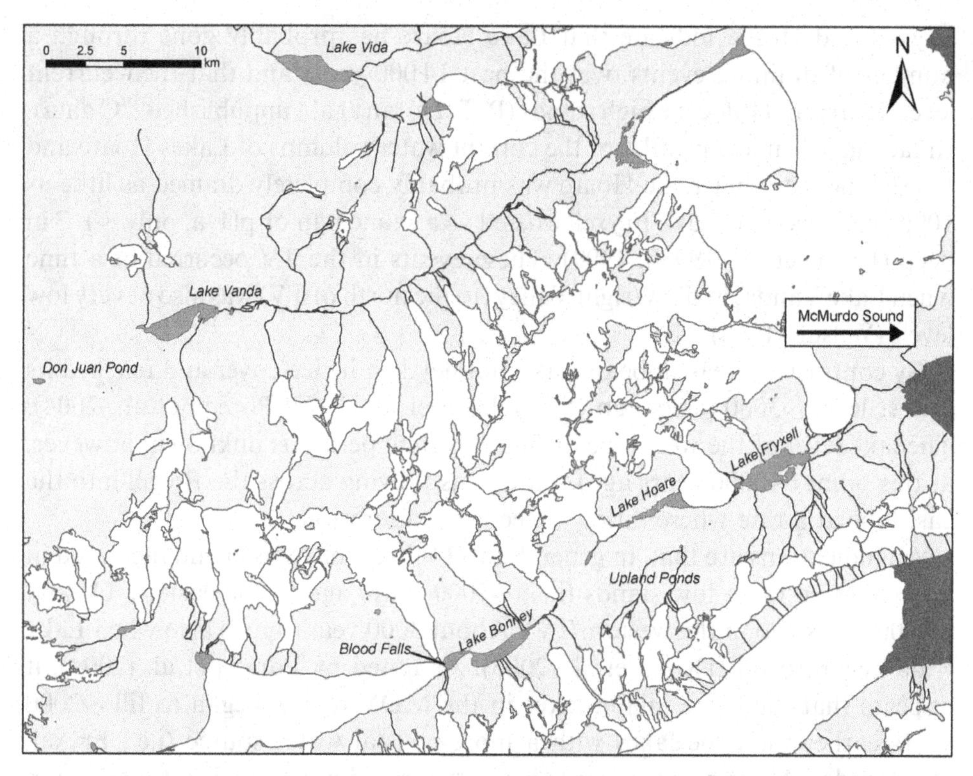

Fig. 6.2. Map of the McMurdo Dry Valleys, Antarctica. Features discussed in this chapter are labeled. Glaciers are white, lakes are dark grey, streams are outlined in black, and sediments are light grey.

Strandlines to 336 m above sea level exist in TV, suggesting that to have lakes this large, drainage of TV was blocked to the east by the Ross Ice Sheet (Hall et al., 2000). The $\delta^{18}O$ of carbonate material deposited during GLW times yields values of −40‰ to −42‰, indicating that the Taylor Glacier supplied water to the glacial lake. The dating of carbonate materials sampled around TV by $^{230}Th/^{234}U$ methods suggests that these carbonates were deposited between 70 000–75 000, 87 000–100 000, and 120 000 years ago. Their $\delta^{18}O$ values indicate that all were deposited from water similar to Taylor Glacier melt (Hendy, 2000). Still older carbonates from 190 000–210 000 and 300 000 ± 40 000 years ago have also been dated (Hendy, 2000). Large lakes also existed at 20 000, 145 000, and ~250 000 years ago during colder, glacial times (Hendy, 2000). Larger lakes were once present in the TV, many of which were generated from Taylor Glacier melt during interglacial times.

Recent analysis from a Lake Fryxell sediment core suggests that GLW was established 42 000 years ago, and that the lake has existed as a water body throughout this period (Wagner et al., 2006). Sediment data from both Lakes

Fryxell and Hoare indicate that Lake Hoare has probably gone through a number of draining events over the past 14 000 years and that their current levels represent Holocene high stands (P. T. Doran et al., unpublished ^{14}C data). In fact, geochemical profiles of the current water columns of Lakes Hoare and Fryxell indicate that Lake Hoare was probably completely drained as little as 1000 years ago and Lake Fryxell existed as a saline pan or playa, only ~1–3 m deep (Lyons et al., 1998b, 1999). These events in the TV occurred at a time when Lake Vanda, in the Wright Valley (to the north of TV) was also at very low levels (Wilson, 1964).

By contrast, the east lobe of Lake Bonney lost its ice cover and reached its lowest level ~3000 years ago (Matsubaya et al., 1977; Poreda et al., 2004). The lake level of the west lobe during this time period is unknown; however, at this point (~3000 years ago), water was flowing across the Riegel into the east lobe at a rate where inflow exceeded evaporation.

All indications are that, in general, the TV lakes have risen and increased in volume since these low stands (i.e., ~1000 years ago in the eastern TV and ~3000 years ago in the western TV). About 3000 years ago is also when Lake Vida began to fill (Doran et al., 2003). As noted by Poreda et al. (2004), it appears that the more inland lakes in the MDV region began to fill ~2000 years earlier than the lakes with a more coastal water source (i.e., Fryxell and Vanda). More recent records indicate a rapid and significant increase in lake levels over the past century (Chinn, 1993; Hood et al., 1998; Bomblies et al., 2001). However, a cooling trend in the TV region led to a small decrease in lake levels from 1985 to 2002 (Doran et al., 2002). This decrease in lake level was interrupted by a ~0.5 m lake level rise during a warm period in the summer of 2001–2002.

Consequences of lake level fluctuations

Ecological studies conducted during the recent warm summer 2001–2002 have shown definitively that increased inflow and lake level rise are associated with increased nutrient input to the euphotic zone, which results in a significant increase in primary production in these lakes (Foreman et al., 2004; Lyons et al., 2006). In contrast, the ecological response to the changes in the lake characteristics during the recent cooling period was mainly a decrease in primary production attributed to the increase in ice-cover thickness (Doran et al., 2002). Thus, the climatic fluctuations of TV lake levels impart a strong influence on the biology of these systems. These changes in lake level and corresponding size of the MDV lakes are due to the response of the TV hydrological regime to both global and perhaps regional/local climate forcings (Hendy, 2000;

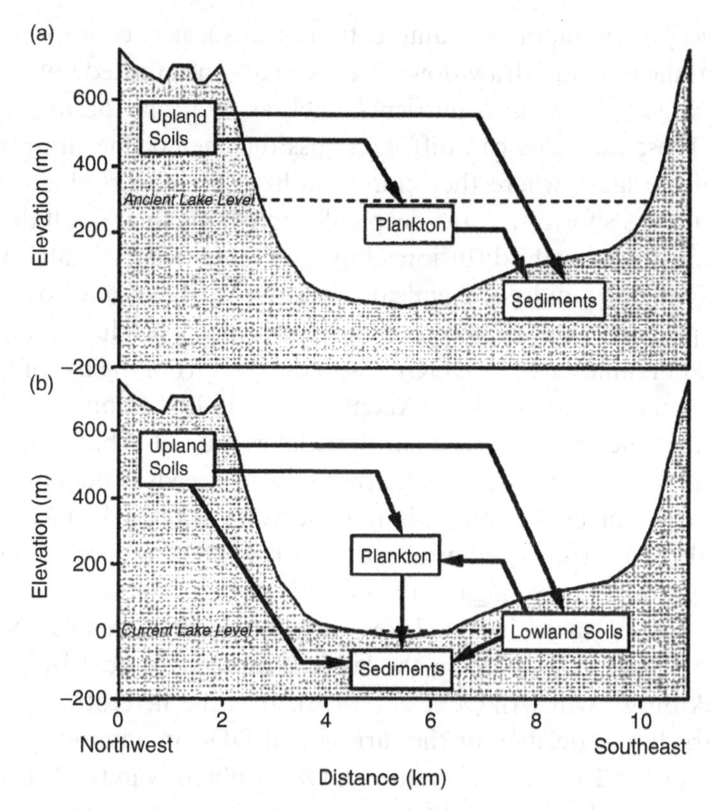

Fig. 6.3. Cross sections of the Lake Fryxell basin, Taylor Valley, Antarctica. (a) Approximate depth of Lake Washburn (approximately 20 000 years ago). (b) Approximate depth of modern Lake Fryxell. Flow diagrams represent primary fluxes of organic carbon. (Reprinted from Moorhead et al., 1999; copyright, American Institute of Biological Sciences.)

Doran et al., 2002; Poreda et al., 2004). Paradoxically, the highest lake levels have occurred during both the coldest and the warmest times (Hendy, 2000).

As the lakes enlarge and expand, soils become subaqueous and water solubilizes salts, such as the abundant nitrate salts found throughout TV soils which are deposited via atmospheric aerosols (Keys and Williams, 1981). Soil nitrates are also mobilized by streamflow. In turn, autochthonous organic matter produced in the littoral zone is deposited at the sediment/soil–lake interface. As climate changes and lake levels decrease, the subaerial portion of the landscape increases and the previously deposited lacustrine organic matter is now incorporated into TV soils. Thus, soil nitrate drives lake primary production and lacustrine soil matter drives heterotrophy in the soils.

This concept – termed legacy by McMurdo Dry Valley LTER scientists – is a key landscape and ecological feature in TV (Priscu, 1995; Burkins et al., 2000; Lyons et al., 2000; Moorhead et al., 2005). Figure 6.3 provides a visual format of this concept. An important ramification of this legacy concept for the TV lakes is that the previous draw-down events have concentrated not only major solutes such as Cl^- but also nutrients, such as NO_3^-, in the hypolimnia of the lakes. These nutrients can diffuse across the chemoclines into the upper portions of the lakes where they can be utilized by autotrophic organisms. Priscu (1995) has shown that the deep chlorophyll maxima of all the lakes in the MDV are driven by the diffusional inputs of these "legacy" nutrients. Thus, the climatic history and geochemical evolution of these lakes overprint an important property that helps determine their primary production.

Dissolved organic carbon (DOC) is also diffused from the hypolimnia into the mixolimnia of these lakes (Aiken et al., 1996). Lyons et al. (2000) suggested that the "legacy" DOC in these lakes may also be responsible for their current metabolic status, where respiration exceeds production (Takacs et al., 2001). Dissolved organic carbon in the MDV lakes is distinctive relative to many other lacustrine systems because the lake DOC is derived solely from microbial production by algae and bacteria in the water columns and in benthic microbial mats in the lakes and in-flowing streams (McKnight et al., 1998, 1999). The DOC in the lakes is thought to be a by-product of phytoplankton growth (Aiken et al., 1996). It is the heterotrophic bacteria that are largely responsible for the turnover of DOC in aquatic systems, and this is true for the TV lakes. Therefore, DOC availability in the MDV lakes is regulated both by the upward diffusion of nutrients and internal production.

Salts are another important legacy component of the TV lakes. Lyons et al. (2005) have demonstrated, based on halogen profiles and ^{129}I and ^{36}Cl measurements, that the source of solutes to Lake Bonney are remnants of ancient marine salts that have been modified by the input of chemical weathering products and periods of cryoconcentration. The source of these marine salts may be Miocene or Pliocene age seawater sequestered in TV when it was a fiord. Blood Falls, a saline discharge at the snout of the Taylor Glacier, is likely a frozen portion of a larger saline lake or even part of the original seawater source (Lyons et al., 2005). Blood Falls currently supplies both solutes and organisms to the west lobe of Lake Bonney (Mikucki et al., 2004; Lyons et al., 2005). This input of ancient salt and organisms to Lake Bonney is another example of a "legacy" effect and reflects the long and rather complicated history of the TV lakes.

Lake Fryxell

Lake Fryxell (LF) is a relatively large and shallow lake; LF has a surface area of $7.08\,m^2$, is $\sim$19 m at its deepest, and has a volume of 4.42×10^{10} L (Green et al., 1988; Spiegel and Priscu, 1998). The position of LF adjacent to the McMurdo Sound leads to increased precipitation within its basin and a resultant higher flux of marine aerosols (Lyons et al., 1998). Thirteen different melt streams emanating from the surrounding glaciers provide fresh water and nutrients to LF (Conovitz et al., 1998). Further discussion of the MDV streams and their relevance as a Mars analog are discussed in more detail in this volume (Gooseff et al., Chapter 5). Lake Fryxell is a well-stratified meromictic lake with relatively fresh water on top of brackish bottom waters that are ~about one fifth the salinity of seawater (Aiken et al., 1996).

Lake Fryxell, like the other lakes in the MDV, supports algal, bacterial, and microbial mat communities. In terms of bacterial production, LF is highest among the TV lakes (Takacs and Priscu, 1998). The geochemical stratification, specifically the interplay between light, oxygen, and sulfide gradients, results in the distinct positioning of microbial populations in the water column. Light penetration through the ice cover is attenuated to less than 10% of incoming surface radiation; despite the low light levels, a diversity of photosynthetic organisms are found in the lake environment. Fifty-six species of algae were identified in LF; these phytoplankton were primarily cryptophyte and chlorophyte flagellates, and filamentous cyanobacteria (Laybourn-Parry et al., 1996). Laybourn-Parry et al. (1996) found vertical stratification in some taxa (*Oscillatoria limnetica*, *Phormidium angustissimum*, *Pyramimonas* sp., *Oscillatoria* sp.), while others showed no distinct positioning (*Chlamydomonas subcaudata*, *Cryptomonas* sp.). The observed stratification of these species is attributed to nutrient and light gradients and the distinct requirements of each species (Spaulding et al., 1994). Despite the limitations on life under permanent ice, the MDV microorganisms have acclimated to extremely low nutrient and light levels. For example, recent physiological studies on a select species of *Chlamdyomonas* revealed adaptation to low temperature and low light conditions that was reflected in the unique photosynthetic machinery of these organisms (Morgan-Kiss et al., 2006).

In addition to oxygenic photosynthesis by a diversity of algae, anoxygenic photosynthesis occurs in LF. The light that does penetrate the ice cover, although low, reaches into the anoxic zone (~9 m) where significant H_2S is available to phototrophic purple bacteria (Karr et al., 2005). Chemolithoautotrophic sulfur oxidizing isolates have also been cultivated from LF (Sattley et al., 2006). These organisms dominate at the chemocline where the gradient between sulfide and oxygen is strongest. These novel sulfur oxidizers were shown to be cold adaptive, demonstrating growth at $-2°$ C (Sattley et al., 2006). Below the chemocline, LF becomes anoxic and highly sulfidic due to biological sulfate reduction (Karr et al., 2005). It is the only lake in the TV where *Archaea* have been identified (Karr et al., 2005) and isolated. For example, a relative of a methylotrophic *Methanosarcina* sp. that occupies the surface of LF sediments was detected by both molecular methods (Karr et al., 2005) and isolation (Singh et al., 2005). The overall contribution of the *Archaea* to lake productivity has yet to be determined, as does their presence in other MDV lakes.

Lake Hoare

Lake Hoare (LH) is separated from LF by the Canada Glacier. It has a surface area of $\sim 3 \times 10^6\,m^2$ and a volume of 2.6×10^{10} L (Wharton et al., 1993b). Lake Hoare is the freshest lake in TV with Cl^- values of 5.84 mM at 20 m depth (Fig. 6.4). It has a maximum depth of 34 m, which is intermediate between LF and Lake Bonney (LB), and a depth to the oxycline of 28 m (Wharton et al., 1993b). The salinity structure of LH is similar to that of the other MDV lakes in some respect, but also has distinctive features: a gradient exists from just below the ice to ~13 m while from 13.5 m to 23 m salinity is nearly uniform (Spigel and Priscu, 1998). Lake temperature varies less than 1 °C (Spigel and Priscu, 1998). Tritium, CFC, and noble gas profiles suggest that there has been periodic incursions of dense surface waters to deeper portions of the lake which might be related to downwelling of moat water or waters in contact with the Canada Glacier (Tyler et al., 1998). Because the surface water is so fresh, variations in seasonal inflow from year to year can have a great impact on the salinity and major solute concentrations, with decreases in inflow rapidly leading to surface water salinity increases as solutes are excluded from the ice during ice-cover refreeze and accretion (Welch et al., 2000). NO_3^-, PO_4^3, DOC, dissolved inorganic carbon (DIC), and H_4SiO_4 all increase with depth, with the least increase shown by PO_4^{3-} (Fig. 6.4). Surface water NO_3^- and PO_4^{3-} concentrations are extremely low, with nitrate concentrations being some of the lowest observed in the surface waters of the MDV lakes (Fig. 6.4). Dissolved organic carbon values at depth are the lowest observed in all of the lakes. The H_4SiO_4:Cl and DIC: Cl profiles suggest little overprinting of biogeochemical processes on these constituents (Pugh et al., 2003; Neumann et al., 2004).

Primary production in the lake's water column is very low and the benthic mat production is probably greater than that in the pelagic realm (Moorhead et al., 2005). Still, $\delta^{13}C$ profiles of the DIC do indicate carbon uptake in the euphotic zone (i.e., ^{12}C depletion) and organic matter remineralization (i.e., ^{12}C enrichment) at depth in the water column (Neumann et al., 2004). These data indicate that biogeochemical processes are important to the overall geochemistry of LH and that much of the organic carbon fixed is

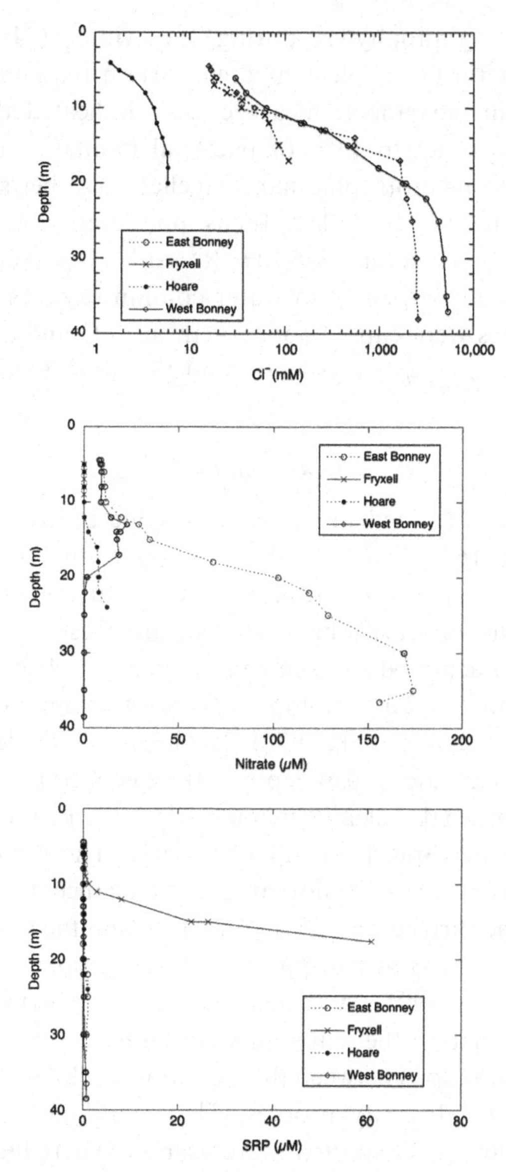

Fig. 6.4. Water column profiles for Lakes Fryxell and Hoare, and east and west lobes of Lake Bonney. Chloride (Cl^-), nitrate, and soluble reactive phosphorus (SRP) are plotted.

later oxidized as it falls through the water column (Wharton et al., 1993a). Cryptophytes are the dominant phytoflagellates in LH and are capable of mixotrophy (i.e., carbon metabolism using both autotrophy and heterotrophy) (Roberts and Laybourn-Parry, 1999). Despite the high photosynthetic efficiency of these cryptophytes, during the winter (24 hour darkness) they consume bacteria to supplement their carbon requirements. During a one-month study in November, measurements indicated that mixotrophic cryptophytes removed up to 13% of bacterial biomass and had a greater grazing impact than heterotrophic nanoflagellates (Roberts and Laybourn-Parry, 1999). There has been less focus on describing the prokaryotic component in the water column of LH. Recently, however, chemoorganotrophic isolates were enriched from water column samples (Clocksin et al., 2007). These isolates were capable of growth at 0 °C and are related to the *Alpha-*, *Beta-*, and *Gammaproteobacteria* and the *Actinobacteria* genera.

Lake Bonney

Lake Bonney (LB) is located in the western edge of TV's closed drainage basin (at 77°43′S, 162°20′E). The lake occupies two steep-sided basins forming an east and west lobe (Spigel and Priscu, 1998) and is situated at 57 m altitude with the east lobe 25 km and the west lobe 28 km from the McMurdo Sound. Each lobe has a maximum depth of 40 m and is separated by a sill at a depth of ~13 m, resulting in distinct chemistries in the bottom waters of each lobe (Hendy et al., 1977). The lobes are divided vertically by a chemocline at approximately 20 m depth in the east lobe and 15 m in the west lobe. The water beneath the chemoclines is cold (< 0 °C), saline (>3 × seawater salinity), and suboxic (Spigel and Priscu, 1998). The gradients in LB are extreme; for example, dissolved inorganic carbon concentrations range from 0.16 mg L^{-1} C in the surface waters to 34 mg L^{-1} C in the bottom waters. The two lobes of LB contain nutrient-poor trophogenic zones and nutrient-rich bottom waters (Priscu, 1995). A large fraction of the biomass in LB is also found in microbial mats at the sediment/water interface.

The physical connections between the east and west lobe of LB are reflected in observed bacterial diversity profiles. The microbial assemblages in the surface waters, which are connected, were found to share high gene sequence similarity between the two lobes. However, the assemblages in the waters below the chemocline, separated by a sill, are distinct and unique to each individual lobe (B. D. Lanoil, unpublished data). The bottom water of the west lobe is derived from the evapoconcentration of marine water and inputs from Blood Falls, whereas the surface water is derived from glacial runoff into seasonal streams. Thus, the biological assemblages in these lakes reflect the effects of landscape position and ecological legacy (Lyons et al., 2000). Glatz et al. (2006) found that, in general, LB was dominated by species in the *Gamma-proteobacteria* and *Bacteroides* (once commonly referred to as the CFB) bacterial groups and to a lesser extent the *Firmicutes* and *Acinetobacter*. Of the clone library from the east lobe at 25 m depth (a hypersaline depth), 22% was dominated by a clone related to *Halomonas*, an organism capable of extreme salt tolerance (Ventosa et al., 1998; Kaye and Baross, 2000; Glatz et al., 2006). These results reflect how, in the MDV, geochemistry and microbial diversity are intimately linked.

Lake Vanda

Lake Vanda (LV) is located in the Wright Valley which is due north of TV. LV has a surface area of ~7.8 km^2 and a maximum depth of 68 m (Canfield and Green, 1985). The lake receives input from the Onyx River, the longest river in the dry valleys (~30 km in length) which typically flows for 6–10 weeks in the austral summer (data from MCM LTER website www.mcmlter.org). Lake Vanda is one of the clearest lakes in the world, and has a unique temperature profile. Surface waters are on average 7 °C; below 48 m depth a steep thermalcline develops with water rapidly increasing to 25 °C. The deeper portions of the lake are hypersaline, slightly acidic, and anoxic. These deep waters are a Ca–Cl brine that formed from a cryoconcentration event when LV lost its ice cover during a colder climate earlier in LV's history (Green and Canfield, 1984). The lake is ultra-oligotrophic with a peak in photosynthetic activity that occurs at 63 m or just above the anoxic zone, here photosynthesis was measured at ~0.2 μg C L^{-1} h^{-1} (Priscu et al., 1987). This pattern is similar to that of other MDV lakes which have a deep chlorophyll maximum where nutrients diffuse upward from the chemocline and sustain photosynthetic and bacterial production (Priscu et al., 1987).

Lake Vanda supports luxuriant microbial mat communities (as described below and in Fig. 6.5), as well as a diverse but small assemblage of planktonic microorganisms. Cell counts from the water column of LV revealed low bacterial density, 1–5 × 10^3 cells ml^{-1} (Bratina et al., 1998). Sulfate reduction, which occurs only in the anoxic zone, is the dominant process responsible for organic matter oxidation in LV (Canfield and Green, 1985). Significant attention has been paid to metal cycling in LV (e.g., Green et al., 1993, 1998). Particulate manganese oxides (MnO_2) play an important role in scavenging, transporting, and releasing metals; crucial to the manganese cycle is microbially mediated manganese reduction (Bratina et al., 1998). Bratina et al. (1998) isolated and characterized manganese reducers from LV that belonged to the genus *Carnobacterium*. Their exact *in situ* physiology is unknown but initial experiments suggest that the *Carnobacterium* reduces MnO_2, not in a dissimilatory fashion, but rather as a consequence of heterotrophic

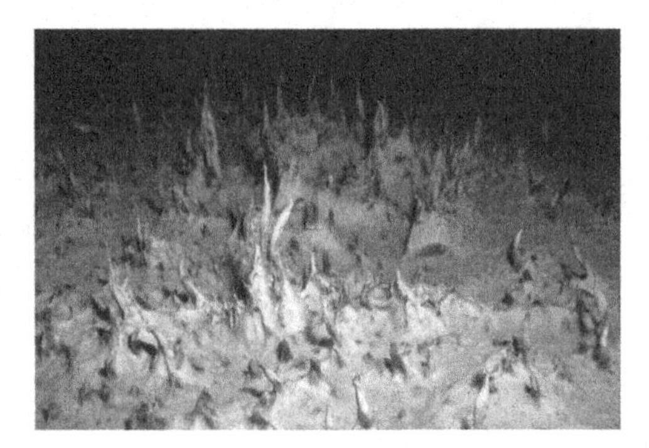

Fig. 6.5. Pinnacle mats at 16 m depth in Lake Vanda.

metabolism (Bratina et al., 1998). Regardless, these findings implicate the microbial inhabitants of LV as playing an active role in *in situ* metal cycling.

Lake Vida

Lake Vida is located in the Victoria Valley and is one of the largest lakes in the Dry Valleys (surface area = 6.8 km^2). Until the mid 1990s, Lake Vida was presumed to be frozen solid. This presumption was largely based on a drill and dynamite survey by Calkin and Bull (1967) in which liquid water was never found. However, using ground penetrating radar, Doran et al. (2003) discovered a large brine pocket (~2 km × 1 km) of undetermined thickness under 19 m of ice in the middle of the lake. During a 1996 drilling expedition into the lake, an ice core was collected which showed that the ice is wet (i.e., contains brine inclusions) below about 16 m (Doran et al., 2003). The brine is dominantly NaCl with major ion ratios very similar to those of Lake Bonney. Based on measurements of extracted ice cores and the long-term deployment of ice thermisters in the borehole, the temperature of the wet ice at 16 m is about −12° C and, by extrapolation, the year-round temperature of the top of the main brine body is −10° C. Victoria Valley has extremely cold winters because its location shelters it from the warming effect of katabatic winds; this attribute serves to maintain Lake Vida's thick ice cover.

Although the Victoria Valley is generally colder than the Wright Valley, summers in Victoria Valley are usually warm enough to produce significant stream melt (Doran et al., 2002, 2003). This stream melt pools on the surface of Lake Vida's ice cover and accumulates layers as it freezes annually (as opposed to a process of ice-cover "growth" via accretion of ice from below). Radiocarbon ages from trapped organic material suggest that the ice cover has been in place, isolating the brine from contact with the atmosphere, for more than 2800 years before present. Prior to this time,

Lake Vida may have been open to the atmosphere. Studies of ice-core profiles have indicated that the microorganisms (primarily filamentous cyanobacteria) associated with sedimentary material within the ice (Mosier et al., 2006) have retained metabolic potential (measured via the incorporation of radio-labeled CO_2, thymidine, and leucine) and are capable of growth if liquid water becomes available within the permanent ice environment (Fritsen and Priscu, 1998).

Dry valley ponds

Shallow ponds are numerous in the McMurdo Dry Valleys region. There is estimated to be over 300 ponds along the Victoria Land coast of Antarctica, occupying a diversity of landscapes and covering $\sim$500 000 m^2 (Hage et al., 2007). Surveys of pond geochemistry and organic matter production have been published over the past 20 years (e.g., Torii et al., 1989; Webster et al., 1994; Moorhead et al., 2003; Healy et al., 2006; Hage et al., 2007; Moorhead, 2007). Most of the MDV ponds are closed basin, but several have direct hydrological connections with other ponds, thus producing flow-through systems. Depending on their location and elevation, the input of liquid water into these ponds can be rather irregular, with some austral summers having no inflow at all (Moorhead et al., 2003). Thus the size and depth of these features can vary dramatically over time. Ponds in close proximity can show quite distinct major solute geochemistries (Lyons et al., unpublished data). The geochemistry of the ponds depends in part on their location and their hydrologic connectivity, with the coastal ponds being Cl^- rich and ponds further inland being NO_3^- dominated (Healy et al., 2006). Water flowing from one pond to another tends to deplete upslope ponds in the most soluble salts, concentrating solutes in downslope ones (Timperley, 1997). Detailed studies of ponds in western Wright Valley indicate that many are thermally stratified in the summer and are chemically stratified during the austral winter (Healy et al., 2006). The basal brines in these ponds are Na–Cl dominated as $CaSO_4{\cdot}2H_2O$ precipitation occurs during the early stages of cryoconcentration. Chemical weathering and possible input of permafrost melt can also influence the geochemistry of the ponds (Healy et al., 2006).

The MDV ponds provide an important habitat for robust microbial mat communities (Hage et al., 2007). Most ponds are highly productive with an average organic matter content of $\sim$260 g C m^{-2} in soils around the ponds in the higher elevation areas of Taylor Valley (Moorhead et al., 2003). As the ponds evaporate and contract in size, the organic matter from the abundant algal mats in many of these ponds can be exposed subaerially and dispersed by the winds. Moorhead (2007) has demonstrated via short-term experiments that from 0.22 to 2.9 g C m^{-2} can be lost from the ponds by this process. As the ponds wax and wane due to variations in climate and glacier melt, their geochemistries change from year to year as well (data from MCM LTER website www.mcmlter.org). Thus both seasonal and annual variations occur in these ponds, making them hydrologically and geochemically dynamic yet capable of sustaining abundant microbial mats.

Description of dry valley microbial mats

In the MDV, microbial mats are widespread and abundant almost wherever liquid water is available. Microbial mats are present in the lake ice covers, at the sediment–water boundary of shallow, saline ponds, and cover large areas of the floors of the deep, perennially ice-covered lakes. Microbial mats are consortia of *Bacteria*, *Archaea*, and simple *Eukarya*, such as diatoms, that together form a laminar structure typically at the sediment–water boundary. The ability to proliferate under extreme temperature, low light level, and other conditions make mats particularly relevant to exobiological research. Two other attributes – their lineage back to the earliest complex communities on Earth (Paerl et al., 2000) and their tendency to self-organize into laminated structures that can leave a recognizable fossil signature (Wharton, 1994) – are also compelling reasons to consider them potential models for recognizable extraterrestrial life forms.

Microbial mats contribute significantly to the productivity of the MDV lakes and ponds. The best understood mats are photosynthetic, with the dominant mat-forming organisms being filamentous cyanobacteria. In terms of biomass, these microbial mats are often the dominant photosynthetic community in both MDV pond and lake systems (Hawes et al., 1993; Moorhead et al., 2005; Howard-Williams and Hawes, 2007). The rapid attenuation of light within the microbial mat results in a strong light resource gradient from excess at the surface, through an optimal zone, the location of which varies over time as incident irradiance varies, and into a suboptimal region over a very short spatial scale (Stal, 1995; Paerl et al., 2000). Extensive metabolic diversity can be found within a single mat consortium; in fact, most of the microbially mediated pathways found on Earth can be found in or associated with these compressed ecosystems. Thus, within the microbial mat, strong redox and biogeochemical gradients exist over short spatial scales based on the distribution of metabolic activities.

Microbial mats appear to survive long periods of extreme conditions, such as prolonged darkness and desiccation with lake level fluctuation, with little lasting damage. They are able to recommence metabolic activity quickly when growth-favorable conditions return (Hawes et al., 1992, 1999, 2001). This strategy allows microbial mats to accumulate biomass over many years; the absence of bioturbation allows microbial mats to gradually accumulate to dominance (e.g., Vincent, 1988, 2000).

Microbial mats growing under perennial ice cover

Mats in deep MDV lakes grow in an extraordinarily quiescent environment. The meters-thick perennial ice cover, in conjunction with the density-stabilized water column, ensures that there is virtually no water movement. This lack of physical disturbance, combined with the lack of bioturbation, allows MDV microbial mat communities to develop macroscopic features. Filamentous cyanobacteria, primarily species of *Lyngbya*, *Leptolyngbya*, *Oscillatoria*, and *Phormidium*, dominate the matrix of these microbial mats. A range of pennate diatoms are also present and increase in relative abundance with increasing depth (Hawes and Schwarz, 1999). As with the cyanobacteria, all of the dominant diatoms are motile, including the commonly encountered species of *Diadesmis*, *Achnanthes*, *Navicula*, *Muelleria*, *Luticola*, and *Stauroneis*. These microbes coalesce to form macroscopic structures, variously categorized by their gross morphology as pinnacle mats, lift-off mats, columnar or prostrate mats. There seems to be a strong tendency for mats to form peak and ridge morphologies, clearly expressed in the pinnacles of Lake Vanda (Fig. 6.5), which extend up to 200 mm from the lake floor.

The alternation of prolonged dark periods, when photosynthetic growth is not possible, and light periods when it is possible, results in annual lamination of lake mats. This growth lamination is on a micrometer to millimeter scale and is manifested in alternating bands of dense and less dense material (Hawes et al., 2001). The motility of the cyanobacteria and diatoms in the microbial mats allows them to move and grow upwards within an exopolysaccharide matrix during summer. Winter darkness interrupts this process, and a dense band is formed that includes a dusting of very fine, sedimenting material. A second scale of lamination (centimeter scale) can result from episodic events when pockets of coarse sediment accumulated within the lake ice cover find their way through the ice and to the lake floor (Squyres et al., 1991).

Analogous growth laminations are found in temperate and tropical microbial mats, where growth and sedimentation are temporally uncoupled. It is the laminated nature of MDV microbial mats that has drawn comparison with ancient laminated fossils, notably stromatolites and microbialites (Parker et al., 1981; Love et al., 1983). In particular, the tendency for MDV microbial mats to form laminated ridge/peak structures, in both shallow and deep communities (Wharton et al., 1983; Vincent, 1988), is similar to fenestrate microbialites, which are common in 3.0–2.5-Ga-old carbonates (Sumner, 1997, 2000). It appears that on Earth, microbial mats that may have shared features with mats extant in MDV lakes occurred very early in the development of biotic structures. If evolution on other planets went through similar stages to those on Earth, residual evidence of early microbial communities, in the presence of biogenic laminated material, may be present as fossils even when conditions cease to be conducive for life.

Microbial mats in shallow water environments

Cyanobacterial mats are also widespread in shallow MDV ecosystems that typically thaw in summer. Here again the mats develop luxuriant biomass (Rautio and Vincent, 2006), with pigment arrays of similar density to eutrophic temperate systems (Hawes et al., 1993). While biomass is high, net productivity of shallow water mats during summer open water periods, as measured by uptake of CO_2 from, or release of O_2 to, the water column frequently approaches zero. This apparent anomaly is largely explained by a close balance between photosynthesis and respiration, indicative of intensive internal reprocessing of resources (Hawes et al., 1993). Dry valley ponds with thick and extensive microbial mats typically have very high pH – often pH 10 or higher – indicative of photosynthetic stripping of DIC to near zero and very low concentrations of dissolved inorganic nutrients (Howard-Williams and Hawes, 2007). This supports speculation that the recycling of materials within the mats and from sediments below results in a highly efficient resource-trapping mechanism. An example is the nitrogen cycling in the mats. McMurdo Dry Valleys microbial mats frequently contain organisms that can mediate nitrogen fixation (Fernández-Valiente et al., 2001) as well as other steps of the nitrogen cycle (Howard-Williams and Hawes, 2007), despite the fact that these processes have different redox requirements. The presumption is that the processes are layered with the oxygen-redox gradient within the mats.

Whereas in permanently ice-covered lakes lack of light energy is a limitation for photosynthesis, in shallow systems, organisms are exposed to high and potentially damaging irradiances, including high UV-A and UV-B, for 24 hours a day in summer. The combined effects of high irradiance, high dissolved oxygen, and low DIC concentrations (from photosynthesis) at high pH and near zero temperature create an extreme photosynthetic environment. This environment threatens both direct photodamage to photosystem components and indirect photodamage through production of highly oxidizing free radicals. Benthic communities in shallow ecosystems employ a number of strategies to adapt to such conditions. As in the deep lakes, these mats are laminated, though in shallow mats the surface layers, rather than being the most productive, show very low rates of oxygen evolution and are usually composed of cyanobacteria very rich in carotenoids and sheath pigments, indicating a possible protective role (Hawes et al., 1993; Quesada and Vincent, 1993; Vincent et al., 1993; Quesada et al., 1999). These upper, relatively inactive layers give the characteristic brown color to shallow-water microbial mats from the *Oscillatoriaceae* and the black color of those due to the *Nostocales*. The capability of carotenoid pigments to contribute towards quenching of free-radical species, coupled with the presence of superoxide dismutase, further alleviates the harmful effects of high radiation (Vincent and Quesada, 1994).

Shallow water MDV mats must contend with a long period of encapsulation in ice. In some cases, this may be for >9 months each year. In addition to low temperature, ice formation imposes strong osmotic and ionic stress on encased organisms (Hawes, 1990; Hawes et al., 1999). The ability to tolerate prolonged freezing is clearly a prerequisite for survival in shallow MDV waters, and the limited literature available shows that antarctic microbial mats are able to recover very quickly from such an excursion (e.g., Davey, 1989; Hawes et al., 1992, 1993). However, the cost appears to be that freeze–thaw cycles result in the loss of substantial amounts of organic solutes from cells (Howard-Williams et al., 1998). Indeed, many shallow water habitats contain exceptional high concentrations of dissolved organics (Howard-Williams and Hawes, 2007), possibly a legacy of freeze–thaw leakage.

Features of the growth of cyanobacterial mats in perennially ice-covered antarctic lakes, including the annual growth, recorded by banding in at least some locations, and the production of persistent, biogenic structures, points to the kind of legacy that simple microbial communities will have left in planets where life began in a similar way. The fact that such traces are found on Earth in Precambrian rocks suggests that this simple growth form is persistent, and may be found in extraterrestrial ecosystems at many stages in their evolution.

Unique McMurdo Dry Valleys features

Permanent lake ice covers

Active microbial communities are also found in the permanent ice covers of the MDV lakes. The MDV ice covers range in thickness from 3 m to 20 m and collect aeolian-derived sediments that are transported through the dry valleys on katabatic winds (Fritsen et al., 1998). These sediments, once deposited on the ice, absorb heat from the sun, melt into the ice cover, and rest at a depth that is in dynamic equilibrium between the downward melting of sediments and upward movement of the ice cover that results from ablation at the surface and accretion or freezing at the base (Priscu et al., 1998). Sediment particles, once in the ice cover, absorb heat in the summer months, melting out pockets of liquid water around the particles. These liquid water inclusions form for ~150 days during the summer when solar radiation is continuous; more than 40% of the ice cover volume can be liquid during this time (Fritsen et al., 1998). The transported sediments effectively bring inorganic and organic nutrients as well as microbial cells to this icy habitat (Priscu et al., 1998). These physical and chemical processes acting in the lake ice covers create yet another niche for life in the polar deserts of the MDV.

Microorganisms that are transported to the ice cover with wind-blown sediments find a suitable habitat for growth within the meltwater inclusions. A diversity of bacteria and cyanobacteria were identified from lake ice samples using molecular techniques (Gorden et al., 2000). Cyanobacteria were the dominate phylotype and were closely related to terrestrial mat samples from the adjacent sediments. These lake ice microcosms contain species that originate from the surrounding terraform and "opportunistically" colonize the lake ice once deposited there (Gorden et al., 1998; Priscu et al., 1998). Organisms that take up residence in the ice cover are metabolically active; primary production can be as high as 7.8 $\mu g\,C\,d^{-1}$; and bacterial activity was measured to be 0.4 nM thymidine d^{-1} (Priscu et al., 1998). Genes that specifically code for the process of nitrogen fixation (the *nifH* gene) were detected in samples from the Lake Bonney ice cover and N_2-fixation activity has been measured using the acetylene reduction assay (Olson et al., 1998). The lake ice microbial assemblage is thus a contained ecosystem with active photosynthesis, nitrogen fixation, and decomposition. The tight coupling of nutrient cycles in the permanent ice covers of the MDV lakes suggests a consortial relationship among the bacterial inhabitants. Here photoautotrophs supply fixed carbon and nitrogen to the heterotrophic organisms in the assemblage; the heterotrophs in turn remineralize organic matter (Paerl and Priscu, 1998; Priscu et al., 1998). The tight coupling of metabolic processes observed in the MDV lake ice may be an important key to survival in icy habitats.

Description of Blood Falls

Taylor Glacier, located at the western end of the Taylor Valley, is an outlet glacier that drains from the polar plateau (Fountain et al., 1999). The glaciers

Fig. 6.6. Scientists collect samples from Blood Falls outflow at the terminus of the Taylor Glacier. (Image credit: W. H. Keys.)

of the Taylor Valley are cold based, with their beds frozen to the underlying rock; however Taylor Glacier is a unique exception. Ice-penetrating radar data indicate water or slush below the glacier corresponding to an 80-m depression in the bedrock topology at ~4 km up-glacier from the terminus (Hubbard et al., 2004). This depression is below sea level and forms what is believed to have been a third lobe of Lake Bonney (Lyons et al., 2005). When the chemically reduced, iron-rich subglacial brine flows from beneath the glacier and is exposed to the atmosphere, it becomes oxidized, and a red salt cone, known as Blood Falls (Fig. 6.6), precipitates at the northern end of the glacier terminus (Black et al., 1965; Keys, 1979). The presence of this subglacial brine near the terminus may cause Taylor Glacier's geomorphic behavior to be more like that of a temperate or polythermal glacier (Robinson, 1984).

The brine below the glacier is thought to originate from the Pliocene Epoch (~2–5 Ma ago). During the Pliocene, the Earth was warmer than at present (Kennett and Hodell, 1993). This warming event resulted in the incursion of marine waters from the Ross Sea Embayment (Denton et al., 1993). When these marine waters retreated from the valley network, a remnant sea remained near the Taylor Glacier terminus. The $\delta^{37}Cl$ composition of Blood Falls (BF) outflow (when mixing of glacial melt is accounted for) was ~0‰, essentially that of seawater (Lyons et al., 1999). These data strongly support a marine origin of BF with its bulk chemistry resembling that of cryoconcentrated seawater, much like west lobe Lake Bonney (Lyons et al., 1999, 2005). A modest expansion of the Taylor Glacier during the late Pliocene or Pleistocene (Marchant et al., 1993) likely covered this concentrated sea, leaving behind a liquid brine of Pliocene marine waters sealed beneath the Taylor Glacier.

Release events from BF are episodic in nature. When brine is released, it is chemically reduced. The outflow is rapidly oxygenated as it exits the subglacial environment at BF and spreads across the lake ice surface, mixing with the Lake Bonney moat. Various minerals, such as mirabilite, halite, gypsum, and goethite (Black et al., 1965; Lyons et al., 2005) precipitate at the surface and then dissolve in glacier melt or streamflow from Santa Fe Creek, and eventually enter the moat waters of the west lobe of Lake Bonney. Keys (1979) estimated that BF discharges an average of $2000\,m^3\,y^{-1}$ of saline water (episodic discharges reach $6000\,m^3\,y^{-1}$).

The subglacial brine contains numerous cells (1×10^5 cells ml^{-1}), including metabolically active and cultivable cells (Mikucki et al., 2004; Mikucki and Priscu, 2007). Isolates obtained from BF outflow revealed similar organisms both below the Taylor Glacier and in the water column of west lobe Lake Bonney. One of these isolates clusters within the *Marinobacter* genus. A clone from BF was most closely related to isolate ELB-17, previously identified throughout the water column of the east and west lobes of Lake Bonney (Ward and Priscu, 1997; Glatz et al., 2006). A clone library from BF outflow showed the presence of *beta-*, *gamma-*, and *deltaproteobacteria* and *Bacteroides* (once commonly referred to as the CFB) groups. The most abundant clone detected in BF samples (Mikucki and Priscu, 2007) was closely related (99% sequence similarity based on 1420 bp) to an obligate chemolithoautotrophic sulfur-oxidizing bacterium, *Thiomicrospira arctica*, which was originally isolated from arctic marine sediments (Knittel et al., 2005). Approximately 11% of the library was most similar based on gene sequence identity (98% identity) to *Geopsychrobacter electrodiphilus*, a psychrotolerant bacterium capable of growth using Fe(III), Mn(IV), and elemental sulfur as electron acceptors (Holmes et al., 2004). The inferred physiologies of the organisms detected in BF brine suggest that sulfur cycling may be an important process in BF (Mikucki and Priscu, 2007). The majority of clones and isolates from BF outflow (>75%) shared high gene sequence similarity to isolates from other cold marine environments, implying that BF has retained its marine legacy within its gene pool.

Microbial dynamics in the subglacial setting would be distinct from that of ice-covered lake systems. Subglacial ecosystems must support growth in the absence of sunlight. As such, the bedrock lithology, preglacial ecosystem and glacier hydrology impart a strong control on subglacial electron flow. In the hydrologic regime of the Taylor Glacier, anoxia also is likely to be an important regulator of microbial energetics. Organisms below the Taylor Glacier must also contend with elevated salinities and high iron concentrations. Biogeochemical data imply that the bacterial assemblage below the

Taylor Glacier can grow chemoautotrophically or chemorganotrophically by harvesting energy from bedrock minerals, or may grow heterotrophically on ancient marine organics by respiring Fe(III) or SO_4^{2-} (Mikucki and Priscu, 2007).

Blood Falls demonstrates microbial survival and growth under isolation and in the absence of sunlight for extended periods of time (~2 Ma). In such a sub-ice scenario, energy for growth would be harnessed from the bedrock chemistry. Metabolism based on subglacial lithology has been proposed for other subglacial environments on Earth (Skidmore et al., 2005; Tranter et al., 2005). However, due to the high iron and salt content of BF and its location under a cold-based glacier, this particular subglacial feature provides a unique earthly analog for putative habitats that may exist below the martian polar caps. A clear benefit of living subglacially on Mars would be protection from harsh surface radiation. Such an ecosystem could persist even under current martian conditions.

Description of Don Juan Pond

Don Juan Pond (DJP) is a small (~100 m × 300 m × 10 cm), hypersaline pond in the Wright Valley at 162 m above sea level. Its salinity ensures that the pond remains liquid year round, despite winter temperatures approaching −40 °C (Meyer et al., 1962; Cartwright and Harris, 1981). The pond stays liquid even in the deep of austral winter: the eutectic point for this water was estimated to be −51.8 °C (Marion, 1997). Don Juan Pond is host to one of the few natural occurrences of the mineral antarcticite ($CaCl_2 6{\cdot}H_2O$) (Marion, 1997). Groundwater flow from the melt of a nearby rock glacier is the only water source and evaporation is the sole water loss mechanism at this site (Harris and Cartright, 1981; Tomiyama and Kitano, 1985; Carlson et al., 1990). The primary solute in DJP is $CaCl_2$, which is solubilized from surrounding evaporite rocks by groundwater and deposited in the basin by illuvation. The concentration of $CaCl_2$ varies from year to year due to variation in groundwater flow; rarely, the pond disappears completely (Harris and Cartright, 1981). Because the groundwater flow is determined by the number of days that the air temperature exceeds the freezing point, salinity is directly dependent upon climate in DJP. $CaCl_2$ concentrations are quite high, approaching saturation (>60% w/v). The water activity in DJP is exceptionally low (0.3 – 0.6) because the primary solute is a divalent rather than a monovalent cation. This water activity is exceedingly low – well below the lowest demonstrated limit for life, and even below the estimated maximum lower limit for life (Group, 2006; Tosca et al., 2008). The solution is so hydroscopic that the sediments of the pond have large void spaces where no water is present, despite lying immediately beneath DJP.

The extremely low water activity, temperatures, and extremely high $CaCl_2$ concentrations seen in DJP are a significant challenge for life (Grant, 2004; Tosca et al., 2008) and mirror the expected chemistry in martian groundwater (Burt and Knauth, 2003). It remains unclear whether life exists in DJP. Several reports (Meyer et al., 1962; Siegel et al., 1979) indicate the presence of life in the water or on exposed sediments; however, it is difficult to say unequivocally that these organisms are autochthonous (Horowitz et al., 1972). Theoretical arguments seem to indicate that life should not occur in DJP (Oren, 1992). In another example of a divalent cation dominated system, DNA and RNA signatures of life were found in a deep hypersaline anoxic basin (Discovery Basin) with concentrations of $MgCl_2$ approaching saturation (5 M) (Van der Wielen et al., 2005; Van der Wielen, 2006). In that case, chaotropicity (the tendency to destabilize and destroy complex molecules such as nucleic acids or proteins), not low water activity, was demonstrated to be the limiting factor in the $MgCl_2$-dominated system (Hallsworth et al., 2007). It is unclear whether a $CaCl_2$-dominated system, such as DJP, would give similar results.

Attempts to obtain any extractable DNA or RNA from DJP water or sediments have been unsuccessful to date (B. D. Lanoil, unpublished data). Preliminary results do indicate the presence of culturable bacteria in DJP waters; however, these isolates will not grow in DJP-based media, but will only grow at lower salt concentrations. It is likely that these are allochthonous microbes, perhaps entering the lake via aeolian deposition, which were "rescued" by cultivation. Very low, but statistically significant, biological activity based on radiolabeled substrate uptake and turnover in DJP water was observed in experiments that were several months long (J. C. Priscu, unpublished data).

The jury is still out regarding whether life can or does exist in DJP. It may be that in years with high flow, inoculation by aeolian deposition allows some growth of cells during summer months; however, during years with low flow, these cells die. Thus there is not an autochthonous microflora present in DJP, but only a transient assemblage that has minimal impact on the surrounding environment. If such is the case for DJP, it may be the only system (to date) found on Earth with liquid water but no indigenous microflora. As such, it provides a constraining "boundary condition" for determining where to look for life on extraterrestrial planets – only those with sufficient water activity (not just the presence of liquid water) at the proper temperature will be likely to support or have supported life.

Summary

Paleolakes on Mars would have provided an environment where liquid water could have existed in sufficient quantities to provide for the needs of biological systems (Carr, 1983). The ice-covered lake and pond habitats of the McMurdo Dry Valleys present a legitimate earthly analog to putative martian ecosystems. Our discussion of these systems and their inhabitants illustrates that, despite the close proximity of the lakes and ponds in the MDV, each system possesses its own unique geochemical and biological signatures. Putative ecosystems on Mars would also have had the tendency to be spatially and temporally diverse – a notion that should be kept in mind when exploring exobiological targets on Mars. The excitement of recent missions that have clearly identified evidence for liquid water on the surface, including a recent report that identified a variety of hydrated minerals over a broad range of landscape features, further supports the notion that a diversity of water habitats once existed on Mars (Mustard et al., 2008). Lake deposits from impact craters are thought to be among the best targets for future exobiology investigations or sample return missions (Newsom et al., 1996; Jakosky et al., 2007). Regardless of whether analyses are preformed *in situ* or samples are returned to Earth, the quantity of sample will likely be small. Particularly for biological analyses, careful consideration should be made of the most prudent measurements to detect life or its signature. For these reasons, thoughtful examination and understanding of the ecology of analogous environments here on Earth may enhance future exploration missions to Mars. The ice-covered lakes and ponds of the MDV can help inform such future investigations.

References

Aiken, G., McKnight, D., Harnish, R., and Wershaw, R. (1996). Geochemistry of aquatic humic substances in the Lake Fryxell Basin, Antarctica. *Biogeochemistry*, **34**, 157–188.

Baker, V. R., Strom, R. G., Gulick, V. C., et al. (1991). Ancient oceans, ice sheets and the hydrological cycle on Mars. *Nature*, **352**, 589–594.

Barrett, P. J. and Hambrey, M. J. (1992). Plio-Pleistocene sedimentation in Ferrar Fiord, Antarctica. *Sedimentology*, **39**, 109–123.

Black, R. F., Jackson, M. L., and Berg, T. E. (1965). Saline discharge from Taylor Glacier, Victoria Land, Antarctica. *Journal of Geology*, **74**, 175–181.

Bomblies, A., McKnight, D., and Andrews, E. D. (2001). Retrospective simulation of lake level rise in Lake Bonney based on recent 21-year record: indication of recent climate change in the McMurdo Dry Valleys, Antarctica. *Journal of Paleolimnology*, **25**, 477–492.

Bratina, B. J., Stevenson, B. S., Green, W. J., and Schmidt, T. M. (1998). Manganese reduction by microbes from oxic regions of Lake Vanda (Antarctica) water column. *Applied and Environmental Microbiology*, **64**, 3791–3797.

Burkins, M. B., Virginia, R. A., Chamberlain, C. P., and Wall, D. H. (2000). Origin and distribution of soil organic matter in Taylor Valley, Antarctica. *Ecology*, **81**, 2377–2391.

Burt, D. M. and Knauth, L. P. (2003). Electrically conducting, Ca-rich brines, rather than water, expected in the Martian subsurface. *Journal of Geophysical Research*, **108**, 1–6.

Cabrol, N. A. and Grin, E. A. (1999). Distribution, classification, and ages of Martian impact crater lakes. *Icarus*, **V142**, 160–172.

Cabrol, N. A. and Grin, E. A. (2001). The evolution of lacustrine environments on Mars: is Mars only hydrologically dormant? *Icarus*, **V149**, 291–328.

Cabrol, N. A. and Grin, E. A. (2005). Ancient and Recent lakes on Mars. In *Water on Mars and Life: Advances in Astrobiology and Biogeophysics*, ed. T. Tokano. Berlin: Springer Verlag, pp. 235–259.

Calkin, P. E. and Bull, C. (1967). Lake Vida, Victoria Valley, Antarctica. *Journal of Glaciology*, **6**, 833–836.

Canfield, D. E. and Green, W. J. (1985). The cycling of nutrients in a closed-basin Antarctic lake: Lake Vanda. *Biogeochemistry*, **1**, 233–256.

Carlson, C. A., Phillips, F. M., Elmore, D., and Bentley, H. W. (1990). Chlorine-36 tracing of salinity sources in the Dry Valleys of Victoria Land, Antarctica. *Geochimica et Cosmochimica Acta*, **54**, 311–318.

Carr, M. H. (1983). Stability of streams and lakes on Mars. *Icarus*, **56**, 476–495.

Cartwright, K. and Harris, H. J. H. (1981). Hydrogeology of the Dry Valley region, Antarctica. *Antarctic Research Series*, **33**, 193–214.

Chinn, T. J. (1993). Physical hydrology of the dry valley lakes. In *Physical and Biogeochemical Processes in Antarctic Lakes*, ed. W. J. Green, and E. I. Freidmann. Washington, D.C.: American Geophysical Union, pp. 1–52.

Clocksin, K. M., Jung, D. O., and Madigan, M. T. (2007). Cold-active chemoorganotrophic bacteria from permanently ice-covered Lake Hoare, McMurdo Dry Valleys, Antarctica. *Applied and Environmental Microbiology*, **73**, 3077–3083.

Conovitz, P. A., McKnight, D. M., MacDonald, L. H., Fountain, A. G., and House, H. R. (1998). Hydrologic processes influencing streamflow variation in Fryxell Basin, Antarctica. In *Ecosystem Dynamics in a Polar Desert: The McMurdo Dry Valleys, Antarctica*, ed. J. C. Priscu. Antarctic Research Series 72. Washington, D.C.: American Geophysical Union, pp. 93–108.

Davey, M. C. (1989). The effects of freezing and desiccation on photosynthesis and survival of terrestrial Antarctic algae and cyanobacteria. *Polar Biology*, **10**, 29–36.

Denton, G. H., Sugden, D. E., Marchant, D. R., Hall, B. L., and Wilch, T. I. (1993). East Antarctic Ice Sheet sensitivity to Pliocene climatic change from a Dry Valleys perspective. *Geografiska Annaler*, **75A**, 155–204.

Doran, P. T., Priscu, J. C., Lyons, W. B., et al. (2002). Recent climate cooling and ecosystem response in the McMurdo Dry Valleys, Antarctica. *Nature*, **415**, 517–520.

Doran, P. T., Fritsen, C. H., McKay, C. P., Priscu, J. C., and Adams, E. E. (2003). Formation and character of an ancient 19-m ice cover and underlying trapped brine in an "ice-sealed" east Antarctic lake. *PNAS*, **100**, 26–31.

Doran, P. T., Priscu, J. C., Lyons, W. B., et al. (2004). Paleolimnnology of extreme cold terrestrial and extraterrestrial environments. In *Long-Term Environmental*

Change in Arctic and Antarctic Lakes, ed. R. Pienitz, M. S. V. Douglas, and J. P. Smoll. Amsterdam, Netherlands: Kluwer Academic Publishers, pp. 475–507.

Fassett, C. I. and Head, III, J. W. (2008). Open-basin lakes on Mars: implications of valley network lakes for the nature of Noachian hydrology. *Lunar and Planetary Science*, **39**, Abstract 1139.

Fernández-Valiente, E., Quesada, A., Howard-Williams, C., and Hawes, I. (2001). N_2-fixation in cyanobacterial mats from ponds on the McMurdo Ice Shelf, Antarctica. *Microbial Ecology*, **42**, 338–349.

Fishbaugh, K., Poulet, F., Langevin, Y., Chevrier, V., and Bibring, J.-P. (2007). On the origin of gypsum in the Mars North Polar Region. *Journal of Geophysical Research*, **112**(E07002), doi: 10.1029/2006JE002862.

Foreman, C. M., Wolf, C. F., and Priscu, J. C. (2004). Impact of episodic warming events on the physical, chemical, and biological relationships of lakes in the McMurdo Dry Valleys, Antarctica. *Aquatic Geochemistry*, **10**, 239–268.

Fountain, A. G., Lyons, W. B., Burkins, M. B., et al. (1999). Physical controls on the Taylor Valley ecosystem, Antarctica. *BioScience*, **4**, 961–973.

Fritsen, C. H. and Priscu, J. C. (1998). Cyanobacterial assemblages in permanent ice covers on Antarctic lakes: distribution, growth rate, and temperature response of photosynthesis. *Journal of Phycology*, **34**, 587–597.

Fritsen, C. H., Adams, E. E., McKay, C. P., and Priscu, J. C. (1998). Permanent ice covers of the McMurdo Dry Valleys Lakes, Antarctica: liquid water contents. In *Ecosystem Dynamics in a Polar Desert: The McMurdo Dry Valleys, Antarctica*, ed. J. C. Priscu. Antarctic Research Series 72. Washington, D.C.: American Geophysical Union, pp. 269–280.

Glatz, R. E., Lepp, P. W., Ward, B. B., and Francis, C. A. (2006). Planktonic microbial community composition across steep physical/chemical gradients in permanently ice-covered Lake Bonney, Antarctica. *Geobiology*, **4**, 53–67.

Goldspiel, J. M. and Squyres, S. W. (1991). Ancient aqueous sedimentation on Mars. *Icarus*, **89**, 392–410.

Gorden, D. A., Priscu, J. C., and Giovannoni, S. (2000). Origin and phylogeny of microbes living in permanent Antarctic lake ice. *Microbial Ecology*, **39**, 197–202.

Grant, W. D. (2004). Life at low water activity. *Philosophical Transactions of the Royal Society of London, Series B*, **359**, 1249–1267.

Green, W. J. and Canfield, D. E. (1984). Geochemistry of the Onyx River (Wright Valley, Antarctica) and its role in the chemical evolution of Lake Vanda. *Geochimica et Cosmochimica Acta*, **48**, 2457–2468.

Green, W. J., Anglem, P., and Chavek, E. (1988). The geochemistry of Antarctic streams and their role in the evolution of four lakes of the McMurdo Dry Valleys. *Geochimica et Cosmochimica Acta*, **52**, 1265–1274.

Green, W. J., Canfield, D. E., Shengsong, Y., et al. (1993). Metal transport and release processes in Lake Vanda: the role of oxide phases. In *Physical and Biogeochemical Processes in Antarctic Lakes*, ed. W. J. Green and E. I. Freidmann. Washington, D.C.: American Geophysical Union, pp. 145–163.

Green, W. J., Canfield, D. E., and Nixon, P. (1998). Cobalt cycling and fate in Lake Vanda. In *Ecosystem Dynamics in a Polar Desert: The McMurdo Dry Valleys, Antarctica*, ed. J. C. Priscu. Antarctic Research Series 72. Washington, D.C.: American Geophysical Union, pp. 205–215.

Group, MSR–SA (2006). Findings of the Mars Special Regions Science Analysis Group. *Astrobiology*, **6**, 677–732.

Hage, M. M., Uhle, M. E., and Macko, S. (2007). Biomarker and stable isotope characterization of coastal pond-derived organic matter, McMurdo Dry Valleys, Antarctica. *Astrobiology*, **7**, 645–661.

Hall, B. L. and Denton, G. H. (2000). Radiocarbon chronology of Ross Sea Drift, Eastern Taylor Valley, Antarctica: evidence for a grounded ice sheet in the Ross Sea at the last glacial maximum. *Geografiska Annaler*, **82**, 305–336.

Hall, B. L., Denton, G. H., and Hendy, C. H. (2000). Evidence from Taylor Valley for a grounded ice sheet in the Ross Sea, Antarctica. *Geografiska Annaler*, **82**, 275–303.

Hallsworth, J. E., Yakimov, M. M., Golyshin, P. N., et al. (2007). Limits of life in $MgCl_2$-containing environments: chaotropicity defines the window. *Environmental Microbiology*, **9**, 801–803.

Harris, H. J. H. and Cartright, K. (1981). Hydrology of the Don Juan basin, Wright Valley, Antarctica. *Antarctic Research Series*, **33**, 161–184.

Hauber, E., Van Gasselt, S., Ivanov, B., et al. (2005). Discovery of a flank caldera and very young glacial activity at Hecates Tholus, Mars. *Nature*, **434**, 356–361.

Hawes, I. (1990). Effects of freezing and thawing on a species of *Zygnema* (Chlorophyta) from the Antarctic. *Phycologia*, **29**, 326–331.

Hawes, I. and Howard-Williams, C. (1998). Primary production processes in streams of the McMurdo Dry Valleys, Antarctica. *Antarctic Research Series*, **72**, 129–140.

Hawes, I. and Schwarz, A. M. (1999). Photosynthesis in an extreme shade environment, benthic microbial mats from Lake Hoare, a permanently ice-covered Antarctic lake. *Journal of Phycology*, **35**, 448–459.

Hawes, I., Howard-Williams, C., and Vincent, W. F. (1992). Desiccation and recovery of Antarctic cyanobacterial mats. *Polar Biology*, **12**, 587–594.

Hawes, I., Howard-Williams, C., and Pridmore, R. D. (1993). Environmental controls on microbial biomass in the ponds of the McMurdo Ice Shelf, Antarctica. *Archive fur Hydrobiologie*, **127**, 271–287.

Hawes, I., Smith, R., Howard-Williams, C., and Schwarz, A. M. (1999). Environmental conditions during freezing, and response of microbial mats in ponds of the McMurdo Ice Shelf, Antarctica. *Antarctic Science*, **11**, 198–208.

Hawes, I., Moorhead, D. L., Sutherland, D., Schmeling, J., and Schwarz, A. M. (2001). Benthic primary production in two perennially ice-covered Antarctic lakes, patterns of biomass accumulation with a model of community metabolism. *Antarctic Science*, **13**, 18–27.

Head, J. W., Neukum, G., Jaumann, R., et al. (2005). Tropical to mid-latitude snow and ice accumulation, flow, and glaciation on Mars. *Nature*, **434**, 346–351.

Healy, M., Webster-Brown, J. G., Brown, K. L., and Lane, V. (2006). Chemistry and stratification of Antarctic meltwater ponds. II. Inland ponds in the McMurdo Dry Valleys, Victoria Land. *Antarctic Science*, **18**, 525–533.

Hendy, C. H. (2000). Late Quaternary lakes in the McMurdo Sound region of Antarctica. *Geografiska Annaler*, **82**, 411–432.

Hendy, C. H., Wilson, A. T., Popplewell, K. B., and House, D. A. (1977). Dating of geochemical events in Lake Bonney, Antarctica and their relation to glacial and climatic changes. *New Zealand Journal of Geology and Geophysics*, **20**, 1103–1122.

Holmes, D. E., Nicoll, J. S., Bond, D. R., and Lovley, D. R. (2004). Potential role of a novel psychrotolerant member of the family Geobacteraceae, *Geopsychrobacter electrodiphilus* gen. nov., sp. nov., in electricity production by a marine sediment fuel cell. *Applied Environmental Microbiology*, **70**, 6023–6030.

Hood, E. M., Howes B. L., and Jenkins, W. J. (1998). Dissolved gas dynamics in perennially ice-covered Lake Fryxell, Antarctica. *Limnology and Oceanography*, **43**, 265–272.

Horowitz, N. H., Cameron, R. E., and Hubbard, J. S. (1972). Microbiology of the Dry Valleys of Antarctica. *Science*, **176**, 242–245.

Howard-Williams, C. and Hawes, I. (2007). Ecological processes in Antarctic inland waters: interactions between physical processes and the nitrogen cycle. *Antarctic Science*, **19**, 205–217.

Howard-Williams, C., Schwarz, A. M., Hawes, I., and Priscu, J. C. (1998). Optical properties of lakes of the McMurdo Dry Valley region, Antarctica. In *The McMurdo Dry Valleys, Antarctica: A Cold Desert Ecosystem*, ed. J. C. Priscu. Washington, D.C.: American Geophysical Union, pp. 189–203.

Hubbard, A., Lawson, W., Anderson, B., Hubbard, B., and Blatter, H. (2004). Evidence for subglacial ponding across Taylor Glacier, Dry Valleys, Antarctica. *Annals of Glaciology*, **39**, 79–84.

Jakosky, B. M., Amend, J. P., Berelson, W. M., et al. (2007). *An Astrobiology Strategy for the Exploration of Mars*. Report by Space Studies Board, National Research Council. Washington, D.C.: National Academies Press.

Karr, E. A., Sattley, W. M., Jung, D. O., et al. (2003). Remarkable diversity of phototrophic purple bacteria in a permanently frozen Antarctic lake. *Applied Environmental Microbiology*, **69**, 4910–4914.

Karr, E. A., Sattley, W. M., Rice, M. R., et al. (2005). Diversity and distribution of sulfate-reducing bacteria in permanently frozen Lake Fryxell, McMurdo Dry Valleys, Antarctica. *Applied and Environmental Microbiology*, **71**, 6353–6359.

Kaye, J. Z. and Baross, J. A. (2000). High incidence of halotolerant bacteria in Pacific hydrothermal-vent and pelagic environments. *FEMS Microbiology Ecology*, **32**, 249–260.

Kennett, J. P. and Hodell, D. A. (1993). Evidence for relative climatic stability of Antarctica during the early Pliocene: a marine perspective. *Geografiska Annaler*, **75**, 204–220.

Keys, J. R. (1979). Saline discharge at the terminus of the Taylor Glacier. *Antarctic Journal of the United States*, **14**, 82–85.

Keys, J. R. and Williams, K. (1981). Origin of crystalline, cold desert salts in the McMurdo region, Antarctica. *Geochimica et Cosmochimica Acta*, **45**, 2299–2309.

Knittel, K., Kuever, J., Meyerdierks, A., et al. (2005). *Thiomicrospira arctica* sp. nov. and *Thiomicrospira psychrophila* sp. nov., psychrophilic, obligately chemolithoautotrophic, sulfur-oxidizing bacteria isolated from marine Arctic sediments. *International Journal of Systematic and Evolutionary Microbiology*, **55**, 781–786.

Laybourn-Parry, J., James, M. R., McKnight, D. M., et al. (1996). The microbial plankton of Lake Fryxell, southern Victoria Land, Antarctica during the summers of 1992 and 1994. *Polar Biology*, **17**, 54–61.

Love, F. G., Simmons, Jr., G. M., Parker, B. C., Wharton, Jr., R. A., and Seaburg, K. G. (1983). Modern conophyton-like microbial mats discovered in Lake Vanda, Antarctica. *Geomicrobiology*, **3**, 33–48.

Lyons, W. B., Tyler, S. W., Wharton, Jr., R. A., McKnight, D. M., and Vaughn, B. H. (1998). A Late Holocene desiccation of Lake Hoare and Lake Fryxell, McMurdo Dry Valleys, Antarctica. *Antarctic Science*, **10**, 247–256.

Lyons, W. B., Frape, S. K., and Welch, K. A. (1999). History of McMurdo Dry Valley lakes, Antarctica, from stable chlorine isotope data. *Geology*, **27**, 527–530.

Lyons, W. B., Fountain, A., Doran, P., et al. (2000). Importance of landscape position and legacy: the evolution of the lakes in Taylor Valley, Antarctica. *Freshwater Biology*, **43**, 355–367.

Lyons, W. B., Welch, K. A., Snyder, G., et al. (2005). Halogen geochemistry of the McMurdo Dry Valleys lakes, Antarctica: clues to the origin of solutes and lake evolution. *Geochimica et Cosmochimica Acta*, **69**, 305–323.

Lyons, W. B., Laybourn-Parry, J.,Welch, K. A., and Priscu, J. C. (2006). Antarctic lake systems and climate change. In *Trends in Antarctic Terrestrial and Limnetic Ecosystems: Antarctica as a Global Indicator*, ed. D. M. Bergstrom, P. Convey, and A. H. L. Huiskes. Dordrecht: Springer, pp. 273–295.

Malin, M. C., Edgett, K. S., Posiolova, L. V., McColley, S. M., and Dobrea, E. Z. N. (2006). Present-day impact cratering rate and contemporary gully activity on Mars. *Science*, **314**, 1574–1577.

Marchant, D. R., Swisher, III, C. C., Lux, D. R., West, Jr., D. P., and Denton, G. H. (1993). Pliocene paleoclimate and East Antarctic Ice Sheet history from surficial ash deposits. *Science*, **260**, 667–670.

Marion, G. M. (1997). A theoretical evaluation of mineral stability in Don Juan Pond, Wright Valley, Victoria Land. *Antarctic Science*, **9**, 92–99.

Matsubaya, O., Sakai, H., Torii, T., Burton, H., and Kerry, K. (1977). Antarctic saline lakes – stable isotope ratios, chemical compositions, and evolution. *Geochimica et Cosmochimica Acta*, **43**, 7–25.

McKnight, D. M., Alger, A., Takte, C. M., Shupe, G., and Spaulding, S. (1998). Longitudinal patterns in algal abundance and species distribution in meltwater streams in Taylor Valley, Southern Victoria Land, Antarctica. In *Ecosystem Dynamics in a Polar Desert: The McMurdo Dry Valleys, Antarctica*, ed. J. C. Priscu. Antarctic Research Series 72. Washington, D.C.: American Geophysical Union, pp. 153–188.

McKnight, D. M., Niyogi, D. K., Alger, A. S., et al. (1999). Dry valley streams in Antarctica: ecosystems waiting for water. *BioScience*, **49**, 985–995.

Meyer, G. H., Morrow, M. B., Wyss, O., Berg, T. E., and Littlepage, J. L. (1962). Antarctica: the microbiology of an unfrozen saline pond. *Science*, **138**, 1103–1104.

Mikucki, J. A. and Priscu, J. C. (2007). Bacterial diversity associated with Blood Falls, a subglacial outflow from the Taylor Glacier, Antarctica. *Applied Environmental Microbiology*, **73**, 4029–4039.

Mikucki, J. A., Foreman, C. M., Sattler, B., Lyons, W. B., and Priscu, J. C. (2004). Geomicrobiology of Blood Falls: an iron-rich saline discharge at the terminus of the Taylor Glacier, Antarctica. *Aquatic Geochemistry*, **10**, 199–220.

Moorhead, D. L. (2007). Mesoscale dynamics of ephemeral wetlands in the Antarctic Dry Valleys: implications to production and distribution of organic matter. *Ecosystems*, **10**, 87–95.

Moorhead, D. L., Doran, P. T., Fountain, A. G., et al. (1999). Ecological legacies: impacts on ecosystems of the McMurdo Dry Valleys. *BioScience*, **49**(12), 1009–1019.

Moorhead, D. L., Barrett, J. E., Virginia, R. A., et al. (2003). Organic matter and soil biota of upland wetlands in Taylor Valley, Antarctica. *Polar Biology*, **26**, 567–576.

Moorhead, D. L., Schmeling, J., and Hawes, I. (2005). Modelling the contribution of benthic microbial mats to net primary production in Lake Hoare, McMurdo Dry Valleys. *Antarctic Science*, **17**, 33–45.

Morgan-Kiss, R. M., Priscu, J. C., Pocock, T., Gudynaite-Savitch, L., and Huner, N. P. A. (2006). Adaptation and acclimation of photosynthetic microorganisms to permanently cold environments. *Microbiology and Molecular Biology Reviews*, **70**, 222–252.

Mosier, A. C., Murray, A. E., and Fritsen, C. H. (2006). Microbiota within the perennial ice cover of Lake Vida, Antarctica. *FEMS Microbiology Ecology*, **59**, 274–288.

Mustard, J. F, Murchie, S. L., Pelkey, S. M., et al. (2008). Hydrated silicate minerals on Mars observed by the Mars Reconnaissance Orbiter CRISM instrument. *Nature*, **454**, 305–309.

Neumann, K., Lyons, W. B., Priscu, J. C., Des Marais, D., and Welch, K. A. (2004). The carbon isotopic composition of dissolved inorganic carbon in perennially ice-covered Antarctic lakes: searching for a biogenic signature. *Annals of Glaciology*, **39**, 518–524.

Newsome, H. E., Brittelle, G. E., Hibbitts, C. A., Crossey, L. J., and Kudo, A. M. (1996). Impact crater lakes on Mars. *Journal of Geophysical Research*, **101**, 951–955.

Olson, J. B., Steppe, T. F., Litaker, R. W., Paerl, H. W. (1998). N_2-fixing microbial consortia associated with the ice cover of Lake Bonney, Antarctica. *Microbial Ecology*, **36**, 231–238.

Oren, A. (1992). Ecology of extremely halophilic microorganisms. In *The Biology of Halophilic Bacteria*, ed. R. H. Vreeland and L. I. Hochstein. Boca Raton, FL: CRC Press, pp. 25–54.

Paerl, H. W. and Priscu, J. C. (1998). Microbial phototrophic, heterotrophic and diazotrophic activities associated with aggregates in the permanent ice cover of Lake Bonney, Antarctica. *Microbial Ecology*, **36**, 221–230.

Paerl, H. W., Pinckney, J. L., and Steppe, T. F. (2000). Cyanobacterial-bacterial mat consortia: examining the functional unit of microbial survival and growth in extreme environments. *Environmental Microbiology*, **2**, 11–26.

Parker, B. C., Simmons, Jr., G. M., Love, F. G., Wharton, Jr., R. A., and Seaburg, K. G. (1981). Modern stromatolites in Antarctic Dry Valley lakes. *BioScience*, **31**, 656–661.

Poreda, R. J., Hunt, A. G., Lyons, W. B., and Welch, K. A. (2004). The helium isotopic chemistry of Lake Bonney, Taylor Valley, Antarctica. *Aquatic Geochemistry*, **10**, 353–371.

Priscu, J. C. (1995). Phytoplankton nutrient deficiency in lakes of the McMurdo Dry Valleys, Antarctica. *Freshwater Biology*, **34**, 215–227.

Priscu, J. C., Priscu, L. R., Vincent, W. F., and Howard-Williams, C. (1987). Photosynthate distribution by microplankton in permanently ice-covered Antarctic desert lakes. *Limnology and Oceanography*, **32**, 260–270.

Priscu, J. P., Fritsen, C. H., Adams, E. E., et al. (1998). Perennial Antarctic lake ice: an oasis for life in a polar desert. *Science*, **280**, 2095–2098.

Priscu, J. C., Wolf, C. F., Takacs, C. D., et al. (1999). Carbon transformations in a perennially ice-covered Antarctic lake. *BioScience*, **49**(12), 997–1008.

Pugh, H. E., Welch, K. A., Lyons, W. B., Priscu, J. C., and McKnight, D. (2003). Biochemistry of Si in the McMurdo Dry Valley lakes, Antarctica. *International Journal of Astrobiology*, **1**, 737–749.

Quesada, A. and Vincent, W. F. (1993). Adaptation of cyanobacteria to the light regime within Antarctic microbial mats. *Verhangen International Verein Limnology*, **25**, 960–965.

Quesada, A., Vincent, W. F., and Lean, D. R. S. (1999). Community and pigment structure of Arctic cyanobacterial assemblages: the occurrence and distribution of UV absorbing compounds. *FEMS Microbiology Ecology*, **28**, 315–323.

Rautio, M. and Vincent, W. F. (2006). Benthic and pelagic food resources for zooplankton in shallow high-latitude lakes and ponds. *Freshwater Biology*, **51**, 1038–1052.

Roberts, E. C. and Laybourn-Parry, J. (1999). Mixotrophic cryptophytes and their predators in the Dry Valley lakes of Antarctica. *Freshwater Biology*, **41**, 737–749.

Robinson, P. H. (1984). Ice dynamics and thermal regime of Taylor Glacier, South Victoria Land, Antarctica. *Journal of Glaciology*, **30**, 153–159.

Sattley, W. M. and Madigan, M. T. (2006). Isolation, characterization, and ecology of cold-active, chemolithotrophic, sulfur-oxidizing bacteria from perennially ice-covered Lake Fryxell, Antarctica. *Applied Environmental Microbiology*, **72**, 5562–5568.

Scott, D. H., Rice, J. W., and Dohm, J. M. (1991). Paleolakes and lacustrine basins on Mars. *Lunar and Planetary Science*, **22**, Abstract 1203–1204. Lunar and Planetary Institute, Houston.

Siegel, B. Z., McMurty, G., Siegel, S. M., Chen, J., and LaRock, P. (1979). Life in the calcium chloride environment of Don Juan Pond, Antarctica. *Nature*, **280**, 828–829.

Singh, N., Kendall, M. M., Liu, Y., and Boone, D. R. (2005). Isolation and characterization of methylotrophic methanogens from anoxic marine sediments in Skan Bay, Alaska, description of *Methanococcoides alaskense* sp. nov., and emendation of *Methanosarcina baltica*. *International Journal of Systematic and Evolutionary Microbiology*, **55**, 2531–2538.

Skidmore, M., Anderson, S. P., Sharp, M., Foght, J. M., and Lanoil, B. D. (2005). Comparison of microbial community compositions of two subglacial environments reveals a possible role for microbes in chemical weathering processes. *Applied and Environmental Microbiology*, **71**, 6986–6997.

Solomon, S. C., Aharonson, O., Aurnou, J. M., et al. (2005). New perspectives on ancient Mars. *Science*, **307**, 1214–1220.

Spaulding, S. A., McKnight, D. M., Smith, R. L., and Dufford, R. (1994). Phytoplankton population dynamics in perennially ice-covered Lake Fryxell, Antarctica. *Journal of Plankton Research*, **16**, 527–541.

Spigel, R. H. and Priscu, J. C. (1998). Physical limnology of the McMurdo Dry Valley lakes. In *Ecosystem Dynamics in a Polar Desert: The McMurdo Dry Valleys, Antarctica*, ed. J. C. Priscu. Antarctic Research Series 72. Washington, D.C.: American Geophysical Union, pp. 153–188.

Squyres, S. W., Anderson, D. W., Nedell, S. S., and Wharton, Jr., R. A. (1991). Lake Hoare, Antarctica, sedimentation through a thick perennial ice cover. *Sedimentology*, **38**, 363–379.

Squyres, S. W., Knoll, A. H., Arvidson, R. E., et al. (2006). Two years at Meridiani Planum: results from the Opportunity Rover. *Science*, **313**, 1403–1407.

Stal, L. J. (1995). Physiological ecology of cyanobacteria in microbial mats and other communities. Tansley Review 84. *New Phytologist*, **131**, 1–32.

Sumner, D. Y. (1997). Carbonate precipitation and oxygen stratification in late Archean seawater as deduced from facies and stratigraphy of the Gemohaan and Frisco formations, Transvaal Supergroup, South Africa. *American Journal of Science*, **297**, 455–487.

Sumner, D. Y. (2000). Microbial versus environmental influences on the morphology of late Archean fenestrate microbialites. In *Microbial Sediments*, ed. R. Riding and S. M. Awramik. Berlin: Springer, pp. 307–314.

Takacs, C. D. and Priscu, J. C. (1998). Bacterioplankton dynamics in the McMurdo Dry Valley lakes, Antarctica: production and biomass loss over four seasons. *Microbial Ecology*, **36**, 239–250.

Takacs, C. D., Priscu, J. C., and McKnight, D. M. (2001). Bacterial dissolved organic carbon demand in McMurdo Dry Valley lakes, Antarctica. *Limnology and Oceanography*, **46**(5), 1189–1194.

Timperley, M. H. (1997). A simple temperature-based model for the chemistry of melt-water ponds in the Darwin Glacier area, 80° S. In *Ecosystem Processes in Antarctic Ice-Free Regions*, ed. W. B. Lyons, C. Howard-Williams, and I. Hawes. Rotterdam, Netherlands: AA Balkema, pp. 221–230.

Tomiyama, C. and Kitano, Y. (1985). Salt origin in the Wright Valley, Antarctica. *Antarctic Record*, **86**, 17–27.

Torii, T., Nakaya, S., Matsubaya, O., et al. (1989). Chemical characteristics of pond waters in the Labyrinth of southern Victoria Land, Antarctica. *Hydrobiology*, **172**, 255–264.

Tosca, N. J., Knoll, A. H., and McLennan, S. M. (2008). Water activity and the challenge for life on early Mars. *Science*, **320**, 1204–1207.

Tranter, M., Skidmore, M., and Wadham, J. L. (2005). Hydrological controls on microbial communities in subglacial environments. *Hydrological Processes*, **19**, 995–998.

Tyler, S. W., Cook, P. G., Butt, A. Z., et al. (1998). Evidence of deep circulation in two perennially ice-covered Antarctic lakes. *Limnology and Oceanography*, **43**, 625–635.

Van der Wielen, P. W. J. J. (2006). Diversity of ribulose-1,5-bisphosphate carboxylase/oxygenase large-subunit genes in the $MgCl_2$-dominated deep hypersaline anoxic basin Discovery. *FEMS Microbiology Letters*, **259**, 326–331.

Van der Wielen, P. W. J. J., Bolhuis, H., Borin, S., and the BioDeep Scientific Party (2005). The enigma of prokaryotic life in deep hypersaline anoxic basins. *Science*, **307**, 121–123.

Ventosa, A., Nieto, J. J., and Oren, A. (1998). Biology of moderately halophilic aerobic bacteria. *Microbiology and Molecular Biology Reviews*, **62**, 504–544.

Vincent, W. F. (1988). *Microbial Ecosystems of Antarctica*. Cambridge, UK: Cambridge University Press.

Vincent, W. F. (2000). Cyanobacterial dominance in the polar regions. In *Ecology of the Cyanobacteria: Their Diversity in Space and Time*, ed. B. Whitton and M. Potts. Dordrecht, Netherlands: Kluwers Academic Press, pp. 321–340.

Vincent, W. F. and Quesada, A. (1994). Ultraviolet radiation effects on cyanobacteria: implications for Antarctic microbial ecosystems. In *Ultraviolet Radiation in the Antarctic Environment: Measurements and Biological Effects*, ed. S. Weiler and P. Penhale. Washington, D.C.: American Geophysical Union, pp. 111–124.

Vincent, W. F., Downes, M. T., Castenholz, R. W., and Howard-Williams, C. (1993). Community structure and pigment organisation of cyanobacteria-dominated microbial mats in Antarctica. *European Journal of Phycology*, **28**, 213–221.

Wagner, B., Melles, M., Doran, P., et al. (2006). Glacial and postglacial sedimentation in the Fryxell basin, Taylor Valley, southern Victoria Land, Antarctica. *Palaeogeography Palaeoclimatology Palaeoecology*, **241**, 320–337.

Ward, B. B. and Priscu, J. C. (1997). Detection and characterization of denitrifying bacteria from a permanently ice-covered Antarctic lake. *Hydrobiology*, **347**, 57–68.

Webster, J. G., Brown, K. L., and Vincent, W. F. (1994). Geochemical processes affecting meltwater chemistry and the formation of saline ponds in the Victoria Valley and Bull Pass Region, Antarctica. *Hydrobiology*, **281**, 171–186.

Welch, K. A., Lyons, W. B., McKnight, D., et al. (2000). Climate and hydrologic variations and implications for lake and stream ecological response in the McMurdo Dry Valleys, Antarctica. In *Climate Variability and Ecosystem Response at Long Term Ecological Research Sites*, ed. D. Greenland, D. G. Goodin, and R. C. Smith. Oxford, UK: Oxford University Press, pp. 174–195.

Wharton, Jr., R. A. (1994). Stromatolitic mats in Antarctic lakes. In *Phanerozoic Stromatolites*, Vol. II, ed. J. Bertrand-Sarfati and C. Monty. Dordrecht, Netherlands: Kluwer Academic Publishers, pp. 53–70.

Wharton, Jr., R. A., Parker, B. C., and Simmons, Jr., G. M. (1983). Distribution, species composition, and morphology of algal mats in Antarctic dry valley lakes. *Phycologia*, **22**, 355–365.

Wharton, Jr., R. A., Lyons, W. B., and Des Marais, D. (1993a). Stable isotope biogeochemistry of carbon and nitrogen in a perennially ice-covered Antarctic lake. *Chemical Geology*, **107**, 159–172.

Wharton, Jr., R. A., McKay, C. P., Clow, G. D. and Andersen, D. T. (1993b). Perennial ice covers and their influence on Antarctic lake ecosystems. In *Physical and Biogeochemical Processes in Antarctic Lakes*, ed. W. J. Green and E. I. Freidmann. Washington, D.C.: American Geophysical Union, pp. 53–70.

Wharton, Jr., R. A., Crosby, J. M., McKay C. P., and Rice, J. W. (1995). Paleolakes on Mars. *Journal of Paleolimnology*, **13**, 267–283.

Wilson, A. T. (1964). Evidence from chemical diffusion of a climatic change in the McMurdo Dry Valleys 1,200 years ago. *Nature*, **201**, 176–177.

7

The biogeochemistry and hydrology of McMurdo Dry Valley glaciers: is there life on martian ice now?

MARTYN TRANTER, LIZ BAGSHAW, ANDREW G. FOUNTAIN, AND CHRISTINE FOREMAN

Introduction

Microbial life on Earth usually requires at least five prerequisites: innoculi, liquid water, and sources of energy, carbon, and nutrients (Rothschild and Manicelli, 2001). One of the major advances in the cryospheric sciences during the last decade is the realization that microbial life or innoculi are found in a whole spectrum of environments throughout glacier ice masses of all scales, from the snow cover, through ice surface (or supraglacial) environments, within ice (or englacial) environments through to ice bed (or subglacial) environments (Hodson et al., 2008). A remarkable observation is that apparently viable microbes can be found throughout the whole 4 km of ice column found near the center of the East Antarctic Ice Sheet above subglacial Lake Vostok (Priscu et al., 2008). Hence, glaciers on Earth can now be regarded as biomes or ecotomes, and the question arises whether or not glaciers on other celestial bodies have the potential to act as ecotomes. This chapter begins to provide an answer by first describing how microbial life exists in the cold glaciers of the McMurdo Dry Valleys, and second, by speculating on whether or not there is the chance of life in the glaciers and ice caps of Mars. We make the assumption that potential microbial life on Mars is carbon based and requires the same five prerequisites for microbial life as on Earth (Rothschild and Manicelli, 2001).

Dry Valley glaciers

The McMurdo Dry Valleys comprise the majority of the 2% of ice-free land in Antarctica (Cowan and Tow, 2004), and have long been recognized as one of the best terrestrial analogs for conditions on Mars (e.g., McKay et al., 2005;

Life in Antarctic Deserts and Other Cold Dry Environments: Astrobiological Analogs, ed. Peter T. Doran, W. Berry Lyons and Diane M. McKnight. Published by Cambridge University Press.

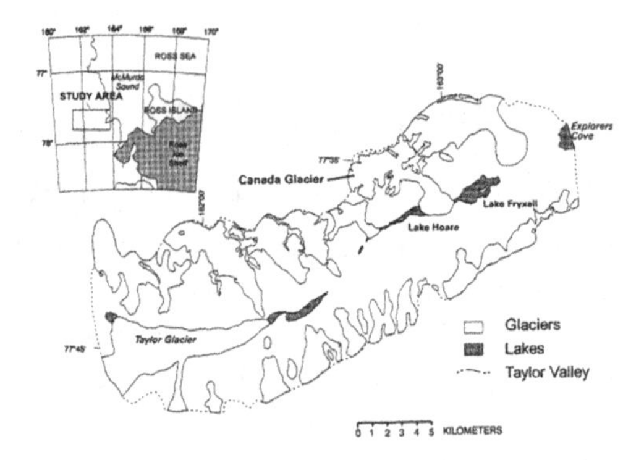

Fig. 7.1. Taylor Valley, adjacent to McMurdo Sound and the East Antarctic Ice Sheet. Major features include glaciers and perennially ice-covered lakes. (From Lewis et al., 1999.)

Head and Marchant, this volume, Chapter 2). The Dry Valleys are situated on the edge of the East Antarctic Ice Sheet (EAIS), within Southern Victoria Land. The Dry Valley network consists of Taylor, Wright, and Victoria Valleys, with Garwood, Marshall, and Miers Valleys further south. The Dry Valleys are ice free because the Trans-Antarctic Mountains block the flow of the EAIS into the Ross Sea, maintaining perennial ice-free conditions (Fountain et al., 1999).

Taylor Valley (Fig. 7.1) is the location of a U.S. National Science Foundation (NSF) Long Term Ecological Research (LTER) site, established in 1993. Taylor Valley was chosen because of its position as an "end-member" of the LTER sites. Life is found here despite Taylor Valley being one of the harshest environments on Earth (Fountain et al., 1999). A prerequisite for life in a cold polar desert is liquid water (Kennedy, 1993). The small amount of snow ($<10\,\mathrm{cm\,y^{-1}}$ water equivalent) which reaches the valley bottom sublimates quickly (Fountain et al., 1999), due to the high wind speeds and the low relative humidity of the frequent katabatic winds (Nylen et al., 2004), and so makes little hydrologic contribution. So, paradoxically, the 15 cold, polar glaciers which cover ~40% of the valley floor (Lyons et al., 2003) prove to be the key water source. The glaciers themselves are host to a range of biota which also depend on meltwater for their viability. We describe below how climate, topography, wind, and debris are key factors in the presence and supply of liquid water, innoculi, and nutrients.

Climate

The Dry Valleys have one of the coldest and driest climates on Earth. Mean annual temperatures range from −16 °C to −30 °C, with average summer temperatures of −6 °C, and may vary by more than 10 °C in less than 40 km and 300 m elevation, due to topographic effects and wind patterns (Doran et al., 2002). The local meteorology is dominated by opposing winds, which have a large impact on the prevailing climate. Cool, easterly sea breezes bring moist air from the Ross Sea into the Dry Valleys, while eastward katabatic winds are warm and dry whilst descending from the EAIS. Sea breezes dominate the summer regime, reaching speeds of $4\,m\,s^{-1}$. Eastward locations are subject to a greater influence of the sea breezes and precipitation, and hence there is a pronounced east–west precipitation gradient up-valley. This is heightened by the presence of the Nussbaum Riegel, which prevents the moist air from moving up-valley (Fountain et al., 1999).

The westerly katabatic winds are stronger and more frequent in winter (Nylen et al., 2004). They form as a consequence of net longwave radiation losses cooling near-surface air on the Antarctic Plateau, so increasing the air density and causing it to flow downhill. The winds displace less dense air at lower elevations, and channeling by the steep slopes and topographic features of the Dry Valleys causes increasing speeds, which typically reach 20 to $35\,m\,s^{-1}$. The katabatic winds have a significant effect on the temperature in Taylor Valley. Adiabatic warming as the winds descend from the Plateau results in a local temperature increase of up to 4 °C in summer, and up to 30 °C in winter, for periods of to several days. Nylen et al. (2004) suggest that average annual temperatures would be up to 2.3 °C cooler without the effect of the katabatic winds. Both their onset and termination is usually abrupt, but the elevated temperatures may persist for days afterwards.

The katabatic winds transport large amounts of debris around the valleys (Lancaster, 2002), which is often sand sized (Fig. 7.2). The aeolian debris also contains microorganisms and higher animals in their resting (or anhydrobionic) states (Nkem et al., 2006). Hence, the winds, including the sea breezes, are key components in the transfer of innoculi around the Dry Valleys (see below). Debris deposited on the glacier surfaces also promotes the production of water during the summer, and may help sculpture certain topographical features on some of the glaciers (see below).

The flux of solar and photosynthetically available radiation (PAR) varies according to cloudiness and topography (Nylen et al., 2004), but light is seldom limiting on the potential for photosynthesis on glacier surfaces during the summer.

Physical characteristics

Glaciers in the Dry Valleys are generally of the polar-alpine type, often being valley glaciers which are frozen to the underlying substrate (Fountain

Fig. 7.2. Sand-sized debris that has been blown onto Canada Glacier and subsequently redistributed by the hydrological system. The stream is ~30–50 cm wide.

et al., 1999). They are typically small, being on the order of 1–10 km in length and ~1–3 km wide (Fig. 7.1). The smaller glaciers are fed by accumulation zones some 2000 m above the main valley floor, but larger glaciers at the head of Victoria, Wright, and Taylor Valleys are outlet lobes of the East Antarctic Ice Sheet. Those smaller glaciers flowing from the Asgard Range to the north of Taylor Valley are generally three times the area of those flowing from the southerly Kukri Hills (Fig. 7.1). They have larger and higher accumulation basins, and less direct exposure to solar radiation due to their south-facing aspect. The temperature of the ice is typically on the order of −20 °C. They flow slowly, up to a few millimeters per year, since they are largely frozen to the substrate (Fountain et al., 1998). They often fan out from the side-valley confines into distinctive terminal lobes on the main valley floor. These lobes have characteristic vertical ice cliffs, some tens of meters high (Fig. 7.3). Small radial crevasses are a common feature of these lobes. The crevasses often anneal or become full of water, which quickly refreezes, during the height of the ablation season (Fig. 7.4).

The glaciers in Taylor Valley generally receive 10–30 cm accumulation annually in their upper zones, and lose between 6 cm and 15 cm per year in their ablation zones.

Only some 30% of this ablation is due to direct melting, whereas 70% is due to evaporation and sublimation (Fountain et al., 1999). Ice lenses within the snowpack indicate that there is some melting in the accumulation zone, but the majority of the snowmelt refreezes within the snowpack and so makes little contribution to runoff. The katabatic winds erode significant quantities

Fig. 7.3. The ice cliffs at the eastern terminus of Canada Glacier. The cliffs are ~20 m high. Snow picks out the frozen supraglacial channels, and the dust that is transported through these channels in the near-surface drainage system is draped around the top of the cliffs.

Fig. 7.4. Expansion cracks on the surface of Canada Glacier, intersecting and slowly draining a small, semi-thawed surface pond.

of debris from the arid soils on and near the valley floors and glacial debris from the upper valley slopes. Deposition of the debris onto the glacier reduces the albedo of the ice, and this enhanced melting due to dust deposition imparts a distinctive, valley-like geomorphology on the glacier surfaces, which is roughly parallel to the main terrestrial valley walls, marking the dominant wind direction (Fig. 7.5). These valleys become important areas of meltwater-flow focus on the glacier surface during the ablation season.

The low angle of incident solar radiation in the Dry Valleys promotes the more rapid ablation of vertical relative to horizontal surfaces. Vertical surfaces, such as ice cliffs and channels, tend to become exaggerated if there are suitable positive feedbacks, such as the rotting and melting of ice cliff bases and meltwater export to the subsurface drainage system (Lewis et al., 1999). Ablation is typically an order of magnitude higher on vertical surfaces than

Fig. 7.5. Up- and down-valley (west–east) orientation of supraglacial drainage channels on the southern terminus of Canada Glacier.

on the horizontal ice tongues. In particular, there is higher specific humidity and increased sun exposure on north-facing cliffs, and hence higher sensible and latent heat fluxes produce increased melting on these cliffs. Wind speeds are half of those across the horizontal ice surface and so sublimation is significantly reduced (Lewis et al., 1999), and hence almost 100% of summer ablation from the ice cliffs is via melting. Lewis et al. (1999) show that the terminus cliffs on Canada Glacier make up 2% of the ablation area, but contribute 15–20% of overall summer runoff. This increases to ~40% in cool summers, such as 1994–95. The cliffs may produce melt even when the air temperatures are below freezing due to their favorable radiation budget. They are unaffected by summer snow storms which otherwise shut down meltwater production on the glaciers due to increased albedo (Lewis et al., 1999).

A notable feature on Dry Valley glaciers are cryoconite holes (Fig. 7.6), which cover at least 4.5% of the ablation zones of glaciers in Taylor Valley (Fountain et al., 2004). Frozen ponds and lakes (cryolakes) are also found on certain glaciers. Cryoconite holes are common to glaciers worldwide, and are small water-filled cylindrical holes found in the surface of glacier ice (Hodson et al., 2008). Each hole has a thin layer of sediment at the bottom. Debris and sediment patches residing on the glacier surface absorb more solar radiation than the surrounding ice and melt into the ice, eventually forming a cylindrical water-filled hole with a layer of sediment on the bottom (Gribbon, 1979). The cryoconite holes on Dry Valley glaciers are unique in that they have ice lids and remain largely isolated from the atmosphere throughout the year. Cryoconite holes only form on the lower, ice-covered, and snow-free parts of glaciers. Higher on glaciers, the net accumulation of snow rapidly buries any surficial sediment.

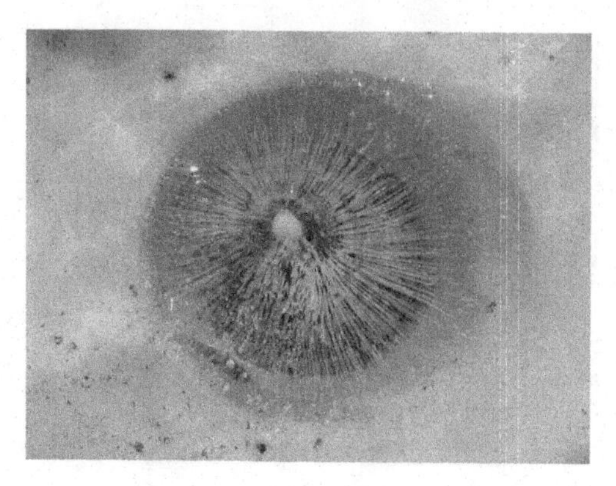

Fig. 7.6. Cryoconite hole in the surface of a Dry Valley glacier. The clear ice is refrozen meltwater, which strongly contrasts with the opaque glacier ice. The prominent debris layer some 30 cm below is clearly visible, as is some recently deposited aeolian dust on the ice lid. The diameter of the hole is ~20 cm. The color version of this figure can be found at http://www.cambridge.org/uk/catalogue/catalogue.asp?isbn=9780521889193.

Fig. 7.7. A frozen pool and riffle sequence on Canada Glacier.

Certain glaciers, such as Canada and Taylor Glaciers, have prominent surface channels, up to tens of meters wide and deep, in the ice slopes leading towards the terminal cliffs. The frozen channel floor morphology resembles the pool and riffle sequences of upland temperate streams (see Fig. 7.7). Some of the frozen ponds are large, with diameters >10 m, and are better described as cryolakes (see Fig. 7.8). The cryolakes are prominent and important features within the glacier valley floors. They act as sediment traps from upstream environments, often have steep ice cliffs along their margins and, being topographic lows, act as atmospheric dust traps. Hence, these environments have a tendency to grow once they are initiated and are inoculated by microorganisms in the atmospheric dust which, when heated by solar

Fig. 7.8. A cryolake on Canada Glacier, with prominent ice cliffs in the background and a basal layer of debris clearly visible.

radiation, melts through their frozen ice surface (Fountain et al., 2004, 2008). Cryoconite holes, frozen ponds, and cryolakes are important elements of the hydrological system of Dry Valley glaciers, and we show below that they are important semisealed oases of life near the glacier surface.

Hydrology

Runoff from glaciers in the McMurdo Dry Valleys is generally very low. The total runoff volume over the two summers of 1994–95 and 1995–96 was 198×10^3 m^3 (Lewis et al., 1999). However, warm summers have been recorded periodically, with a return time of up to a decade or so, and high flow events are then recorded. For example, warm summers occurred three times in the last 20 years, in 1985, 1990, and 2001 (Foreman et al., 2004). These high flow years completely change the nature of the glacier surface, creating large supraglacial streams, similar to those found on more temperate glaciers (Fig. 7.9).

Meltwater cannot penetrate far into the surface of cold glacier ice, and so is usually confined to the surface or very near surface. The exception is if a water flow path intersects an expansion crack, a narrow crevasse on the order of millimeters to a few centimeters wide, which forms when the lobes slowly form by lateral motion in the valley flow (Fig. 7.4). Meltwater draining into these cracks freezes on prolonged contact with the cold glacier ice and forms vertical bands of clear ice. Otherwise, most meltwater drains just below the ice surface in a near-surface drainage system, which is maintained by the solar heating of subsurface debris in otherwise frozen supraglacial channels, lakes, and cryoconite holes. The evolution and current configuration of the drainage system is probably dependent on the high rates of melting of the

Fig. 7.9. Supraglacial streams on Canada Glacier during the warm summer of 2001. Sediment is redistributed along the watercourses, forming a sediment trail that forms the basis of the near-surface drainage system in subsequent cooler summers.

glacier surface during the relatively warm summers. Then, the subsurface drainage network is exposed to form the basis of a supraglacial drainage system (Fountain et al., 2004). These events transport significant quantities of sediment into the channels, ponds, and cryolakes, so providing a sediment trail that is the basis of the near-surface drainage system in subsequent cold years (Fig. 7.9).

Hence, the main drainage routes on the glaciers of the Dry Valleys are largely subsurface during most summers, consisting of channels beneath approximately 10–30 cm of ice (Tranter et al., 2005). Much of the water is sourced from internal melting of the surface ice, by solar heating of debris. Internal melting occurs within cryoconite holes, channel beds, frozen pools, and cryolakes. Major subsurface channels develop preferentially along the base of the major pool and riffle sequence, and smaller flow paths connect ~50% of the cryoconite holes which occupy 1–3% of the surface of the upper, flatter ablation zone (Fountain et al., 2004). These connected cryoconite holes may produce up to 20% of the runoff from Canada Glacier, and their connection to the major channels is most likely via subsurface fissures (Fountain et al., 2008). The cryoconite holes deliver a supply of relatively solute-rich water to the subsurface drainage system, and help explain why waters near the top of the major pool and riffle systems may be relatively solute rich. Ice cliffs along the main channels (Figs. 7.7–7.9) grow larger as the glacier ice moves towards the terminus, and contribute increasingly significant quantities of runoff. Water from the ice cliffs seeps into the subsurface drainage system via rotten ice at the base of the cliffs. Continual leaching of solute from the ice cliffs during the ablation season depletes the

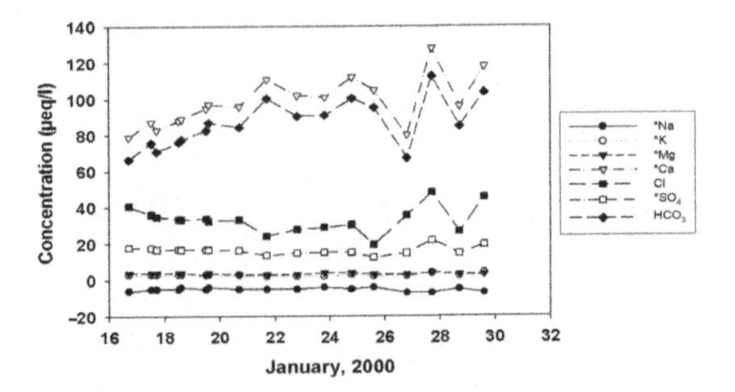

Fig. 7.10. The temporal variation in the concentration of major ions in the main outflow from Canada Glacier during January 2000. An asterisk * denotes that ion concentrations are sea-salt corrected with respect to Cl^-. The drainage system begins to freeze on January 27. (After Tranter et al., 2005. Courtesy of the International Glaciological Society.)

ice of solute, and hence the more solute-rich meltwaters from further up the glacier are diluted during progression towards the glacier terminus (Tranter et al., 2005). Ice cliff melt is sensitive to incoming solar radiation, and is much reduced during cold spells when there is freeze-up of the glacier surface. By contrast, water may continue to be produced in the subsurface via solar heating. Hence, ion concentrations in runoff increase during the freeze (Fig. 7.10), both due to decreased dilution by ice cliff melt because ions are preferentially rejected from growing ice crystals and remain in solution (Killawee et al., 1998).

Cryoconite holes and cryolakes: biogeochemical reactors

Cryoconite holes are an element of glacier surfaces that have been the subject of much recent interest. Cryoconite holes are small, cylindrical depressions which occur on the surface ice of glaciers throughout the world (Hodson et al., 2008). They form when solar radiation heats debris resting on the ice surface, so melting the debris into the ice to form a small, water-filled depression (Fig. 7.11). The debris melts into the ice surface until a steady-state depth is reached, when downward melt-in rate equals the ablation rate of the glacier surface (Gribbon, 1979). Consequently, these isolated cryoconite holes may sequester solute and nutrient from glacier ice and debris over time. Wholescale flushing of the glacier surface to downstream ecosystems during warm summers stimulates significant increases in the primary production of the ice-covered lakes (Foreman et al., 2004).

Cryoconite holes in Taylor Valley are typically 30 to 80 cm deep, and 20 to 120 cm in diameter (Tranter et al., 2004). An ice lid between 10 and 30 cm may

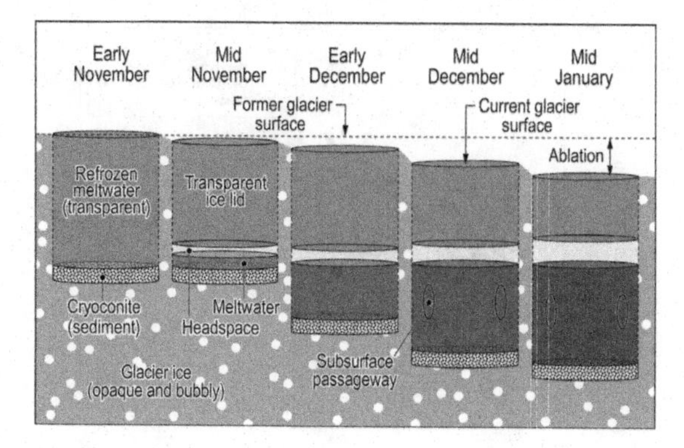

Fig. 7.11. Evolution of cryoconite holes in the Dry Valleys over the duration of an ablation season. (From Fountain et al., 2008.)

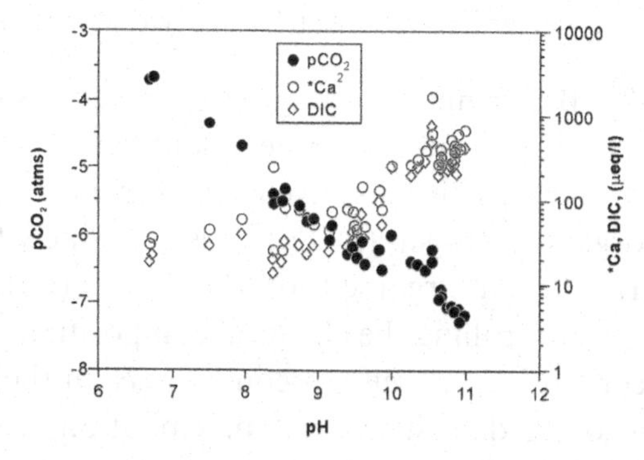

Fig. 7.12. Chemical characteristics of waters in cryoconite holes on Canada Glacier, 2000. (From Tranter et al., 2004.)

form over the hole, below which is a head space (Fig. 7.11). Water in summer is generally clear, but can be tea colored. Cryoconite holes are refugia for life in an otherwise extreme glacial environment (Wharton et al., 1985). The relatively simple systems which operate in the holes also provide an opportunity to study ecosystem dynamics on a simplified scale, and provide an analog for potential habitats on other icy terrestrial bodies (Priscu and Christner, 2002). The cryoconite material at the base of the hole is derived from aeolian sources (Lyons et al., 2003), and acts as a substrate for microorganisms such as heterotrophic bacteria, fungi, cyanobacteria, green algae, rotifers, and tardigrades (Mueller and Pollard, 2004; Porazinska et al., 2004). The cyanobacteria, in particular, photosynthesize when sunlight illuminates the holes (Tranter et al., 2004). The holes freeze solid over winter and reform during the following spring. Heterotrophy and photoautotrophy by biota in the holes, coupled with weak chemical buffering, can cause extreme chemical conditions to develop: pH values approach 11 (Fig. 7.12) and O_2 saturation

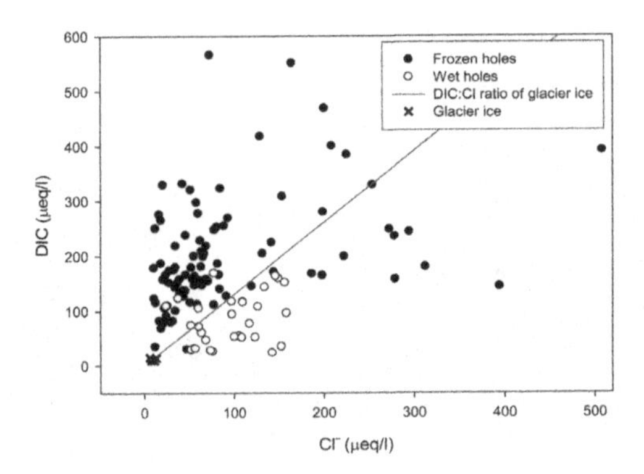

Fig. 7.13. The DIC (dissolved inorganic carbon) content of waters in cryoconite holes (wet holes) and frozen holes relative to a conservative ion, Cl^-. Samples were collected from Canada Glacier in 2004. Note that wet holes largely lie below the glacier ice line, due to photosynthesis in the summer.

may approach 160%. Biochemical cycling between periods of net photosynthesis and net respiration produces large quantities of dissolved organic carbon (DOC) and much of the dissolved nitrogen pool (>99%) is in the form of dissolved organic nitrogen (DON). It is very likely that much of the dissolved phosphorus is in an organic form (DOP) too (Stibal et al., 2008).

The main processes controlling the chemical composition of solutes in both frozen and water-containing or wet cryoconite holes in the Dry Valleys are dissolution of cryoconite debris, cyclical precipitation and dissolution of carbonates, net photosynthesis over summer, and net respiration during the autumnal freeze (Bagshaw et al., 2007; Fountain et al., 2008). Dissolution of primary carbonate and sulfate minerals in the debris, sourced from Dry Valley soils of variable salinity, which forms the hole produces a water chemistry that reflects the heterogeneous nature of the debris. Thereafter, slower dissolution of silicates adds ions such as K^+ to the holes. Net photosynthesis occurs during the summer months, and DIC (dissolved inorganic carbon) may be lost from the waters (Fig. 7.13). Net respiration is likely to occur during the autumnal freeze, and possibly over winter. DOC is regenerated during the thaw, and the mass of DOC accumulates annually. The isolation age, based on the mass of excess Cl^- in the holes, appears to be crudely indicative of the extent of isolation of these simple, closed biogeochemical systems from the near-surface drainage system. Our best estimate of DOC generation in the cryoconite holes is on the order of $7.5\,\mu g\,C\,cm^{-2}\,y^{-1}$, and the holes progressively deepen over time, perhaps as a consequence of the waters becoming darker.

Porazinska et al. (2004) hypothesize that the isolation caused by the ice lids may result in the development of unique biotic communities. If the hole

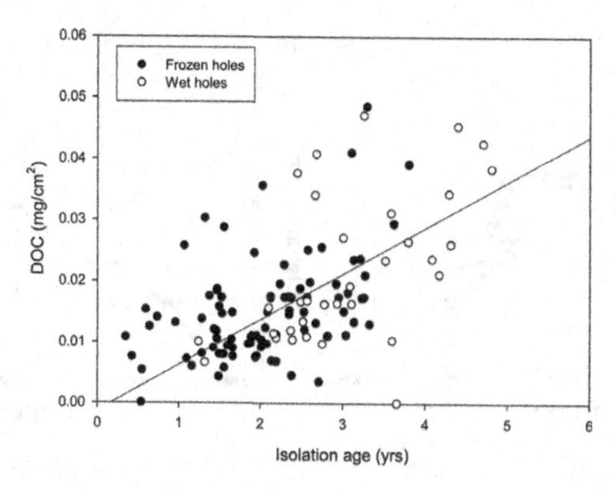

Fig. 7.14. Increase in the DOC (dissolved organic carbon) content of wet and frozen cryoconite holes over time at Canada Glacier, 2004. The best fit linear regression line is shown for aid of interpretation only. (From Bagshaw et al., 2007.)

remains isolated, primary production will progressively use up the more readily available nutrients and drive up the pH of the holes (Fig. 7.12). Feedbacks between physical, chemical, and biological processes are necessary to continue primary production in the closed system once the initial inorganic supplies of nutrient are exhausted (Tranter et al., 2004). For example, the ecosystem in a number of holes appears to have adapted to deriving nitrogen from dissolved organic nitrogen compounds. Foreman et al. (2007) suggested that the greater β-glucosidase activity in certain holes was indicative of organisms producing enzymes to acquire additional dissolved organic carbon within the holes.

Cryolakes are important supraglacial environments because they act as temporary reservoirs of ice melt and host a spectrum of photoautotrophic cyanobacteria and other organisms, potentially including N-fixing bacteria. These ice-topped reservoirs, whose volume, porosity, and hydraulic connectivity are climatically sensitive, have the capacity to enhance the flux of nutrients from the glaciers to downstream ecosystems (Tranter et al., 2005). They also have the capacity to moderate the dissolved phase in which nutrient N and P are found. Our contention is that biogeochemical reactions in cryolakes are similar to those occurring in cryoconite holes, since both contain high quantities of debris in their bases, receive innoculi from similar sources, and contain liquid water for significant periods during the summer ablation season. Cryolakes differ from cryoconite holes in terms of size, having length scales of tens of meters, and because there is at least partial flushing during most melt years. However, they are important components of the near-surface drainage system as most meltwater flow is focused through them because of their location in the bottom of the glacier valleys.

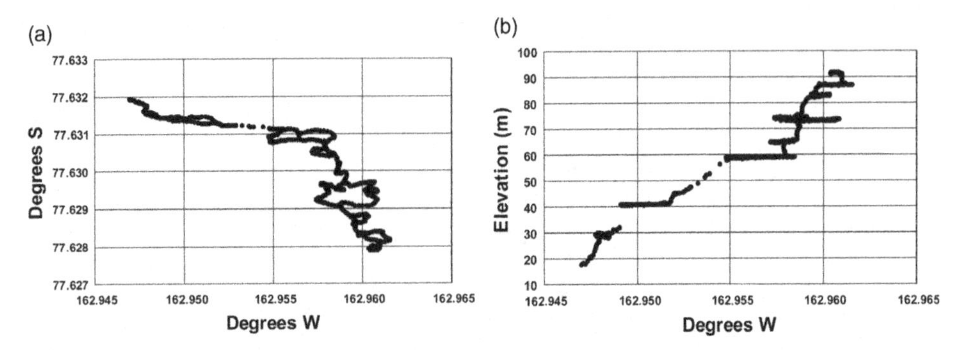

Fig. 7.15. (a) Plan view of cryolakes within the main glacier valley at the western margin of Canada Glacier. Three larger lakes and four smaller lakes can be seen in outline. The lines joining the lakes are the bottom of the glacier valley. The main outlet is to the top left of the diagram (and so water drains from right to left). Data were obtained from a high precision GPS survey. (b) Elevation of the cryolakes (open areas) within the main glacier valley at the western margin of Canada Glacier. The cryolakes are the flat horizontal features. Connections between the cryolakes are via steep largely frozen waterfalls or riffles, which often freeze. Hence, water ponds in the cryolakes and is hydrologically isolated until warmer periods during each ablation season.

Figure 7.15 shows the plan and elevation of the cryolake complex on the western margin of Canada Glacier. There are three larger cryolakes and four smaller pools within this glacier valley. The lakes occupy prominent depressions in the down-glacier profile, and are connected by a series of steeper slopes. This frozen pool and riffle system is often poorly connected during colder weather, so flow between the cryolakes is periodically prohibited (Tranter et al., 2005). Meltwater ponded within the cryolakes during colder weather does not completely freeze because solar radiation continues to heat debris in the cryolake beds. Hence, biogeochemical reactions continue to take place within them, as in unconnected cryoconite holes.

Evidence for this assertion can be found in Table 7.1, which shows the chemical composition of glacier ice, waters in cryoconite holes, and outflow from the cryolake system in January 2000. The cryolake system is a clear source of $*Ca^{2+}$, DIC, $*SO_4^{2-}$, and $*K^+$, since enrichment factors with respect to glacier ice are >1. The "*" denotes that a sea salt has been made to the ion concentration, so that the source of the ion is weathering of debris in this context. By contrast, the cryolake system is a clear sink for DIN (dissolved inorganic nitrogen) and DIP (dissolved inorganic phosphorus) with enrichment factors of $\ll 1$. Cryoconite holes, which supply $\sim 13\%$ of the water to the cryolakes, have broadly similar enrichment factors, in support of our hypothesis that cryolakes have similar biogeochemical properties to cryoconite

Table 7.1. *The mean composition of glacier ice and waters within cryoconite holes on Canada Glacier,*[a] *in comparison with cryolake outflow* ($\mu eq\, l^{-1}$)

	Cl^-	$*Ca^{2+}$	DIC	$*SO_4^{2-}$	$*K^+$	NH_4^+	NO_3^-	PO_4^{3-}
Mean glacier ice	32	36	35	7.8	2.4	7.1	7.1	0.1
Cryolake outflow	33	105	83	18	3.0	0.05	0.43	0.0
	(1)	(2.8)	(2.3)	(2.2)	(1.2)	(0.007)	(0.06)	(0)
Mean cryoconite holes	106	230	150	63	7.6	0.43	0.23	0.08
	(1)	(1.9)	(1.3)	(2.4)	(0.96)	(0.02)	(0.01)	(0.24)

Notes: Enrichment factors are shown within brackets beneath the mean concentrations, and are defined with respect to glacier ice, using Cl^- as a normalizing ion.

DIC, dissolved inorganic carbon.

[a] After Tranter et al. (2005).

holes. However, there are some differences. Cryolake outflow is proportionally higher in DIC, which may be testimony to the more open system nature of the hydrological flow paths to the cryolakes. It could also be the result of higher respiration relative to primary production, as would be anticipated if the cryolakes are more mature ecosystems than cryoconite holes and are consequently more heterotrophic (Bardgett and Walker, 2004). In contrast to DIC, the outflow is devoid of DIP, has very low concentrations of NH_4^+, and is not as efficient at removing NO_3^- from the subsurface waters. This suggests that subtly different biogeochemical reactions may occur within cryolakes. For example, N fixation could be taking place, as has been found in nearby lake ice cover (Paerl and Priscu, 1998). This would require the cryolakes to have an alternative source of P to DIP, most likely DOP, since DIP is limiting in lake ice cover. Recent investigations indicate that P may be the limiting nutrient in arctic cryoconite holes and proglacial lakes (Mindl et al., 2007; Säwström et al., 2007; Stibal et al., 2008). We hypothesize that microbial activity in cryolakes is also P limited, since DIP concentrations are below detection in the outflow (Table 7.1), but that both autotrophic and heterotrophic microbes may be able to obtain P through phosphatase activity.

Summary of prerequisites for microbial life in Dry Valley glaciers

Microbial life can exist on the surface of Dry Valley glaciers because the five prerequisites for life – innoculi, liquid water, and sources of energy, carbon and nutrients (Rothschild and Manicelli, 2001) – are found. The wind is a vital agent of transport for innoculi, which are deposited onto the glacier surface, along with dust, and melt into the glacier surface. Feedbacks exist that tend to enhance debris capture in valleys and cryolakes as these topographic depressions flow slowly down-glacier. The presence of liquid water is essential for microbial life, and is generated within shallow, subsurface environments by the adsorption of solar radiation by inorganic and organic debris of low albedo. Surface meltwater may be added to the subsurface environment via rotten ice at the base of ice cliffs and via crevasses. The legacy of debris trails from aquatic sediment transport and deposition in supraglacial transport during warm years serves as a template for the subsurface drainage system during colder years, when melting occurs beneath the surface despite the subzero ambient air temperatures. Hence, a secondary means of innoculi transport is via the hydrological drainage system even during cold years. The principal source of energy to organisms in the subsurface drainage system is solar radiation, fixed via photosynthesis. However, other energy sources, such as allochthonous organic carbon and sulfides, may be imported via the melt-in of aeolian debris. Nutrient is obtained from ice melt, and microbial processes serve to extract and recycle nutrient from inorganic and organic debris. Much of the N and P may be transformed into organic phases, which in turn maximizes the chemical potential for dissolution of N and P from inorganic phases. Nitrogen-fixing bacteria may occur in suitable elements of the subsurface drainage system, such as the sediments of cryoconite holes, ponds, and cryolakes. Sources of carbon arise from the atmosphere, bubbles in the ice, dissolution of carbonate, degradation of allochthonous organic matter, and recycling of carbon within the drainage system. Microbial communities have adapted to sourcing C, N, and P from particulate and dissolved organic phases.

Microbial life can also be found beneath at least one Dry Valley glacier, and Taylor Glacier in particular (Mikucki et al., 2004; Mikucki and Priscu, 2007). Here, highly saline water is believed to be a Pliocene remnant some 3–5 million years old. The microbes live in a suboxic environment, finding energy, carbon, and nutrient sources from the formerly marine sediments that the Taylor Glacier has now overridden. Subglacial environments are usually colonized by microbes, provided that liquid water is present (Hodson et al., 2008; Priscu et al., 2008). The combination of geothermal heating and energy generated by internal deformation of the ice, when coupled with a salinity that may reach several times that of seawater, enables the remnant water to exist in a nonfrozen state beneath Taylor Glacier.

Analogs with Mars

Mars is currently being explored by the Phoenix mission, and is the subject of great scientific interest (Head, 2007). We show below that the Dry Valleys have certain features in common with those found on Mars, namely a frigid surface environment, water ice that mostly ablates by sublimation, ubiquitous dust transport by the wind, and dust deposition onto ice during the summer. There is therefore the potential for ice melt to occur in subsurface environments and for cryoconite hole formation to occur if there is a suitable energy balance, particularly since the Phoenix Lander has recently discovered water ice close to the martian surface at a depth of ~5 cm (Muir, 2008). We examine these aspects of the martian environment before questioning whether or not there are suitable glacial aquatic environments to act as potential ecosystems.

Mars has a radius just over half of the Earth's, but the geomorphology of the red planet is much more extreme than that of Earth. The northern and southern hemispheres are very different. The northern hemisphere is smooth and flat and ~6 km lower than the rugged highlands of the south.

Fig. 7.16. The rugged geomorphology of Mars is exemplified by the Valles Marineris, a large split in the surface up to 7 km deep, running roughly along the equator. (Image credit: NASA/JPL/Marlin Space Science Services.)

This two-level geomorphology is very likely a relict of a large impact blast which blew away part of the northern hemisphere about 4.4 billion years ago, when the martian crust was first differentiating (Andrews-Hanna et al., 2008; Marinova et al., 2008; Melosh, 2008; Nimmo et al., 2008). The most notable feature on the surface is Valles Marineris, running roughly along the equator, which is a split in the martian crust some 4000 km long, up to 600 km wide, and some 7 km deep (Fig. 7.16). The Hellas Basin in the southern hemisphere is another prominent feature, which is an impact crater 2300 km in diameter and some 9 km deep.

Mars has a very thin atmosphere, consisting predominantly of CO_2. The surface temperature varies from −126 °C to +26 °C annually. The rotation axis of Mars is tilted with respect to the orbital plane by almost 24°, so there are marked seasonal differences in the amount of solar radiation falling on each hemisphere during the year. The difference between winter and summer is more extreme on Mars than on Earth, due to the greater eccentricity of the martian orbit. Some 40% more solar radiation is received during the southern summer, when Mars is nearest the Sun, than during the southern winter, resulting in relatively hot southern summers and mild northern winters, but cool northern summers and cold southern winters. Orbital variations are such that there may have been at least 40 major ice ages over the last five million years (Head et al., 2003; Schorghofer, 2007). For example, whereas the obliquity of the Earth's orbit ranged from 22° to 24.5° over the last 10 million

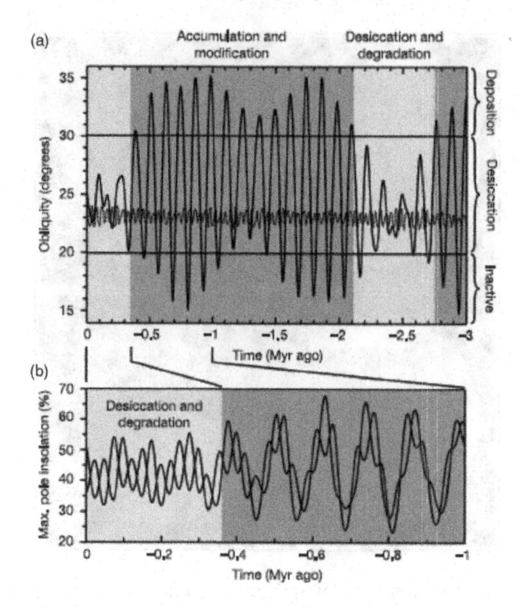

Fig. 7.17. Orbital forcing and ice sheet growth and decay on Mars (after Head et al., 2003). Ice ages correspond to periods of high obliquity (dark grey) and interglacials correspond to periods with lower obliquity (see text for details). The obliquity of Earth is shown for comparison, and is the central low amplitude line (~22–24°) in the upper diagram.

years, that of Mars ranged from 14° to 48° (Head et al., 2003). Those periods of high obliquity, from 2100 to 400 ka ago, correspond to times when there is net water transport from the polar ice caps to mid latitudes (Fig. 7.17). The intervening periods of low obliquity were too short to allow the complete sublimation of the water ice, because the ice was mantled by debris that isolated it from the atmosphere. This combination of high obliquity and dust blanketing of mid-latitude ground ice is the likely cause of martian ice ages (Head et al., 2003). By contrast, the low obliquity of the last 420 ka has resulted in the net transport of water ice from the mid-latitude ground ice, via slow diffusion through the debris mantle, through the atmosphere to the poles, so contributing to their layered structure (see below). These are referred to as interglacial periods (Head et al., 2003). Obliquities below 21° correspond to periods when the atmospheric pressure of CO_2 drops significantly, because of sequestering of CO_2 ice in the polar caps. This retards aeolian erosion of the dust blanket over the ice, so decreasing rates of sublimation of the water ice. These periods are termed inactive as a consequence (Head et al., 2003).

Currently, Mars has ice caps at both its north and south poles. Enough water is locked up in the south pole to cover the planet in a liquid layer ~11 m deep (Plaut et al., 2007). There is significant annual growth and retreat of the

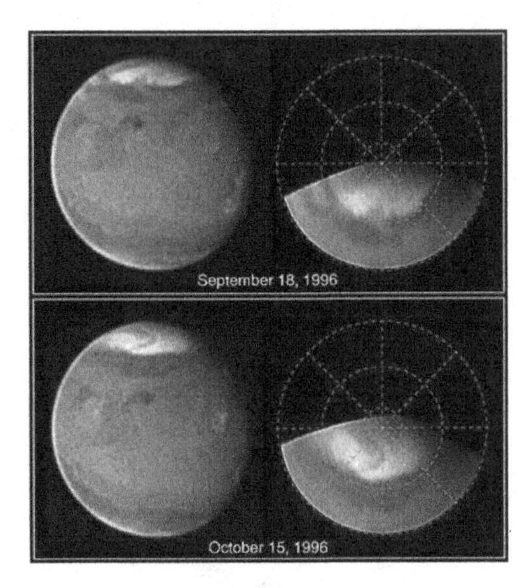

Fig. 7.18. Wispy clouds and the northern ice cap of Mars, as captured by the Hubble telescope on duststorm-free days. (Image credit: James Bell, III (Cornell University), Todd Clancy (Space Science Institute), Phil James (University of Toledo), Steve Lee (University of Colorado), Leonard Martin (Lowell Observatory), Michael Wolff (University of Toledo), and NASA.)

polar caps, which expand to cover ~30% of the planet's surface during the winter and shrink to relatively small caps covering ~1% of the surface in summer. The perennial portion of the north polar cap (Fig. 7.18) consists almost entirely of water ice (Langevin et al., 2005). This gains a seasonal coating of frozen CO_2, which grows to ~1 m thick in winter. The south polar cap consists of two layers. The top layer consists of frozen CO_2 and is ~8 m thick. The bottom layer is very much deeper and is made of water ice (Titus et al., 2003). Flat-floored, circular pits ~8 m deep and 200–1000 m in diameter occur within the upper frozen CO_2 layer during the summer, revealing the water ice below (Byrne and Ingersoll, 2003). Thus, although the two polar caps are similar, the significant difference between them is that the south polar dry ice cover is thicker and does not completely disappear during the summertime.

Reddish atmospheric dust, with a mean diameter of <5 μm, is ubiquitous to the martian atmosphere (Goetz et al., 2005). Dust deposited onto and contained within the ice caps gives them a reddish color, although the aerosol content of the ice is not considered to be high on average (Langevin et al., 2005). Another source of dust to the ice cap surfaces is that locally redistributed by the CO_2 jets (Fig. 7.19) that emerge from the southern ice cap each spring (Kieffer et al., 2006) and which spray fine dark sand over a

Fig. 7.19. An artist's impression of sand-laden CO_2 jets in the southern ice cap during the spring. (Image credit: Ron Miller, Arizona State University.)

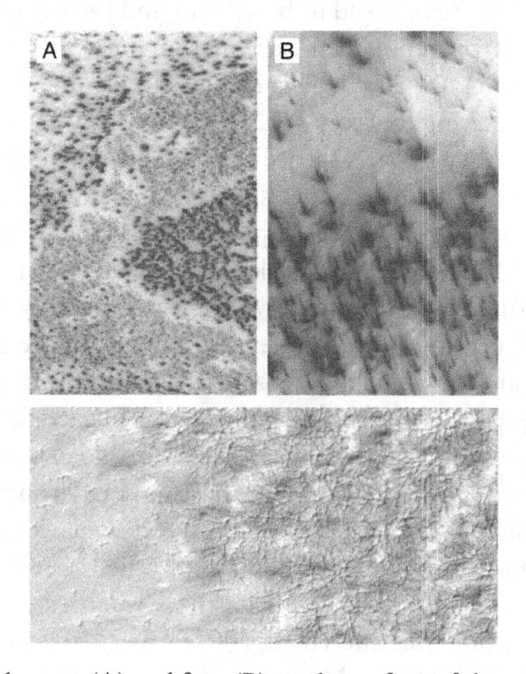

Fig. 7.20. Dark spots (A) and fans (B) on the surface of the south polar cap. Spiders (lower image) on top of the residual polar cap, after the seasonal carbon-dioxide ice slab has disappeared. Next spring, these will likely mark the sites of vents when the CO_2 ice cap returns. Each image is ~3 km wide. (Image credit: NASA/JPL/MSSS.)

hundred meters into the air and around their vent holes through the ice cap surface. These jets are believed to form the dark spot or fan-like features that form on the southern ice cap surface each spring, and which gradually evolve into spider-like features as the summer progresses (Fig. 7.20). The spots are typically 15–50 m wide and a hundred or so meters apart, and usually form in similar locations on an annual basis. The combination of dust-laden water ice would superficially suggest that there is the potential for water production by the heating of subsurface dust within the polar ice caps during the summer. The overlying ice would protect the water from freezing or sublimation into the cold, dry atmosphere, in a similar manner to the protection offered to water produced in the subsurface drainage system of Dry Valley glaciers.

Glaciers were once present and extensively covered the large, high volcanoes of the Tharsis region at low latitudes. Head and Marchant (2003) compare and contrast geomorphological features on the western Arsia Mons with those of the Dry Valleys to demonstrate this point. Glaciers may still be present on the upper slopes of the volcanoes, for example on Olympus Mons above 7 km, but they are believed to be ancient and the water ice is covered by a dust layer which prevents their sublimation into the cold, dry martian atmosphere (Forget et al., 2006). This being the case, the debris is also likely to insulate the underlying ice from solar radiation, and hence the water ice is likely to be cold and dry. Hence, the ice caps are the only likely location where solar heating of debris could produce subsurface melting. Unfortunately, the energy balance conditions in the ice caps are not conducive to the production of liquid water (MEPAG, 2006). Calculations suggest that the maximum possible temperature that subsurface water ice could reach is −43 °C, even with favorable aspect and latitude. Hence, present-day ice cap surface environments are deemed to be solidly frozen and incapable of producing liquid water as a consequence of the solar heating of subsurface debris. By contrast, abundant geomorphological evidence exists for volcano–ice interactions, which have produced meltwater even in these cold environments on Mars in the past (Head and Wilson, 2007). Further, the close association of relatively recent glacial deposits and gullies, interpreted to be the result of local fluvial erosion, has been cited as evidence for localized microenvironments of meltwater generation (Marchant and Head, 2007). Hence, it is possible that liquid water is present periodically on Mars, and there remains the possibility of life in these transient aquatic systems should suitable innoculi be present.

Conclusions

Glaciers of the Dry Valleys and the martian polar ice caps have many similarities, in that both occur in a cold, dry atmosphere, where the main means of ablation is sublimation, and both receive significant annual deposits of aeolian dust which impacts on the surface geomorphology and energy balance. However, a significant difference is that the energy balance of glaciers in the Dry Valleys is such that liquid water is produced just below the surface, so forming a near-surface hydrological system that supports active and diverse microbial communities. By contrast, the surface energy balance of polar ice caps on Mars is currently insufficient to produce subsurface melt, even given optimistic energy balance conditions. Glaciers at low latitude on high volcanic cones are blanketed by dust which prevents sublimation of water vapour, and also prevents subsurface melting. We conclude that glaciers and ice caps on Mars do not routinely contain liquid water, and are therefore unlikely to contain carbon-based microbial life even should suitable innoculi be present. Microbial life is present in aquatic environments beneath Dry Valley glaciers (Mikucki et al., 2004; Mikucki and Priscu, 2007). Glacier beds on Mars at present would require significant sources of geothermal heating to be at the pressure melting point. No compelling case has been made to date for the presence of liquid water beneath martian glaciers, except in instances of ice–volcano interactions and in local microenvironments, and hence the basis of a compelling case for the presence of carbon-based microbial life beneath martian glaciers has yet to be made.

References

Andrews-Hanna, J. C., Zuber, M. T., and Banerdt, W. B. (2008). The Borealis basin and the origin of the martian crustal dichotomy, *Nature*, **453**, 1212–1215.

Bagshaw, E., Tranter, M., Fountain, A., et al. (2007). The biogeochemical evolution of cryoconite holes on Canada Glacier, Taylor Valley, Antarctica. *Journal of Geophysical Research*, **112**, G04S35, doi: 10.1029/2007JG000442.

Bardgett, R. D. and Walker, L. R. (2004). Impact of coloniser plant species on the development of decomposer microbial communities following deglaciation. *Soil Biology and Biochemistry*, **36**, 555–559.

Byrne, S. and Ingersoll, A. P. (2003). A sublimation model for Martian south polar ice features. *Science*, **299**, 1051–1053.

Cowan, D. A. and Tow, L. A. (2004). Endangered Antarctic environments. *Annual Review of Microbiology*, **58**, 649–690.

Doran, P. T., McKay, C. P., Clow, G. D., et al. (2002). Valley floor climate observations from the McMurdo Dry Valleys, Antarctica, 1986–2000. *Journal of Geophysical Research, Atmospheres*, **107**, D24.

Foreman, C. M., Wolf, C. F., and Priscu, J. C. (2004). Impact of episodic warming events on the physical, chemical and biological relationships of lakes in the McMurdo Dry Valleys, Antarctica, *Aquatic Geochemistry*, **10**(3), 239–268.

Foreman, C. M., Sattler, B., Mikucki, J. A., Porazinska, D. L., and Priscu, J. C. (2007). Metabolic activity and diversity of cryoconites in the Taylor Valley, Antarctica. *Journal of Geophysical Research*, **112**, G04S32, doi: 10.1029/2006JG000358.

Forget, F., Haberle, R. M., Montmessin, F., Levrard, B., and Head, J. W. (2006). Formation of glaciers on Mars by atmospheric precipitation at high obliquity. *Science*, **311**, 368–371.

Fountain, A. G., Dana, G. L., Lewis, K. J., Vaughn, B. H., and McKnight, D. (1998). Glaciers of the McMurdo Dry Valleys, Southern Victoria Land, Antarctica. In *Ecosystem Dynamics in a Polar Desert: The McMurdo Dry Valleys, Antarctica*, ed. J. C. Priscu. Antarctic Research Series 72. Washington, D.C.: American Geophysical Union, pp. 65–75.

Fountain, A. G., Lyons, W. B., Burkins, M. B., et al. (1999). Physical controls on the Taylor Valley ecosystem, Antarctica. *BioScience*, **49**(12), 961–971.

Fountain, A. G., Tranter, M., Nylen, T. H., Lewis, K. J., and Mueller, D. R. (2004). Evolution of cryoconite holes and their contribution to meltwater runoff from glaciers in the McMurdo Dry Valleys, Antarctica. *Journal of Glaciology*, **50**(168), 35–45.

Fountain, A. G., Nylen, T. H., Tranter, M., and Bagshaw, E. (2008). Temporal variations in physical and chemical features of cryoconite holes on Canada Glacier, McMurdo Dry Valleys, Antarctica. *Journal of Geophysical Research*, **113**, G01S92, doi: 10.1029/2007JG000430.

Goetz, W. and 19 others (2005). Indication of drier periods on Mars from the chemistry and mineralogy of atmospheric dust. *Nature*, **436**, 62–65.

Gribbon, P. W. (1979). Cryoconite holes on Sermikaysak, West Greenland. *Journal of Glaciology*, **22**, 177–181.

Head, J. W. (2007). The geology of Mars: new insights and outstanding questions. In *The Geology of Mars: Evidence from Earth-Based Analogs*, ed. M. Chapman. Cambridge, UK: Cambridge University Press, pp. 1–46.

Head, J. W. and Marchant, D. R. (2003). Cold-based mountain glaciers on Mars: Western Arsia Mons. *Geology*, **31**, 641–644.

Head, J. W. and Wilson, L. (2007). Heat transfer in volcano–ice interactions on Mars: synthesis of environments and implications for processes and landforms. *Annals of Glaciology*, **45**, 1–13.

Head, J. W., Mustard, J. F., Kreslavsky, M. A., Milliken, R. E., and Marchant, D. R. (2003). Recent ice ages on Mars. *Nature*, **426**, 797–802.

Hodson, A., Tranter, M., Anesio, A. M., et al. (2008). Glacial ecosystems. *Ecological Monographs*, **78**, 41–68.

Kennedy, A. D. (1993). Water as a limiting factor in the Antarctic terrestrial environment: a biogeographical synthesis. *Arctic and Alpine Research*, **25**, 308–315.

Kieffer, H. H., Christensen, P. R., and Titus, T. T. (2006). CO_2 jets formed by sublimation beneath translucent slab ice in Mars' seasonal south polar ice cap. *Nature*, **442**, 793–796.

Killawee, J. A., Fairchild, I. J., Tison, J. -L., Janssens, L., and Lorrain, R. (1998). Segregation of solutes and gases in experimental freezing of dilute solutions: implications for natural glacial systems. *Geochimica et Cosmochimica Acta*, **62**, 3637–3655.

Lancaster, N. (2002). Flux of eolian sediment in the McMurdo Dry Valleys, Antarctica: a preliminary assessment. *Arctic Antarctic and Alpine Research*, **34**, 318–323.

Langevin, Y., Poulet, F., Bibring, J.-P., Schmitt, B., Doute, S., and Gondet, B. (2005). Summer evolution of the north polar cap of Mars as observed by OMEGA/Mars Express. *Science*, **307**, 1581–1584.

Lewis, K. J., Fountain, A. G., and Dana, G. L. (1999). How important is terminus cliff melt?: a study of the Canada Glacier terminus, Taylor Valley, Antarctica. *Global and Planetary Change*, **22**(1–4), 105–115.

Lyons, W. B., Welch, K. A., Fountain, A. G., et al. (2003). Surface glaciochemistry of Taylor Valley, southern Victoria Land, Antarctica and its relationship to stream chemistry. *Hydrological Processes*, **17**(1), 115–130.

Marinova, M. M., Aharonson, O., and Asphaug, E. (2008). Mega-impact formation of Mars hemispheric dichotomy. *Nature*, **453**, 1216–1219.

McKay, C. H., Anderson, D. T., Pollard, W. H., et al. (2005). Polar lakes, streams and springs as analogs for hydrological cycles on Mars. In *Water on Mars and Life, ed.* T. Tokano. Advances in Astrobiology and Biogeophysics. Berlin: Springer-Verlag, pp. 219–233.

Melosh, J. H. (2008). Did an impact blast away half of the martian crust? *Nature Geosciences*, **1**, 412–414.

MEPAG (Mars Exploration Program Analysis Group) (2006). Findings of the Mars Special Regions Science Analysis Group. *Astrobiology*, **6**, 677–732.

Mikucki, J. A. and Priscu, J. C. (2007). Bacterial diversity associated with Blood Falls, a subglacial outflow from the Taylor Glacier, Antarctica. *Applied and Environmental Microbiology*, **73**, 4029–4039.

Mikucki, J. A., Foreman, C. H., Sattler, B., Lyons, W. B., and Priscu, J. A. (2004). Geomicrobiology of Blood Falls: an iron rich saline discharge at the terminus of Taylor Glacier, Antarctica. *Aquatic Chemistry*, **10**, 199–220.

Mindl, B., Anesio, A. M., Meirer, K., et al. (2007). Factors influencing bacterial dynamics along a transect from supraglacial runoff to proglacial lakes of a high Arctic glacier. *FEMS Microbiol Ecology*, **59**, 307–317.

Mueller, D. R. and Pollard, W. H. (2004). Gradient analysis of cryoconite ecosystems from two polar glaciers. *Polar Biology*, **27**(2), 66–74.

Muir, H. (2008). Phoenix lander uncovers ice on Mars. *New Scientist*. www.newscientist.com/article/dn14143.

Nimmo, F., Hart, S. D., Korycansky, D. G., and Agnor, C. B. (2008). Implications of an impact origin for the martian hemispheric dichotomy. *Nature*, **453**, 1220–1223.

Nkem, J. N., Wall, D. H., Virginia, R. A., et al. (2006). Wind dispersal of soil invertebrates in the McMurdo Dry Valleys, Antarctica. *Polar Biology*, **29**(4), 346–352.

Nylen, T. H., Fountain, A. G., and Doran, P. T. (2004). Climatology of katabatic winds in the McMurdo dry valleys, southern Victoria Land, Antarctica. *Journal of Geophysical Research, Atmospheres*, **109**, D3.

Paerl, H. W. and Priscu, J. C. (1998). Microbial phototrophic, heterotrophic, and diazotrophic activities associated with aggregates in the permanent ice cover of Lake Bonney, Antarctica. *Microbial Ecology*, **36**, 221–230.

Plaut, J. J. and 23 others (2007). Subsurface radar sounding of the south polar layered deposits of Mars. *Science*, **316**, 92–95.

Porazinska, D. L., Fountain, A. G., Nylen, T. H., et al. (2004). The biodiversity and biogeochemistry of cryoconite holes from McMurdo Dry Valley glaciers, Antarctica. *Arctic Antarctic and Alpine Research*, **36**(1), 84–91.

Priscu, J. C. and Christner, B. C. (2002). Earth's icy biosphere. In *Microbial Biodiversity and Microprospecting*, ed. A. Bull. Washington, D.C.: ASM Press, pp. 130–145.

Priscu, J. C., Tulaczyk, S., Studinger, M., et al. (2008). Antarctic subglacial water: origin, evolution and microbial ecology. In *Polar Lakes and Rivers: Limnology of Arctic and Antarctic Aquatic Ecosystems*, ed. W. Vincent and J. Laybourn-Parry. Oxford, UK: Oxford University Press, pp. 119–135.

Rothschild, L. J. and Manicelli, R. L. (2001). Life in extreme environments. *Nature*, **409**, 1092–1101.

Säwström, C., Granéli, W., Laybourn-Parry, J., and Anesio, A. M. (2007). High viral infection rates in Antarctic and Arctic bacterioplankton. *Environmental Microbiology*, **9**, 250–255.

Schorghofer, N. (2007). Dynamics of ice ages on Mars. *Nature*, **449**, 192–194.

Stibal, M., Tranter, M., Benning, L. G., and Řehák, J. (2008). Microbial primary production on an Arctic glacier is insignificant in comparison to allochthonous organic carbon input. *Environmental Microbiology*, doi: 10.1111/j.1462–2920.2008.01620.

Titus, T. N., Kieffer, H. H., and Christensen, P. R. (2003). Exposed water ice discovered near the south pole of Mars. *Science*, **299**, 1048–1051.

Tranter, M., Fountain, A. G., Fritsen, C. H., et al. (2004). Extreme hydrochemical conditions in natural microcosms entombed within Antarctic ice. *Hydrological Processes*, **18**(2), 379–387.

Tranter, M., Fountain, A. G., Lyons, W. B., Nylen, T. H., and Welch, K. A. (2005). The chemical composition of runoff from Canada Glacier, Antarctica: implications for glacier hydrology during a cool summer. *Annals of Glaciology*, **40**, 15–19.

Wharton, R. A., McKay, C. P., Simmons, G. M., and Parker, B. C. (1985). Cryoconite holes on glaciers. *BioScience*, **35**(8), 499–503.

8

Factors promoting microbial diversity in the McMurdo Dry Valleys, Antarctica

CRISTINA TAKACS-VESBACH, LYDIA ZEGLIN, J.E. BARRETT, MICHAEL N. GOOSEFF, AND JOHN C. PRISCU

The McMurdo Dry Valleys (MDV) comprise a mosaic of habitats at scales ranging from micrometers to the kilometer scale. The varied landscape of the valleys, combined with strong physical and chemical gradients within and across the terrestrial and aquatic habitats, yields an ecosystem dominated by microbes that is both complex and diverse (Gordon et al., 2000; Smith et al., 2006; Mikucki and Priscu, 2007). The cold desert environment is analogous to icy conditions found on other icy worlds. For example, the low organic carbon, cold, arid soils of the MDV are similar to Mars' present-day terrestrial environment and the glaciers and ice-covered lakes of the MDV are comparable to conditions that existed on Mars in the past (Priscu et al., 1998; Wynn-Williams and Edwards, 2000; McKay et al., 2005). If there are extant or extinct life forms on Mars, they likely experience similar physical constraints and environmental challenges as do microbial communities in the MDV. Therefore, the MDV provide a unique earthly setting to gain insight into the diversity, adaptation, and function of life on other icy worlds. Here we describe the ecological processes and conditions that contribute to the microbial diversity observed in the MDV and relate these to potential life on Mars.

The McMurdo Dry Valley ecosystem

The MDV include a variety of unique habitats that are connected physically, chemically, and energetically (Fig. 8.1). Solar radiation and wind are the underlying forces that determine the existence and distribution of biota throughout the valleys (Dana et al., 1998; Nkem et al., 2006). Incoming solar

Life in Antarctic Deserts and Other Cold Dry Environments: Astrobiological Analogs, ed. Peter T. Doran, W. Berry Lyons and Diane M. McKnight. Published by Cambridge University Press.

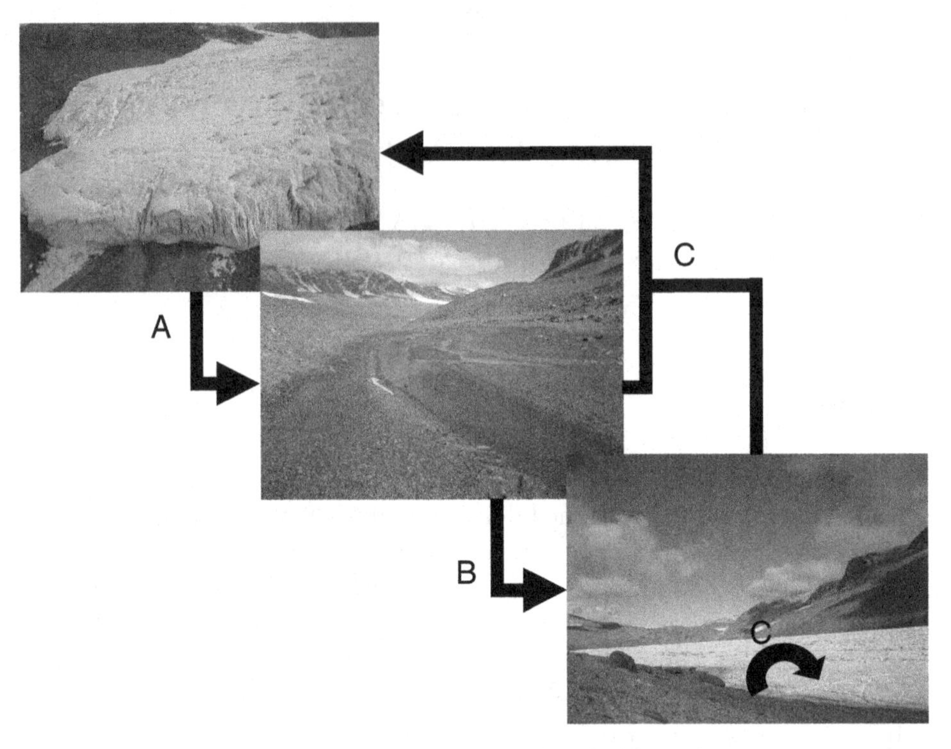

Fig. 8.1. The MDV include a variety of unique habitats that are connected physically, chemically, and energetically. Water and thermal energy are transported down the valleys from the glaciers to streams (from A) and the streams to lakes (B), whereas sediment is transferred up-valley (C, or onto lake ice), and promotes more melt and creates additional microbial habitats in the ice.

radiation is important not only for photosynthesis, but also for melting the glaciers that provide the majority of liquid water to the area (Dana et al., 1998). During 6 to 10 weeks each summer, glacial melt results in ephemeral streams, which harbor photosynthetic cyanobacterial mats and associated heterotrophic microorganisms.

Streams are significant sources of organic carbon and inorganic nutrients to the lakes they feed because of photosynthetic mat productivity and in-stream weathering of rock and sediment, and they are also a critical source of thermal input for the lakes (Vincent, 1988; Lyons et al., 1998; Spigel and Priscu, 1998; Takacs et al., 2001). Water in streams can reach temperatures of 15 °C (Conovitz et al., 1998; Cozzetto et al., 2006), making them important heat sources to the lakes (McKay et al., 2005). Liquid water columns exist within the lakes because of the annual thermal input of the ephemeral streams that essentially accumulate and transfer solar radiation from a wider surface area than the area of the lake. With the exception of Don Juan Pond in the Wright Valley, which has salinities exceeding 600‰ due to high levels of calcium chloride (Horowitz et al., 1972), MDV lakes have permanent ice covers that serve to insulate the water column and prevent wind-driven mixing (Spigel and Priscu, 1998).

Lake ice is formed in a conveyor-belt fashion (Fritsen et al., 1998). Water freezes at the ice–water interface and is ablated at the ice–air interface (Adams et al., 1998). Less than 10% of the incident solar radiation penetrates the lake ice covers and most of this is in the blue-green part of the spectrum (Lizotte and Priscu, 1992; Howard-Williams et al., 1998). Ultraviolet radiation also penetrates the ice covers and can inhibit photosynthesis in certain lakes (Vincent et al., 1998). Wind-blown sediments deposited on the ice melts downward during summer and forms liquid water inclusions that support phototrophic and heterotrophic activity (Priscu et al., 1998). This novel microbial consortium exploits a habitat in an environment that would appear to be inhospitable for life (Paerl and Priscu, 1998).

The lakes are the sites of the highest biological activity in the MDV owing to the accumulation of nutrients, organic matter, and thermal input, and year-round constant temperatures that are above freezing (Hopkins et al., 2006). Strong physical and chemical gradients exist in the water columns, which provide a spectrum of potential habitats for the lake communities (Spigel and Priscu, 1998). In contrast are the inhospitable soils that comprise a majority of the MDV surface area (Burkins et al., 2001). Although the MDV receive up to 10 cm of moisture annually, precipitation is in the form of snow, which tends to sublimate or blow away before it melts and percolates the soils (Campbell et al., 1998). Owing to a lack of moisture and highly variable temperatures, the soils have relatively low biomass per unit area compared with lakes and streambeds of the MDV. The exception is the soils adjacent to streams and lakes which have higher biomass and activity than the surrounding soils (Hopkins et al., 2006; Ayres et al., 2007; Zeglin et al., 2009).

Wind plays an important role in the MDV in that it physically redistributes organic matter and soils (Lancaster, 2002; Nkem et al., 2006). For example, benthic microbial mats in the lakes lift off the bottom of the lakes, are conveyed through the ice cover and released at the surface where wind transports them throughout the valleys (Parker et al., 1982). Similarly, stream microbial mats may be dispersed by wind to new locations in the valleys. Additionally, winds are often strong enough to transport soil particles, which may be important in redistributing the microorganisms associated with them (Nkem et al., 2006). For example, sediment may be deposited by the wind on the surface of glaciers (Porazinska et al., 2004). Because the deposits have a higher albedo than the surrounding ice, they absorb more energy and create vertical melt holes called cryoconites. The debris seeds the cryoconites with microorganisms and provides minute amounts of nutrients that enable biological activity. Biomass synthesized in cryoconites is eventually transferred down-valley during major melt events that flush melt holes out of the glacier and complete the cycle of biomass and energy transfer in the valleys (Wharton et al., 1985; Christner et al., 2003; Fountain et al., 2004; Porazinska et al., 2004; Foreman et al., 2007).

The McMurdo Dry Valley food web

Biodiversity in the MDV is composed entirely of microinvertebrates including prokaryotes (Bacteria and Archaea), protists (phytoplankton and heterotrophic nanoflagellates), and invertebrates (Arthropoda, Tardigrada, Nematoda, and Rotifera). There are no extant vertebrate communities in the MDV (Adams et al., 2006). For many years, the food web of MDV lakes was characterized as relatively simple compared with more temperate ecosystems. This perception arose because the lakes were discovered to contain no crustaceans or fish. Biotic interactions such as predation were considered to be insignificant because protozoa were believed to be relatively sparse and the existence of viruses was unknown (Vincent and James, 1996). It is now clear that biotic interactions are much more complex than originally anticipated (Laybourn-Parry, 1997; Takacs and Priscu, 1998; Roberts et al., 2000) and, at least in the aquatic environment, viral lysis and predation have been detected (Kepner et al., 1998; Lisle and Priscu, 2004; Säwström et al., 2008).

Nematodes, which feed on bacteria and algae, are at the top of the food chain and are represented by only five species (Adams et al., 2006). Tardigrades and rotifers are found in many of the MDV habitats, but have received little attention compared with the other groups of MDV inhabitants. Mosses are found in wetted areas of the terrestrial environment, for example along streams, near glaciers, and in depressions where snow may collect (Schwarz et al., 1992; Dale et al., 1999; Adams et al., 2006). Algae occur throughout the MDV, and as many as 40 unique diatoms have been identified in streams (Esposito et al., 2006). Protists and fungi are most abundant in the aquatic and the terrestrial habitats, respectively (Baublis et al., 1991; Kepner et al., 1999; Adams et al., 2006; Connell et al., 2006; Fell et al., 2006).

Diversity is much higher among the prokaryotes, or members of the Bacteria and Archaea. Bacteria exist throughout the valleys in the soils, lakes, streams, and even the lake and glacier ice (Adams et al., 2006) and include the cyanobacteria, which are the major primary producers in many of these systems. Bacteria have been studied in the MDV since the 1960s, but it is only recently that we have begun to appreciate the diversity of these organisms (Gordon et al., 2000; Cowan and Tow, 2004; Stackebrandt et al., 2004; Barrett et al., 2006; Sattley and Madigan, 2006; Smith et al., 2006).

A major obstacle in studying microorganisms in nature has been methodology. Early studies of prokaryotic diversity relied to a large extent on enrichment cultures and isolating individual species. The first reports of MDV microbial diversity were of bacteria isolated from the lakes; only a handful of species were isolated, leading to a perception that bacteria were not very diverse or important in the MDV (Cameron et al., 1970; Horowitz et al., 1972; Friedmann, 1993). We now know that not all microorganisms are amenable to culturing, making this method highly selective and unrepresentative of actual biodiversity. It is estimated that less than 1% of organisms in the environment are directly cultivable using current techniques (Amann et al., 1995) and the cultivable organisms rarely represent the ecologically dominant members of the community (Ward, 1998). This is true of all environments, and perhaps more so for extreme environments where the isolation of organisms may require creative approaches to approximate the conditions under which life can thrive (Glausiusz, 2007; Stingl et al., 2008). Presently, the most effective means we have of assessing microbial diversity is by surveying community genomic DNA for phylogenetically important genes, such as the gene that codes for the small subunit of the ribosome (SSU rRNA genes or 16S and 18S rDNA in prokaryotes and eukaryotes, respectively), which allows us to make evolutionary comparisons among all domains of life.

Most of the recent prokaryotic microbial diversity assessments in the MDV have focused on the Bacteria and few reports of Archaea exist. The current understanding is that Archaea are not widespread in the MDV, but limited to anoxic or high salinity regions that harbor either methanogens or halophilic Archaea, respectively (Franzmann et al., 1997; Bowman et al., 2000; Brambilla et al., 2001). However, a more complete picture of microbial diversity in the MDV will develop as the region is sampled more extensively and alternative approaches such as environmental metagenomics are applied to the MDV. Given that the MDV environment is analogous to past and present conditions on Mars (McKay et al., 2005), we can use these observations to make inferences about the potential for life and its diversity on Mars.

Microbial biodiversity of MDV habitats

Streams

Streams within the MDV are ephemeral, with seasonal flow driven by solar radiation, geomorphology, and hyporheic exchange (Conovitz et al., 1998; Cozzetto et al., 2006). The annual flow period is on average six weeks, though discharge is highly variable on annual and diel timescales. Streams in the MDV support high levels of biomass compared with the surrounding soils (McKnight et al., 1998), though much of this biomass is thought to be inactive or senescing (Vincent and Howard-Williams, 1989). Biodiversity studies of MDV streams have focused on the algal and cyanobacterial communities and no reports of nonphotosynthetic bacteria exist. Liquid water is present during the austral summer and autumn, which interacts with the sediments and soil, where C and N can be stored from year to year to supply organisms with essential nutrients and a medium for biosynthesis to occur (Runkel et al., 1998). Cyanobacterial mats and associated biota (Vincent, 1988; McKnight et al., 1999) in and along streams are centers of biological activity, including photosynthesis and nitrogen fixation by cyanobacteria, which is transferred to other organisms within the mat.

Given the paucity of grazing organisms in the MDV and the strong physical forcing of streams (e.g., highly variable flow, seasonal solar radiation, freeze–thaw), abiotic factors may be expected to structure microbial assemblages rather than biotic factors such as grazing and competition (Poff and Ward, 1989; Hogg et al., 2006). Geomorphic template and flow regime have been shown to control the distribution of photosynthetic mats in many MDV streams (McKnight et al., 1998, 1999; Esposito et al., 2006). For example, flow in streams with steep gradients is either too high or does not provide an adequate substrate for mat colonization. Mid-gradient reaches with stone pavement substrate support the highest densities of mat biomass. Dominant cyanobacterial species in stream mats vary spatially with flow regime. Mats dominated by *Nostoc* sp. generally line the edges of streams, while mats dominated by *Oscillatoria* sp. and *Phormidium* sp. occur in the middle of streams where flow rates are the highest (McKnight et al., 1998).

Cyanobacterial mats in MDV streams can survive freeze/thaw cycles and extended periods in a lyophilized state during the austral winter. They revive quickly upon exposure to liquid water during the austral summer (Vincent and Howard-Williams, 1986) and begin photosynthesizing in as little as 10 to 20 minutes after rehydration (Vincent and Howard-Williams, 1986). Furthermore, one week after restoring flow to a relict channel that had not received significant flow in approximately 20 years, the cyanobacterial mats became more abundant than in the surrounding streams due to increased solute concentrations in the reactivated stream channel. Cryptobiotic preservation of cyanobacterial mats enables MDV streams to respond rapidly to climatic and geomorphological change, similar to other arid zone stream ecosystems (McKnight et al., 2007).

Lakes

The lakes of the MDV have varying degrees of chemical stratification and, in general, all contain nutrient-rich deep water covered by a relatively nutrient-poor trophogenic zone (Priscu, 1995; Dore and Priscu, 2001). The permanent ice cover of these lakes prevents wind-driven mixing, which coupled with low advective stream input allows vertical chemical and biological gradients to develop and persist (vertical mixing is at the molecular level throughout the water column, Spigel and Priscu, 1998). These physicochemical gradients provide a spectrum of diverse habitats that are represented by the unique microbial communities that inhabit them (Karr et al., 2003; Glatz et al., 2006; Sattley and Madigan, 2006; Clocksin et al., 2007).

The Taylor Valley lakes have received the most attention among lakes of the MDV. Lake Bonney is the most stratified of the lakes, followed by Lake Fryxell, and then Lake Hoare. Lake levels have fluctuated over the past 6000 years, effectively concentrating the solutes in the bottom waters of the lakes by evaporation and sublimation. An exception is Lake Hoare, which is believed to have dried out completely approximately 1200 years ago, whereas Lakes Fryxell and Bonney are believed to have persisted as ice-free brine ponds during the periods of low lake levels (Matsubaya et al., 1979; Lyons et al., 2004). Bottom-water dissolved inorganic carbon ages measured by ^{14}C dating are approximately 1200 years in Lake Hoare and 8000 years in Lake Bonney (Doran et al., 1994). The ^{14}C age of the fulvic acid fraction of Lake Fryxell's bottom waters has been determined to be approximately 3000 years (Aiken et al., 1996).

The lakes were discovered during Scott's exploration of the area in the early 1900s (Scott, 1905), but were not extensively studied until the 1960s when the first quantitative physical, chemical, and biological measurements were made (Armitage and House, 1962; Angino et al., 1964; Goldman, 1964; Goldman et al., 1967). However, a majority of these studies focused on the phytoplankton (Koob and Leister, 1972; Parker et al., 1977; Vincent, 1981; Lizotte et al., 1996). Of the studies that were concerned with the bacterioplankton, biomass and production were largely underestimated. Essentially all of these studies were conducted during a time when bacterial colony plate counts were the standard technique of determining bacterial abundance (Koob and Leister, 1972; Mikell et al., 1986).

The complexity and diversity of the lake microbial communities have only begun to be appreciated in the past decade (Laybourn-Parry, 1997; Ward and Priscu, 1997; Kepner et al., 1998; Takacs and Priscu, 1998; Priscu et al., 1999; Voytek et al., 1999; Roberts et al., 2000, 2004; Lisle and Priscu, 2004;

Mikucki and Priscu, 2007). Bacterial abundance and activity measurements revealed that bacteria play a significant role in carbon transformations in the lakes, and biomass losses indicated that grazing must be occurring (Takacs and Priscu, 1998). Seminal studies describing micro-eukaryote and viral populations expanded our understanding of the MDV lake food webs and provided evidence of an active microbial loop (Kepner et al., 1998; Priscu et al., 1999; Roberts and Laybourn-Parry, 1999; Roberts et al., 2000; Lisle and Priscu, 2004).

An extensive study of nitrogen dynamics in Lake Bonney revealed the importance of nitrification and denitrification in this system (Priscu, 1995; Priscu et al., 1996, 2002, 2008; Ward and Priscu, 1997; Voytek et al., 1999; Lee et al., 2004). A major focus of this research was on the high levels of nitrous oxide (>70 000% air saturation) and apparent lack of denitrification in the suboxic waters of the east lobe of the lake. A thermodynamic study of the biogeochemistry in MDV lakes indicates that present-day conditions in the east lobe of Lake Bonney are not favorable for contemporary formation of the nitrous oxide found there and that it is likely a vestige of some previous microbial community that existed when the redox conditions of the lake were more reducing (Lee et al., 2004). Another study based on metabolic genes and stable isotopes concluded that the gradients in nitrous oxide present in the lake today are relicts of ancient biogeochemical transformations, which owing to the 50 000-year mixing time of the lake, have yet to diffuse to equilibrium (Priscu et al., 2008).

Immunofluorescent probes specific for two denitrifying isolates from Lake Bonney revealed that potential denitrifiers were scarce in the deep waters of the east lobe, relative to the west lobe. Denitrification was proposed to be absent in the east lobe because of inhibition by salts, temperature, or possibly some other chemical limiting or inhibiting factor (Ward and Priscu, 1997). Denitrification could be induced in east lobe lake water that was diluted with ion-free water and incubated at temperatures higher than those *in situ* (Ward et al., 2005).

The importance of nitrifying bacteria in MDV lakes was determined by the development of a polymerase chain reaction (PCR) assay for the detection of ammonium oxidizers (Voytek and Ward, 1995; Voytek et al., 1999; Priscu et al., 2008). Ammonium oxidizers of the β-Proteobacteria were present in all lakes tested, whereas members of the γ-Proteobacteria, which is represented primarily by marine organisms, were detected in the saline Lakes Fryxell and Bonney, but not in Hoare (Voytek et al., 1999). Ammonium oxidizers were most abundant above the chemocline of the lakes and were associated with lower concentrations of ammonium and higher concentrations of nitrate and nitrite.

The microbial diversity of Lake Bonney's water column and ice cover has been assayed by SSU rRNA gene analysis (Gordon et al., 2000; Glatz et al., 2006). The sequences detected were unique among the depths sampled and many of the depths had little overlap in their communities. Diversity is low compared with other extreme aquatic habitats (Hughes et al., 2001), especially at the deepest depth surveyed, 25 m of Lake Bonney's east lobe, where temperatures and salinity are most limiting compared with the other depths sampled (Glatz et al., 2006).

A number of the phylotypes detected in Lake Bonney SSU rRNA gene analyses have now been cultured using a high throughput culturing method by amending filter sterilized lake water with additional carbon and nitrogen and inoculating media with lake water diluted to extinction (Stingl et al., 2008). This approach ensures that organisms that are most numerous in the water column serve as inoculum by diluting out any zymogenous or transient organisms and enriches the native organisms by providing media that closely resemble *in situ* conditions. Based on SSU rRNA gene analysis, 18 unique psychrotolerant bacterial cultures were isolated. In comparison with many standard culture studies, a majority of the isolates were not closely related to previously described bacterial species, and some of the cultures represent deeply branching members of their respective phyla (Stingl et al., 2008).

Lake Fryxell, a meromictic lake, is the most productive of the Taylor Valley lakes per unit volume (Takacs and Priscu, 1998). In a survey of the diversity of *pufM* genes, which encode for a pigment-binding protein in the photosynthetic reaction center of all purple phototrophic bacteria, no purple sulfur bacteria were detected; only purple nonsulfur bacteria were found (Karr et al., 2003), which was surprising given that Lake Fryxell has an active sulfur cycle. However, purple nonsulfur bacteria are metabolically more diverse than purple sulfur bacteria and are capable of chemoorganotrophic and chemolithotrophic growth in the dark. Thus, purple nonsulfur bacteria may maintain a competitive edge owing to the low light levels in Lake Fryxell (Karr et al., 2003).

Of the four groups of purple nonsulfur bacteria found in Lake Fryxell, only one was >90% identical in its *pufM* gene to previously cultured organisms. The four groups were detected in separate areas of the water column, distributed presumably according to oxygen, light, and sulfur concentrations indicating niche differentiation among the organisms represented by the sequences. One representative each from two of the four Lake Fryxell purple nonsulfur bacteria groups have been isolated. One of the isolates, *Rhodoferax antarcticus* (strain Fryx1), is the first member of the purple nonsulfur bacteria isolated to contain gas vesicles (Karr et al., 2003). A separate strain of

R. antarcticus has also been isolated from antarctic mat material (Madigan et al., 2000), but the mat strain does not have gas vesicles, indicating that strain Fryx1 is uniquely adapted to planktonic life within Lake Fryxell (Jung et al., 2004).

Many nonphotosynthetic sulfur bacteria have been identified in Lake Fryxell (Karr et al., 2005; Sattley and Madigan, 2006). The gradients of sulfide and oxygen provide an ideal environment for microorganisms capable of sulfur oxidation. Three strains related to *Thiobacillus thioparus* have been isolated from Lake Fryxell and this population of sulfur-oxidizing chemolithotrophs is most abundant at the oxycline of the lake where they are positioned perfectly to take advantage of ideal gradients of sulfide and oxygen (Sattley and Madigan, 2006). Sulfide is produced below the oxycline by sulfate-reducing bacteria belonging to seven groups of sequences representative of the *Desulfovibrio*, *Desulfonema*, *Desulfobulbus*, *Desulfomusa* (δ-Proteobacteria), the *Desulfotomaculum* (Firmicutes), and two deeply branching groups, one of which could not be phylogenetically constrained. Nucleotide sequences from all of these sulfate-reducing bacteria were only distantly (63–77% identical) related to their closest relatives. Three of the Lake Fryxell sulfate-reducing bacterial groups were found throughout the water column, whereas the remainder were distinctly stratified (Sattley and Madigan, 2006).

Two major phyla of the Archaea, the Euryarchaeota and the Crenarchaeota, have been detected in Lake Fryxell (Karr et al., 2006). Methanogenic archaea were found only in the sediments where acetogens also have been cultured (Sattley and Madigan, 2007). Lake Fryxell acetogens and methanogens may be involved in syntrophic interspecies hydrogen transfer, which is known to occur in laboratory cultures, sludges, and the sediments of freshwater lakes (Jones and Simon, 1985). Acetogenesis, which can be endergonic under fermentative conditions, is thermodynamically favored by the utilization of H_2 by methanogens. Because sulfate is low in Lake Fryxell sediments, sulfate reducers are incapable of competing with methanogens for H_2. The presence of methanotrophic archaea suggests that methane may serve as an electron donor for sulfate reducers above the sediments of Lake Fryxell (Karr et al., 2006). The Crenarcheota sequences detected in Lake Fryxell group with Marine Benthic Group C and are presumed to represent oxygenic sulfide and thiosulfate oxidizing organisms (Karr et al., 2006).

The microbial community of Lake Hoare, which is the least stratified and productive lake of the Taylor Valley, has received little attention. Based on culture dependent studies, a diverse group of chemoorganotrophic bacteria belonging to the Actinobacteria and the α-, β-, and γ-Proteobacteria have been isolated (Clocksin et al., 2007). Although all of the cultures could grow at subzero temperatures, only one of the strains was a true psychrophile incapable of growth at temperatures greater than 20 °C. The lack of psychrophily by these and other MDV microbial isolates (Madigan et al., 2000; Jung et al., 2004; Sattley and Madigan, 2006; Stingl et al., 2008) relative to marine sediment isolates, which are more frequently psychrotolerant, suggests that MDV lake microorganisms have not fully adapted to the cold owing to the young age of the lakes (Clocksin et al., 2007).

A diverse assemblage of bacterial SSU rRNA gene sequences were detected in the sediments that exist within the ice cover of Lake Bonney (Gordon et al., 2000). Sediments are deposited on lake ice in the MDV by wind and represent a dynamic equilibrium between downward movement because of melt balanced with upward movement by sublimation of the ice cover (Priscu et al., 1998). Liquid water has been shown to exist in association with these sediments for up to 150 days during summer (Fritsen et al., 1998). The microbial assemblage is composed of cyanobacterial and heterotrophic bacteria that represent a complete consortium capable of organic carbon production and nutrient cycling (Paerl and Priscu, 1998; Priscu et al., 1998).

Microorganisms have also been discovered in the ice cover of Lake Vida (Fritsen and Priscu, 1998; Mosier et al., 2007), which lies in the Victoria Valley of the MDV. This lake is unique among MDV lakes because it is composed of a 19-m ice cover that has sealed off a cold (−10 °C) highly saline (~245‰) brine from the surrounding environment for approximately 2800 years (Doran et al., 2003). Polymerase chain reaction screening indicated that bacterial and eukaryal SSU rRNA genes were present at all depths of the ice cover tested, but no archaeal genes were detected (Mosier et al., 2007). Distinct communities were found at all depths; the nucleotide sequences detected in the upper ice are representative of organisms from soil and freshwater habitats and are believed to originate primarily from aeolian deposition and glacial meltwater, which is deposited on the ice surface annually. Although the brine was not sampled, evidence exists that the bottom depths of the ice result from the liquid water below the ice cover (Doran et al., 2003). Bacterial SSU rRNA gene sequences detected in this portion of the ice cover were most closely related to sequences that were predominantly (65% identical) of marine origin (Mosier et al., 2007).

Microbial mats exist in the littoral zone of the lakes and harbor a diversity of microorganisms dominated by cyanobacteria. Lake Fryxell microbial mats were estimated to contain up to 20 taxa using Fourier-transform infrared spectroscopy (Tindall et al., 2000). In a separate study, at least 133 unique SSU rRNA gene sequences of bacteria and archaea were identified in microbial mats of Lake Fryxell (Brambilla et al., 2001). Recent phylogenetic analyses using genomic tools have shown that the cyanobacterial taxa in the MDV lake mats not only have higher diversity than expected but also revealed a number of putative endemic taxa (Brambilla et al., 2001; Van Trappen et al., 2002, 2004; Taton et al., 2003; Shivaji et al., 2005).

Soils

Soils represent one of the harshest habitats within the MDV landscape, being extremely dry, low in organic matter and other nutrients, and subject to freeze–thaw action on a regular basis (Campbell et al., 1998). Though these conditions do not appear to be conducive to life, certain MDV soils support life in surprising abundance and diversity (Cowan et al., 2002). Various reports have been published recently detailing the bacterial diversity of MDV soils, which is orders of magnitude higher than eukaryotic diversity (which may be universally true for soils). For example, although only eight soil bacteria were cultured, 61 unique SSU rRNA gene types were identified among four soil types with varying pH, moisture, and level of human disturbance, and from largely separated sites around the MDV and the McMurdo area (Aislabie et al., 2006). Diversity among the four sites was greatest in the wetter soils, but was nearly two times lower than found in similarly arid, but temperate soils from Arizona (Dunbar et al., 1999). Bacterial communities in three of the sites were largely composed of a few dominant sequence types, which is contrary to the more even species distribution of most studied soils (Borneman and Triplett, 1997; Torsvik et al., 2002; Lozupone and Knight, 2007) and was interpreted to indicate that the dominant sequence types represent organisms that are highly adapted to *in situ* conditions (Aislabie et al., 2006). Communities from all four sites were composed of members from the Bacteroidetes, Actinobacteria, Proteobacteria, Deinococcus-Thermus, Acidobacteria, and Cyanobacteria bacterial divisions, but the dominant sequence types differed among the sites, indicating that there may be regional variations in the soil microbial communities. Additionally, many of the SSU rRNA gene sequences were only distantly related to previously reported sequences (88–95% identical), indicating that there may be an endemic MDV soil microbial community. In a similar study aimed at characterizing the bacterial communities associated with five different MDV soils, similarly diverse communities were discovered (Aislabie et al., 2008). The major distinction among the soils was that the less suitable habitats (i.e., lower nutrient and water availability) were composed of novel 16S rRNA gene sequence types assigned to uncultured and unrecognized genera, whereas the community in an ornithogenic soil (high nutrients due to avian inputs) was dominated by well-characterized spore-forming bacteria.

A much higher level of diversity has been found in the wetted soils adjacent to MDV streams and lakes (parafluvial sediments), which have higher levels of organic matter and moisture relative to the surrounding soils (Hopkins et al., 2006). In a soil sample from the edge of the Onyx River (Wright Valley), 376 unique 16S rRNA gene sequence types were detected, but the survey was estimated to be only 53% complete (Zeglin, 2008). This level of diversity is similar to what has been found in the most diverse soils reported, such as from the Amazon rain forest and arid temperate regions (Borneman and Triplett, 1997; Fierer and Jackson, 2006). Overall, the Onyx River parafluvial sediment communities were similar to MDV soils at the phylum level, but contained a higher level of genus-level diversity.

The near-surface permafrost of the MDV, which occurs 10 to 25 cm below the surface soils of the valleys, contains only 10^3 to 10^5 bacterial cells g^{-1}. Although bacterial abundances are low relative to other antarctic habitats and arctic permafrost, these organisms have been shown to remain viable following melt despite the low ambient temperatures (Gilichinsky, 2007). Bacteria, algae, and fungi have been found in the MDV permafrost. For the bacteria there was little difference between the phyla found above ground in the surface soils and throughout the permafrost. Owing to the vast range of soil exposure times and age of permafrost, antarctic permafrost likely contains some of the oldest individuals on Earth. Martian permafrost may harbor similarly ancient organisms.

The Beacon Valley sandstones of the MDV can retain water internally for several days after wetting by snowmelt and provide a refugium for terrestrial life in the MDV (de la Torre et al., 2003). Cryptoendolithic microorganisms inhabit the near-surface pore spaces of light transparent exposed rocks and form complex communities that are classified as dominated either by lichens or cyanobacteria (Friedmann et al., 1988). A molecular comparison of the two cryptoendolithic community types detected 51 unique SSU rRNA gene sequences belonging to the Bacteria and Eukarya and no Archaea were detected. Only 12% of the nucleotide sequences were greater than 98% identical to known sequences. Despite the diversity of nucleotide sequences detected, the dominant communities were composed of only a few functional groups. The lichen community was dominated by an ascomycete (Fungi) and a green alga. The cyanobacterial community was dominated by a sequence that was most closely related to cyanobacterial sequences found throughout the MDV and grouped with *Phormidium* spp. Additionally, a *Deinococcus*-like sequence (only 90% identical) and an aerobic anoxygenic phototrophic α-Proteobacteria sequence represented 26% and 31% of the sequences detected, respectively (de la Torre et al., 2003). Whereas the lichen community was composed of Bacteria and Eukarya, no eukaryote sequences were found in the cyanobacterial community.

Cryoconites

Cryoconites are cylindrical melt holes that form on the ablation zone of glaciers worldwide (Wharton et al., 1985). As sediments are deposited on glaciers by wind, they are preferentially heated during the summer owing to their low albedo compared with the surrounding ice. A melt hole forms that contains the deposited sediments at the base and, during summer, liquid water. Antarctic cryoconites are unique in that they contain a thicker ice lid (up to 36 cm) than cryoconite holes from elsewhere. In the MDV, where cryoconite holes cover 4–6% of glacier surfaces, they range from 5 cm to 145 cm in diameter and 4 cm to 56 cm in depth (Fountain et al., 2004). With the formation of water, hole deepening is accelerated by albedo feedback. The water has a lower albedo than the surrounding ice, and the formation of water melts more ice (Wharton et al., 1985). Finally, as the hole becomes deeper, growth slows because less solar radiation reaches the sediment and a steady state is achieved with respect to depth as the amount of melt equals the ablation rate of the glacier surface. Liquid water may persist in the holes for 1–3 months during the summer and individual cryoconites can persist for several years to a decade. Eventually the holes are lost by either hydraulic washout or glacial calving (Fountain et al., 2004) distributing their contents to the surrounding stream, soil, and lake habitats.

Cryoconites have been recognized as biological hot spots in glacial ecosystems for decades (Wharton et al., 1985). In the MDV, these holes harbor a diverse and active community, and the presence of microorganisms appears to increase melt and cryoconite hole development. For example, experiments have shown that the depth of live holes is 10% deeper compared with killed controls. Wind-deposited sediments seed the holes with microorganisms and minerals that are leached into the meltwater and provide nutrients for growth (Wharton et al., 1985). Photosynthesis and nitrogen fixation by cyanobacteria and algae in the summer provide organic carbon and nitrogen and support a food web that is similar in composition to other MDV habitats. Bacteria, fungi, ciliates, rotifers, tardigrades, and nematodes have been detected by amplification of SSU rRNA genes (Christner et al., 2003). Similarities of the detected gene sequences indicated that the cryoconite consortia originated from the surrounding stream, sediment, and lake habitats

(Foreman et al., 2007). Enumeration of rotifers by microscopy suggests that they are more abundant than would be expected of an inactive community that was composed of cells simply deposited by wind (Porazinska et al., 2004). Additionally, bacterial biomass, abundance, and activity were always highest in the sediment of cryoconite holes, rather than the overlying water. The composition of the microbial consortia varied between the sediment and ice layer and among holes from different glaciers (Foreman et al., 2007). These differences point to the importance of sediments as sources of nutrients to the consortium. Based on the density of cryoconite holes on MDV glaciers, it has been suggested that in warm years, flushing could provide significant nutrient and organic carbon to the MDV (Foreman et al., 2007). Chemistry of cryoconite hole water relative to glacier ice suggests that this would influence the chemistry of the glacial melt-stream (Fountain et al., 2004; Barrett et al., 2007).

Another source of runoff from glaciers is the discharge that originates subglacially, such as Blood Falls on the terminus of the Taylor Glacier. The source of Blood Falls is believed to be remnant ancient water from a Pliocence warming event that resulted in ocean level rises that filled the Taylor Valley. When the water receded, a marine sea remained at the Taylor Glacier terminus that was covered during a subsequent expansion of the glacier's terminus (Mikucki, 2005). Blood Falls waters are saline, iron rich, and contain viable microorganisms that are active and diverse. Microbial community activity and diversity is low, but is comparable to similar polar systems, though the phylogenetic diversity of the organisms detected was broad (Mikucki and Priscu, 2007).

Ecological factors promoting diversity in the MDV

Although the microbial diversity of all known MDV habitats has been described (Vishniac, 1993; Voytek and Ward, 1995; Gordon et al., 2000; Brambilla et al., 2001; Van Trappen et al., 2002; de la Torre et al., 2003; Karr et al., 2003, 2005; Stackebrandt et al., 2004; Adams et al., 2006; Barrett et al., 2006; Glatz et al., 2006; Aislabie et al., 2008; Mikucki and Priscu, 2007; Sattley and Madigan, 2007; Stingl et al., 2008; Zeglin, 2008), very few of these reports explain the factors that contribute to the observed distribution and diversity (but see de la Torre et al., 2003; Karr et al., 2003, 2005, 2006; Glatz et al., 2006; Mikucki and Priscu, 2007; Aislabie et al., 2008) and a broad understanding of the factors structuring MDV microbial diversity is lacking. Here we synthesize published MDV bacterial 16 S rRNA gene sequence data as a basis for inferring the processes that contribute to the microbial diversity patterns that have been observed in the MDV.

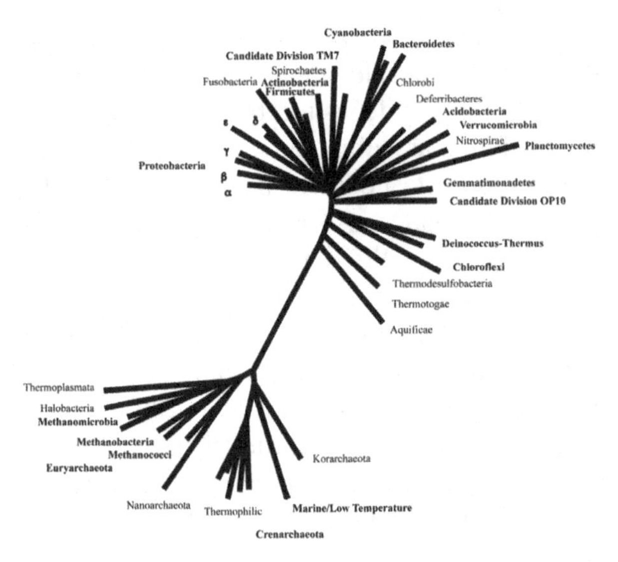

Fig. 8.2. Phylogenetic tree of the Bacteria and Archaea. Phyla for which sequences from the MDV have been retrieved are shown in bold. In addition to the phyla shown, many MDV sequences group with uncultured environmental sequences of unknown phylogeny or candidate divisions, such as TM7 and OP10 (shown) or OD1 and BD.

Surveys of SSU rRNA genes in MDV habitats indicate a broad taxonomic diversity among the bacteria, sequences group with over 15 bacterial phyla (Fig. 8.2) and revealed more diversity than observed among MDV eukaryotes (Brambilla et al., 2001; Smith et al., 2006; Stingl et al., 2008). A simple species richness estimator, such as the proportion of unique sequences (or operational taxonomic units, OTUs) detected per sample, enables comparison of 16S rRNA gene sequence richness from MDV habitats (Fig. 8.3). A similar level of richness was detected among the soil samples (Aislabie et al., 2006, 2008), a Lake Fryxell sample (C. Takacs-Vesbach et al., unpublished data), and from wetted sediments surrounding MDV streams (C. Takacs-Vesbach et al., unpublished data). A recent meta-analysis of over 21 000 16S rRNA gene sequences from diverse habitats worldwide concluded that microbial diversity is highest in sediments (Lozupone and Knight, 2007). Richness in the MDV is lowest in the icy habitats, such as cryoconites (Christner et al., 2003), lake ice (Gordon et al., 2000), and the outflow of Blood Falls (Mikucki and Priscu, 2007). Unfortunately, no SSU rRNA gene sequence data for MDV streams have been published (only algal diversity has been reported, e.g., Esposito et al., 2006).

Here we compare MDV bacterial richness with two similar studies conducted on Crater Lake, Oregon (Urbach et al., 2001, 2007) and soil from the Amazon rain forest (Borneman and Triplett, 1997; Urbach et al., 2001, 2007), and a microbial inventory of Yellowstone thermal features (C. Takacs-Vesbach,

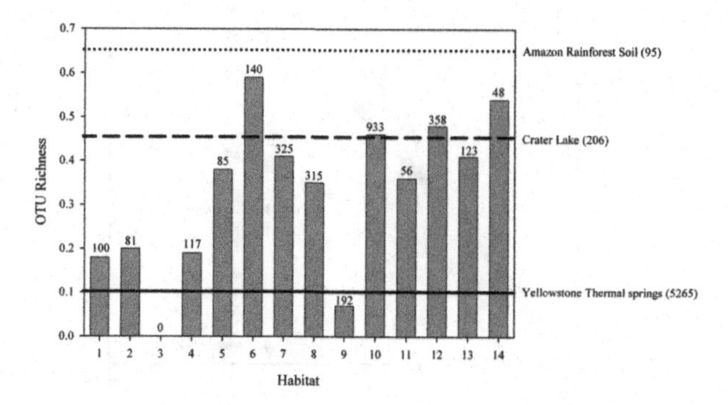

Fig. 8.3. Richness (proportion of unique bacterial sequences in a survey) of MDV habitats. 1, Cryoconite (Christner, 2003); 2, Blood Falls (Mikucki and Priscu, 2007); 3, streams (no data); 4, east lobe Lake Bonney (Glatz et al., 2006); 5, west lobe Lake Bonney (Glatz et al., 2006); 6, Lake Fryxell (C. Takacs-Vesbach, unpublished); 7, Lake Fryxell mat (Brambilla et al., 2001, Taton et al., 2003); 8, Lake Vida (Mosier et al., 2007); 9, Lake Bonney ice (Gordon et al., 2000); 10, wetted sediments along stream margins (C. Takacs-Vesbach, unpublished); 11, soils from Miers Valley (Aislabie et al., 2007); 12, soils from Wright Valley (Aislabie et al., 2006); 13, lichen dominated cryptoendolith (de la Torre et al., 2003); 14, cyanobacterial dominated cryptoendolith (de la Torre et al., 2003). Numbers above each bar represent the number of sequences surveyed in each study. Horizontal lines represent richness detected in a survey of 100 Yellowstone thermal springs (Takacs-Vesbcah, unpublished), Crater Lake (Urbach et al., 2001, 2007), and an Amazon rain forest soil (Borneman and Triplett, 1997). Numbers in parentheses represent the total number of sequences screened in each study.

unpublished data). Crater Lake and the Amazonian soil were reported to have remarkably high levels of microbial diversity and MDV richness is lower, but higher than found in a survey of Yellowstone thermal features, which tend to be dominated by many closely related sequences from only a few genera (Hall et al., 2008; Takacs-Vesbach et al., 2008). Richness was used to compare all of these studies because robust diversity estimators were not reported in all of the studies, some of the studies did not completely survey their sequence libraries, and some reports did not include the number of sequences screened. However, comparison of the more complete data sets revealed that the amount of diversity detected among the sites was similar to the richness results, except that there were several individual sites – for example, sediments along the margins of Taylor Valley streams – which are expected to be as diverse as the Amazon soil sample, based on rarefaction, a more robust diversity estimator (Fig. 8.4).

Three characteristics of MDV diversity that are common among the existing bacterial SSU rRNA gene sequence data are (1) the nature of the abundance of unique SSU rRNA gene sequences, (2) the phylogenetic breadth of the communities, and (3) the novelty of the sequences. In almost all reports of

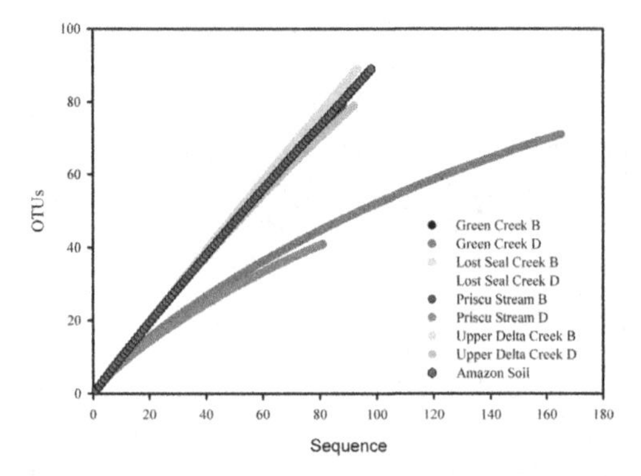

Fig. 8.4. Rarefaction curves of bacterial 16S rRNA gene libraries from margin sediments of MDV streams and an Amazon rain forest soil (Borneman and Triplett, 1997).

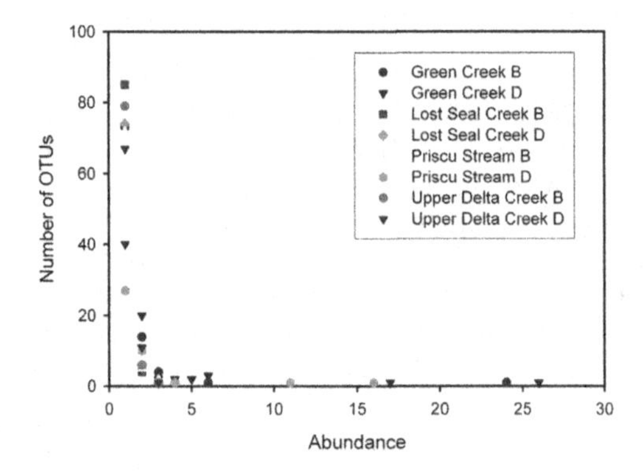

Fig. 8.5. Species abundance curves of the bacterial diversity detected in the wetted sediments found along the margins of Taylor Valley streams (C. Takacs-Vesbach, unpublished). A large proportion of the sequences were unique (<97% identical) and only a few sequences comprised more than five representatives. This type of diversity pattern, described as log series, is common among all the MDV habitats surveyed.

MDV microbial diversity, the SSU rRNA gene libraries are composed of few dominant sequences and a larger proportion of unique sequences that are represented by only one or a few clones (e.g., Brambilla et al., 2001; Mikucki and Priscu, 2007; Aislabie et al., 2008, and see Fig. 8.5). Communities defined by this type of richness abundance pattern are common among all domains of life and similar patterns of diversity have been found by a number of studies in varied MDV and other cold habitats (reviewed in Nemergut et al., 2005). The log series model of species diversity is used to describe communities with a high abundance of rare species and an ecological explanation for this type of diversity distribution is that one or a few factors determine the community assemblage (May, 1975; Magurran, 2004; Mikucki and Priscu, 2007).

Table 8.1. *Phylogenetic breadth of the bacterial sequences detected in MDV habitats*

Habitat	Average branch length[a]
Cryoconite	0.51
Blood Falls	0.36
East lobe Lake Bonney	0.41
Lake Bonney isolates	0.34
West lobe Lake Bonney	0.59
Lake Fryxell	0.64
Lake Fryxell isolates	0.30
Lake Fryxell mat	0.38
Lake Hoare isolates	0.48
Lake Vida	0.39
Lake Bonney Iice	0.48
Stream margins	0.37
Soils	0.38
Cryptoendolith	0.43

Notes: [a] Average branch length calculated as the sum of branch lengths for distance trees constructed for each habitat divided by the number of sequences in the tree (Lozupone and Knight, 2007).

Although species richness may be low in many of the MDV habitats, the phylogenetic breadth of all the communities is notable. A broad phylogenetic swath of microbial diversity is represented in the MDV (e.g., see Fig. 8.2), including sequences and isolates from the Chloroflexi, Deinococcus, Gemmatimonadetes, Planctomyces, Verrucomicrobia, Acidobacteria, Bacteroidetes, Cyanobacteria, Firmicutes and Actinobacteria, and Proteobacteria. Comparison of the average branch length of simple distance trees (Lozupone and Knight, 2007) constructed with the individual data sets reveals that phylogenetic breadth of the sites is similar (Table 8.1), but highest in Lake Fryxell and west Lake Bonney. Finally, many of the sequences detected in the MDV are less than 97% identical (some are as low as 60%) to other known sequences, indicating that MDV microorganisms are novel and may represent species unique to Antarctica.

Principal coordinate analysis (Lozupone and Knight, 2005) of MDV bacterial SSU rRNA gene sequences revealed that a number of the data sets cluster together based on habitat (Fig. 8.6). The bacterial isolates from Lakes Fryxell, Hoare, and Bonney all cluster with the ice habitats. The similarity between uncultured and cultured sea ice microorganisms has been noted before and related to the high success rate of culturing these bacteria (Lozupone and Knight, 2007). Central to the isolate and ice cluster are the soil sequences, which suggests an aeolian origin for the sequences that comprise these

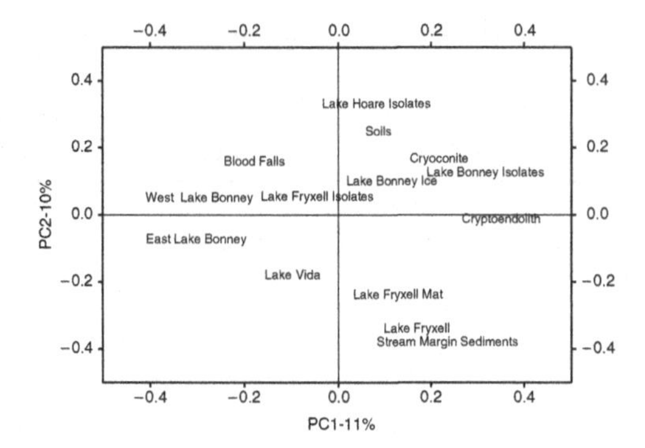

Fig. 8.6. Ordination of bacterial 16S rRNA gene sequences by MDV habitats: cryoconite (Christner et al., 2003); Blood Falls (Mikucki and Priscu, 2007); Lake Bonney (Glatz et al., 2006); Lake Fryxell, Lake Fryxell mat (Brambilla et al., 2001; Taton et al., 2003); Lake Vida (Mosier et al., 2007); Lake Bonney ice (Gordon et al., 2000); soils from Miers and Wright Valleys (Aislabie et al., 2006, 2007); cryptoendolith (de la Torre et al., 2003). Data were aligned using the Greengenes aligner (DeSantis et al., 2006) and analyzed by principal coordinate analysis in Unifrac (Lozupone and Knight, 2005).

habitats. A clear separation among the lake habitats exists, though the two lobes of Lake Bonney are clustered together, as are the Lake Fryxell mat and water column sequences. The sequences from MDV wetted sediments also grouped with Lake Fryxell. In the absence of any obvious similarities between these two habitats, the only explanation that can be provided is that the two habitats share some unidentified environmental factor that correlates with the first axis of the analysis. The cryptoendolithic community was the most divergent data set among the habitats.

That the microbial diversity of MDV habitats shares many patterns of community composition is suggestive that a unifying theme of community assemblage must be at work. However, unlike plant and animal diversity, a comprehensive framework for explaining microbial diversity is largely lacking (Torsvik et al., 2002; Horner-Devine et al., 2003, 2007; Lozupone and Knight, 2007). Because microorganisms and macroorganisms share the same fundamental biology, much of what we already know about plants and animals should also apply to microorganisms. Here we apply ecological explanations developed for eukaryotes to develop a hypothesis that explains the processes promoting microbial diversity in the MDV.

Spatial heterogeneity

In general, species diversity is often higher in communities that inhabit more complex and heterogeneous physical environments (Zhou et al., 2002;

Torsvik et al., 2002; Lozupone and Knight, 2007), because greater spatial heterogeneity contributes to more potential niches (Roth, 1976; Huston, 1979). Spatial heterogeneity occurs over a number of scales in the MDV (Virginia and Wall, 1999). At the regional scale, geographical features (elevation, slopes, and aspect) contribute to differences in microclimate (Dana et al., 1998), while geological variation in soil parent material, mineralogy, and surface exposure age influence nutrient availability in soils, streams, and lakes (Lyons et al., 2004; Barrett et al., 2007; Bate et al., 2007). At meter to kilometer scales, variation in microbial communities (composition and diversity) is influenced by the dominant landforms: glaciers, ephemeral streams, sediments and soils, and lakes. Within each habitat type, we also see differences in biodiversity at finer scales (e.g., micrometer to meter). For example, a synoptic sampling of the soils surrounding the three major lakes of the Taylor Valley, Lakes Fryxell, Hoare, and Bonney, showed that there was a decrease in organic carbon and nitrogen content from east to west, with highest organic matter concentrations in the Fryxell basin and the lowest found in the Bonney basin (Fritsen et al., 2000). These differences in organic matter content broadly correspond to differences in till provenance, surface exposure age, and habitat suitability (Virginia and Wall, 1999; Barrett et al., 2004, 2007; Bate et al., 2007). Spatial heterogeneity in Taylor Valley soils is also hypothesized to affect the distribution of microbial biomass across the valley landscape, which is supported by an assessment of microbial enzymatic potential conducted on MDV soils that revealed similar basin level differences (Zeglin et al., 2009).

On the local scale, within the soils and lakes, gradients of major biotic determinants such as moisture, nutrients, and salt promote diversity because they offer a spectrum of habitat suitability. The water columns of the MDV lakes are hydraulically stable because the permanent ice covers and strong chemical gradients prevent wind-driven mixing (Spigel and Priscu, 1998). The resultant water layers provide a spectrum of microbial habitats. Distinct communities are found at different depths within Lake Fryxell (Fig. 8.7). Likewise, gradients of moisture, salinity, and nutrients in the soils adjacent to streams and lakes are accompanied by changes in community composition within the top 10 cm of sediments (Fig. 8.8).

On the microscale, microhabitats within aquatic systems and on individual grains of soil result from the patchiness found at this scale. For example, in aquatic systems, patches of higher nutrients are hypothesized to occur near a single phytoplankter because of cell exudates (Bell and Mitchell, 1972; Krembs et al., 1998). By clustering around such cells, chemotactic microorganisms capable of attachment have access to elevated nutrient levels

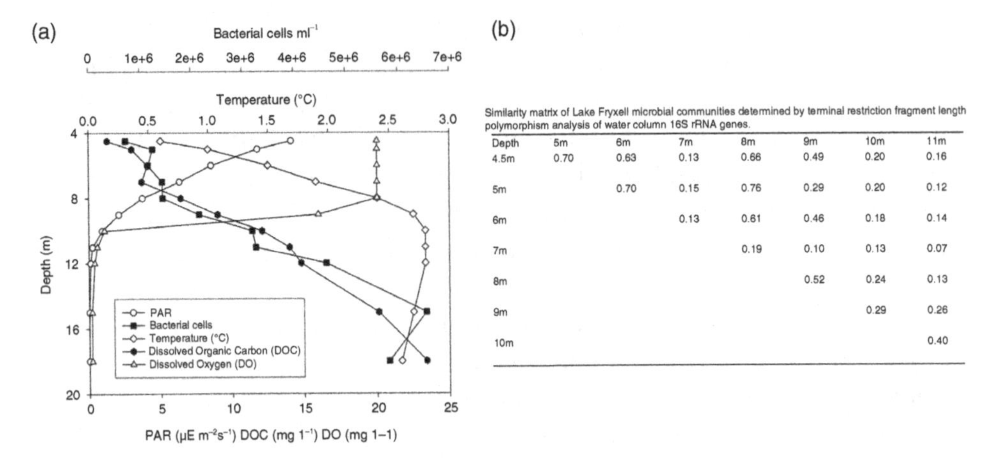

Similarity matrix of Lake Fryxell microbial communities determined by terminal restriction fragment length polymorphism analysis of water column 16S rRNA genes.

Depth	5m	6m	7m	8m	9m	10m	11m
4.5m	0.70	0.63	0.13	0.66	0.49	0.20	0.16
5m		0.70	0.15	0.76	0.29	0.20	0.12
6m			0.13	0.61	0.46	0.18	0.14
7m				0.19	0.10	0.13	0.07
8m					0.52	0.24	0.13
9m						0.29	0.26
10m							0.40

Fig. 8.7. Gradients in Lake Fryxell. (a) Physical, chemical, and biological profiles of Lake Fryxell water column. (b) similarity matrix of Lake Fryxell bacterial communities determined by terminal restriction fragment length polymorphisms of community 16S rRNA genes (C. Takacs-Vesbach, unpublished).

associated with biogeochemical transformations within the aggregate. MDV lakes are high in particulate organic matter where microbial abundance and biogeochemical cycling are elevated (Lisle and Priscu, 2004).

Sediments and soils are heterogeneous on the microscale as well. Steep chemical and physical gradients that affect nutrient concentration, redox conditions, and water availability create a large number of microscale habitats (Young and Ritz, 2000; Torsvik et al., 2002). These microhabitats provide distinct niches, spatially isolating microbial populations and result in a more complex microbial community (Zhou et al., 2002). In the MDV, sediments are biological hot spots in melt holes such as cryoconites and within the lake ice where bacteria and algae are concentrated (Priscu et al., 1998). The sediments contribute to the existence of liquid water and provide increased nutrient concentrations that support a higher abundance of life compared with the surrounding ice. The microbial diversity associated with ice sediments is similar to soil microbial diversity (Gordon et al., 2000) as illustrated in Fig. 8.5, which suggests that ice microbial communities are transplanted by wind, but are capable of growth when liquid water exists (Foreman et al., 2007). Given the wide range of habitat diversity present in the MDV and that microbial activity is detected throughout these habitats, it is not surprising that such a wide range of bacterial sequences have been detected (Voytek and Ward, 1995; Gordon et al., 2000; Madigan et al., 2000; Brambilla et al., 2001; Cowan et al., 2002; de la Torre et al., 2003; Karr et al., 2003, 2005, 2006; Taton et al., 2003; Barrett et al., 2004; Lisle and Priscu, 2004;

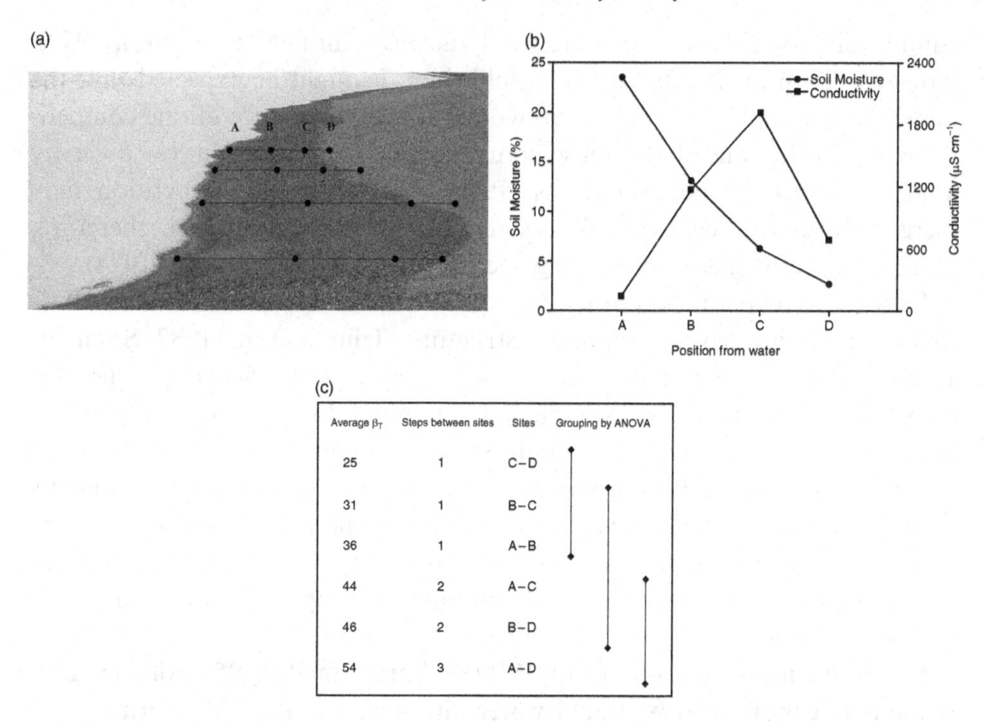

Average β_T	Steps between sites	Sites	Grouping by ANOVA
25	1	C–D	
31	1	B–C	
36	1	A–B	
44	2	A–C	
46	2	B–D	
54	3	A–D	

Fig. 8.8. Spatial heterogeneity of wetted sediments at the hydrologic margins of MDV lakes and streams. (a) Example of the wetted sediments that are clearly visible along MDV lakes and streams in summer. (b) A gradient of moisture and conductivity along the hydrologic margin habitat provides a spectrum of habitable niches. (c) Species turnover (beta diversity) along the wetted margins of Green Creek and Lake Fryxell. Community turnover metric β_T is lowest at 1-step distances between samples; turnover between the two outer sites (C–D) is lowest; the 3-step turnover is highest. Community turnover was determined by using the equation: β_T = (change in number of sequences between two sites/2) * average community richness.

Porazinska et al., 2004; Adams et al., 2006; Aislabie et al., 2006, 2008; Glatz et al., 2006; Smith et al., 2006; Mikucki and Priscu, 2007; Mosier et al., 2007; Stingl et al., 2008).

Competition and predation

Interspecific competition is widely believed to be a key determinant of biodiversity for plants and animals, although contradictory hypotheses exist. The competitive exclusion principle holds that if one competitor is more effective than another, it will eventually overcome the other species (Hardin, 1960). A major criticism of competitive exclusion is the assumption of equilibrium conditions, which are rarely achieved in nature (Huston, 1979). However, even in the absence of equilibrium, it might be expected that the immediate outcome of competition would be the dominance of one competitor over another and that intense competition should result in low diversity among competing species (Levins, 1968). Alternatively, competition may increase diversity because it favors reduced niche breadth and, therefore, more potential niches or coexisting species (May, 1974; Schoener, 1974).

Laboratory experiments of model aquatic microbial communities indicate that competition drives community structure (Tilman et al., 1982; Sommer, 1985), but for more complex communities such as those found in soils, this does not seem to hold. For example, the microbial diversity of heterogeneous versus homogeneous soils under limiting nutrient conditions was investigated by comparing bacterial communities in dry and saturated soils containing low organic carbon. Bacterial diversity was lower in the more homogeneous soil indicating that competitive exclusion was occurring because the species evenness was lower in the saturated soil community compared with the unsaturated soils (Zhou et al., 2002).

Soil moisture is very low in the MDV. Approximately 95% of the landscape is not moistened by liquid water although soil may be intermittently wetted for short periods by snowfall (Campbell et al., 1998). Adjacent to more consistent water sources such as streams and lakes, the moisture content of valley soils/sediments is only 15%. Decreased competition because of spatial separation owing to low moisture can contribute to the diversity observed in MDV soils (Cowan et al., 2002; Connell et al., 2006; Smith et al., 2006). Furthermore, the low growth rates that have been reported from all MDV habitats (Kepner et al., 1998; Priscu et al., 1998; Takacs and Priscu, 1998; Roberts et al., 2000; Foreman et al., 2007; Mikucki and Priscu, 2007; Zeglin et al., 2009) can decrease competitive interactions. An alternative hypothesis of competitive equilibrium focuses on the rate of competitive displacement (Huston, 1979) by comparing the rates at which differences in competitive abilities are expressed. If all competing species are increasing at a very low rate, the better competitor will be slow to predominate. Under fluctuating environmental conditions, competitive equilibrium is not reached and diversity is strongly influenced by the rate of competitive displacement (Huston, 1979). Under conditions where the rate of competitive displacement is low, species evenness will be high and individual species will occur at low density, which is consistent with the patterns of microbial diversity observed in the MDV (see Fig. 8.4). Therefore, spatial and temporal separation of microbial interactions in the MDV likely contributes to the apparent high microbial diversity.

Predation can be important in determining biodiversity in communities with many interactions among members of the food web. Predation is believed to increase biodiversity by maintaining low abundances of individual species, thus reducing competition (Paine, 1966). This is especially true in the tropics where there are many biotic interactions among populations, and has been shown experimentally and theoretically in the seminal works of Connell (1961) and Gause (1934). Historically, predation was not believed to be important in the MDV ecosystem (Vincent and James, 1996; Hogg et al., 2006). However, it is now clear that potential grazers and viruses are abundant in the lakes and bacterial abundances are known to decrease by up to 88% during the summer (Takacs and Priscu, 1998; Lisle and Priscu, 2004). These biotic interactions potentially affect microbial abundance and diversity in the MDV lakes, but less is known about the soils. Nematodes are the major grazers in MDV soils (Treonis et al., 1999), but no relationship between nematode abundances and microbial diversity is apparent in the studies that examined multitrophic relationships (Barrett et al., 2006; Niederberger et al., 2008).

Temporal variability and disturbance

An intermediate level of disturbance maintains spatial and temporal heterogeneity in habitat type and resource availability, and is thus correlated with a high level of diversity (Connell, 1978). This phenomenon is evident in stream ecosystems, which are dynamic by nature (Townsend et al., 1997). In MDV streams, the highest levels of diatom diversity and endemicity are found at moderate flow regime harshness (i.e., length and intensity of stream discharge events, Esposito et al., 2006). Further, some views hold that harsh conditions, periodic disturbances, or resource pulses can promote species coexistence by preventing competitive equilibrium and allowing greater niche overlap (Hutchinson, 1961), while others maintain that harsh, unpredictable conditions drive the evolution of temporally staggered competitive strategies that maintain diversity (Chesson and Huntly, 1997; Chesson et al., 2004). A recent evaluation of factors governing soil invertebrate distribution across the MDV landscape invoked the presence of harshness-tolerant strategies as maintaining regional diversity (Bamforth et al., 2005). Another recent study showed that meltwater pulses occurring during an extreme climate event in the MDV stimulated the abundance and activity of a subordinate soil invertebrate species, and thus may contribute to the maintenance of soil biodiversity (Barrett et al., 2008).

If the lack of competitive displacement explains the persistence of MDV diversity, then how is microbial diversity generated in the MDV? Among naturally occurring populations, there are two opposing forces governing community composition: genetic divergence among populations, and disturbances that purge poorly adapted subpopulations (Thompson et al., 2005). Divergence occurs through genetic mutations that can accumulate and persist because they are neutral or favored by selection (Ward et al., 2008). Although bacterial growth is limited to only a few weeks each year in many MDV habitats, which would limit speciation, the potential exists for heightened mutagenesis due to increased UV at the poles, as has been observed in antarctic mosses (Skotnicki et al., 2000). Despite the broad phylogenetic breadth of microbial sequences detected in the MDV, a striking feature is that within MDV phyla, there are many unique, but closely related 16S rRNA gene sequences (Christner et al., 2003; Karr et al., 2003, 2005; Aislabie et al., 2006, 2008; Glatz et al., 2006). In a survey of bacterial diversity associated with soils collected along the margin of the Onyx River (Wright Valley), 16% of the unique sequences (defined as <97% identical) grouped within a single genus of the Bacteroidetes (*Niastella*, Family Flexibacter) (Zeglin, 2008). This variation may represent accumulated neutral mutations (Thompson et al., 2005) or the persistence of closely related species adapted to different niches (Lozupone and Knight, 2007; Ward et al., 2008).

Disturbances capable of purging genetic variation occur over various timescales in the MDV: for example, the highly variable stream hydrograph in summer (hours to days, Conovitz et al., 1998), the diphotic cycle (months), and cooling and warming trends that may last on the order of decades to millennia (Doran et al., 2002). However, these disturbances presumably are rarely catastrophic and because decomposition rates are so low (Takacs and Priscu, 1998; Priscu et al., 1999; Takacs et al., 2001; Zeglin et al., 2009), any diversity that is generated in the MDV can persist in the environment and be detected by modern molecular diversity assessments.

Conclusions

Although a considerable amount of work has focused on microbial activity and function of the MDV ecosystem, our understanding of microbial biodiversity in this unique environment remains incomplete. However, there are notable similarities regarding the bacterial diversity among MDV habitats that are also characteristic of other cold environments (Nemergut et al., 2005). The varied landscape of the MDV provides a spectrum of habitats that support a broad bacterial richness. We hypothesize that similar to other microbial populations worldwide, diversity is generated by the accumulation of mutations (neutral and adaptive) that persist in the environment because disturbance is rarely catastrophic for these slow-growing populations. At the same time, low bacterial growth rates result in a community where competitive displacement is infrequent.

Physical conditions on Mars have been cold, and in some habitats arid, for at least 3 billion years. Therefore, many contemporary MDV habitats may share similarities with conditions on Mars. Water on Mars must be frozen and solar heating of the subsurface of the ice that covers the planet may result in liquid inclusions similar to lake ice and cryoconites, which could serve as refugia for microorganisms in the hostile glacial surface environment, and may have analogs in the polar ice caps of Mars (Wharton et al., 1985). Antarctic permafrost is of significance because although these organisms are believed to be dormant, if Mars was inhabited in the past, then remnants of this life might be found entrained in Mars permafrost as well. Inspection by electron microscopy of cells defrosted from antarctic permafrost revealed that the cells were largely intact. Indeed, because many cultures have been isolated from the permafrost, this would suggest that perhaps some primitive life may some day be isolated from martian permafrost (Gilichinsky, 2007). Cryptoendolithic communities would be ideally suited to the low water, high ionizing conditions found on Mars. Presumably, life there would have been (or is presently) subjected to similar environmental pressures that would limit biotic interactions and produce similar patterns of microbial diversity.

Examination of the nature of MDV diversity, overwhelmed by rare and divergent sequences, indicates that this is a resource of tremendous genomic potential that may prove to be of applied and medical importance. It will be important in the future to achieve a thorough understanding of this diversity, and the factors promoting and limiting it, because it will provide us with major insights into fundamental ecological processes in the MDV, on Earth, and potentially other planets with similar ecological conditions.

References

Adams, B. J., Bardgett, R. D., Ayres, E., et al. (2006). Diversity and distribution of Victoria Land biota. *Soil Biology and Biochemistry*, **38**, 3003–3018.

Adams, E. E., Priscu, J. C., Fritsen, C. H., Smith, S. R., and Brackman, S. L. (1998). Permanent ice covers of the McMurdo Dry Valley lakes, Antarctica: bubble formation and metamorphism. In *Ecosystem Dynamics in a Polar Desert: The McMurdo Dry Valleys, Antarctica*, ed. J. C. Priscu. Antarctic Research Series 72. Washington, D.C.: American Geophysical Union, pp. 281–296.

Aiken, G., McKnight, D., Harnish, R., and Wershaw, R. (1996). Geochemistry of aquatic humic substances in the Lake Fryxell Basin, Antarctica. *Biogeochemistry*, **34**, 157–188.

Aislabie, J. M., Chhour, K. L., Saul, D. J., et al. (2006). Dominant bacteria in soils of Marble Point and Wright Valley, Victoria Land, Antarctica. *Soil Biology and Biochemistry*, **38**, 3041–3056.

Aislabie, J. M., Jordan, S., and Barker, G. M. (2008). Relation between soil classification and bacterial diversity in soils of the Ross Sea region, Antarctica. *Geoderma*, **144**, 9–20.

Amann, R. I., Ludwig, W., and Schleifer, K.-H. (1995). Phylogenetic identification and in situ detection of individual microbial cells without cultivation. *Microbiological Reviews*, **59**, 143–169.

Angino, E. E., Armitage, K. B., and Tash, J. C. (1964). Physicochemical limnology of Lake Bonney, Antarctica. *Limnology and Oceanography*, **9**, 207–217.

Armitage, K. B. and House, H. B. (1962). A limnological reconnaissance in the area of McMurdo Sound, Antarctica. *Limnology and Oceanography*, **7**, 36–41.

Ayres, E. B., Adams, B. J., Barrett, J. E., Virginia, R. A., and Wall, D. H. (2007). Unique similarity of faunal communities across aquatic-terrestrial interfaces in a polar desert ecosystem: soil-sediment boundaries and faunal community. *Ecosystems*, **10**, 523–535.

Bamforth, S. S., Wall, D. H., and Virginia, R. A. (2005). Distribution and diversity of soil protozoa in the McMurdo Dry Valleys of Antarctica. *Polar Biology*, **28**, 756–762.

Barrett, J. E., Virginia, R. A., Wall, D. H., et al. (2004). Variation in biogeochemistry and soil biodiversity across spatial scales in a polar desert ecosystem. *Ecology*, **85**, 3105–3118.

Barrett, J. E., Virginia, R. A., Wall, D. H., et al. (2006). Co-variation in soil biodiversity and biogeochemistry in northern and southern Victoria Land, Antarctica. *Antarctic Science*, **18**, 535–548.

Barrett, J. E., Virginia, R. A., Lyons, W. B., et al. (2007). Biogeochemical stoichiometry of Antarctic Dry Valley ecosystems. *Journal of Geophysical Research*, **112**, G01010.

Barrett, J. E., Virginia, R. A., Wall, D. H., et al. (2008). Persistent effects of a discrete climate event on a polar desert ecosystem. *Global Change Biology*, doi: 10.1111/j.1365–2486.2008.01641.x.

Bate, D. B., Barrett, J. E., Poage, M. A., and Virginia, R. A. (2007). Soil phosphorus cycling in an Antarctic polar desert. *Geoderma*, **144**, 21–31.

Baublis, J. A., Wharton, R. A., and Volz, P. A. (1991). Diversity of micro-fungi in an Antarctic dry valley. *Journal of Basic Microbiology*, **31**, 3–12.

Bell, W. and Mitchell, R. (1972). Chemotactic and growth responses of marine bacteria to extracellular products. *Biological Bulletin*, **143**, 265.

Borneman, J. and Triplett, E. W. (1997). Molecular microbial diversity in soils from eastern Amazonia: evidence for unusual microorganisms and microbial population shifts associated with deforestation. *Applied and Environmental Microbiology*, **63**, 2647–2653.

Bowman, J. P., McCammon, S. A., Rea, S. M., and McMeekin, T. A. (2000). The microbial composition of three limnologically disparate hypersaline Antarctic lakes. *FEMS Microbiology Letters*, **183**, 81–88.

Brambilla, E., Hippe, H., Hagelstein, A., Tindall, B. J., and Stackebrandt, E. (2001). 16S rDNA diversity of cultured and uncultured prokaryotes of a mat sample from Lake Fryxell, McMurdo Dry Valleys, Antarctica. *Extremophiles*, **5**, 23–33.

Burkins, M. B., Virginia, R. A., and Wall, D. H. (2001). Organic carbon cycling in Taylor Valley, Antarctica: quantifying soil reservoirs and soil respiration. *Global Change Biology*, **7**, 113–125.

Cameron, R. E., King, J., and David, C. N. (1970). Soil microbial ecology of Wheeler Valley, Antarctica. *Soil Science*, **109**, 110–120.

Campbell, I. B., Claridge, G. G., Campbell, D. I., and Balks, M. R. (1998). The soil environment of the McMurdo Dry Valleys, Antarctica. In *Ecosystem Dynamics in a Polar Desert: The McMurdo Dry Valleys, Antarctica*, ed. J. C. Priscu. Antarctic Research Series 72. Washington, D.C.: American Geophysical Union, pp. 297–322.

Chesson, P. and Huntly, N. (1997). The roles of harsh and fluctuating conditions in the dynamics of ecological communities. *American Naturalist*, **150**, 519–553.

Chesson, P., Gebauer, R. L. E., Schwinning, S., et al. (2004). Resource pulses, species interactions, and diversity maintenance in arid and semi-arid environments. *Oecologia*, **141**, 236–253.

Christner, B. C., Kvitko, 2nd, B. H., and Reeve, J. N. (2003). Molecular identification of bacteria and eukarya inhabiting an Antarctic cryoconite hole. *Extremophiles*, **7**, 177–183.

Clocksin, K. M., Jung, D. O., and Madigan, M. T. (2007). Cold-active chemoorganotrophic bacteria from permanently ice-covered Lake Hoare, McMurdo Dry Valleys, Antarctica. *Applied and Environmental Microbiology*, **73**, 3077–3083.

Connell, J. H. (1961). The influence of interspecific competition and other factors on the distribution of the barnacle *Chthamalus stellatus*. *Ecology*, **42**, 710–723.

Connell, J. H. (1978). Diversity in tropical rain forests and coral reefs: high diversity of trees and corals is maintained only in a non-equilibrium state. *Science*, **199**, 1302–1310.

Connell, L., Redman, R., Craig, S., and Rodriguez, R. (2006). Distribution and abundance of fungi in the soils of Taylor Valley, Antarctica. *Soil Biology and Biochemistry*, **38**, 3083–3094.

Conovitz, P. A., McKnight, D. M., MacDonald, L. H., Fountain, A. G., and House, H. R. (1998). Hydrological processes influencing streamflow variation in Fryxell Basin, Antarctica. In *Ecosystem Dynamics in a Polar Desert: The McMurdo Dry Valleys, Antarctica*, ed. J. C. Priscu. Antarctic Research Series 72. Washington, D.C.: American Geophysical Union, pp. 93–108.

Cowan, D. A. and Tow, L. A. (2004). Endangered Antarctic environments. *Annual Review of Microbiology*, **58**, 649–690.

Cowan, D. A., Russell, N. J., Mamais, A., and Sheppard, D. M. (2002). Antarctic dry valley mineral soils contain unexpectedly high levels of microbial biomass. *Extremophiles*, **6**, 431–436.

Cozzetto, K., McKnight, D., Nylen, T., and Fountain, A. (2006). Experimental investigations into processes controlling stream and hyporheic temperatures, Fryxell Basin, Antarctica. *Advances in Water Resources*, **29**, 130–153.

Dale, T. M., Skotnicki, M. L., Adam, K. D., and Selkirk, P. M. (1999). Genetic diversity in the moss *Hennediella heimii* in Miers Valley, southern Victoria Land, Antarctica. *Polar Biology*, **21**, 228–233.

Dana, G. L., Wharton, Jr., R. A., and Dubayah, R. (1998). Solar radiation in the McMurdo Dry Valleys, Antarctica. In *Ecosystem Dynamics in a Polar Desert: The McMurdo Dry Valleys, Antarctica*, ed. J. C. Priscu. Antarctic Research Series 72. Washington, D.C.: American Geophysical Union, pp. 39–64.

de la Torre, J. R., Goebel, B. M., Friedmann, E. I., and Pace, N. R. (2003). Microbial diversity of cryptoendolithic communities from the McMurdo Dry Valleys, Antarctica. *Applied and Environmental Microbiology*, **69**, 3858–3867.

DeSantis, T. Z., Hugenholtz, P., Larsen, N., et al. (2006). Greengenes, a chimera-checked 16S rRNA gene database and workbench compatible with ARB. *Applied and Environmental Microbiology*, **72**, 5069–5072.

Dore, J. E. and Priscu, J. C. (2001). Phytoplankton phosphorus deficiency and alkaline phosphatase activity in the McMurdo Dry Valley lakes, Antarctica. *Limnology and Oceanography*, **46**, 1331–1346.

Doran, P. T., Wharton, R. A., and Lyons, W. B. (1994). Paleolimnology of the McMurdo Dry Valleys, Antarctica. *Journal of Paleolimnology*, **10**, 85–114.

Doran, P. T., Priscu, J. C., Lyons, W. B., et al. (2002). Antarctic climate cooling and terrestrial ecosystem response. *Nature*, **415**, 517–520.

Doran, P. T., Fritsen, C. H., McKay, C. P., Priscu, J. C., and Adams, E. E. (2003). Formation and character of an ancient 19-m ice cover and underlying trapped brine in an "ice-sealed" east Antarctic lake. *Proceedings of the National Academy of Sciences of the United States of America*, **100**, 26–31.

Dunbar, J., Takala, S., Barns, S. M., Davis, J. A., and Kuske, C. R. (1999). Levels of bacterial community diversity in four arid soils compared by cultivation and 16S rRNA gene cloning. *Applied and Environmental Microbiology*, **65**, 1662–1669.

Esposito, R. M. M., Horn, S. L., McKnight, D. M., et al. (2006). Antarctic climate cooling and response of diatoms in glacial meltwater streams. *Geophysical Research Letters*, **33**, 1–4.

Fell, J. W., Scorzetti, G., Connell, L., and Craig, S. (2006). Biodiversity of micro-eukaryotes in Antarctic Dry Valley soils with <5% soil moisture. *Soil Biology and Biochemistry*, **38**, 3107–3119.

Fierer, N. and Jackson, R. B. (2006). The diversity and biogeography of soil bacterial communities. *Proceedings of the National Academy of Sciences of the United States of America*, **103**, 626–631.

Foreman, C. M., Sattler, B., Mikucki, J. A., Porazinska, D. L., and Priscu, J. C. (2007). Metabolic activity and diversity of cryoconites in the Taylor Valley, Antarctica. *Journal of Geophysical Research*, **112**, 1–11.

Fountain, A., Tranter, M., Nylen, T., Lewis, K., and Mueller, D. (2004). Evolution of cryoconite holes and their contribution to meltwater runoff from glaciers in the McMurdo Dry Valleys, Antarctica. *Journal of Glaciology*, **50**, 35–45.

Franzmann, P. D., Liu, Y., Balkwill, D. L., et al. (1997). *Methanogenium frigidum* sp. nov., a psychrophilic, H_2-using methanogen from Ace Lake, Antarctica. *International Journal of Systematics and Bacteriology*, **47**, 1068–1072.

Friedmann, E. I. (1993). *Antarctic Microbiology*. New York: Wiley-Liss.

Friedmann, E. I., Hua, M., and Ocampo-Friedmann, R. (1988). Cryptoendolithic lichen and cyanobacterial communities of the Ross Desert, Antarctica. *Polarforschung*, **58**, 251–259.

Fritsen, C. H. and Priscu, J. C. (1998). Cyanobacterial assemblages in permanent ice covers on Antarctic lakes: distribution, growth rate, and temperature response of photosynthesis. *Journal of Phycology*, **34**, 587–597.

Fritsen, C. H., Adams, E. E., McKay, C. P., and Priscu, J. C. (1998). Permanent ice covers of the McMurdo Dry Valleys lakes, Antarctica: liquid water contents. In *Ecosystem Dynamics in a Polar Desert: The McMurdo Dry Valleys, Antarctica*, ed. J. C. Priscu. Antarctic Research Series 72. Washington, D.C.: American Geophysical Union, pp. 269–280.

Fritsen, C. H., Grue, A. M., and Priscu, J. C. (2000). Distribution of organic carbon and nitrogen in surface soils in the McMurdo Dry Valleys, Antarctica. *Polar Biology*, **23**, 121–128.

Gause, G. F. (1934). *The Struggle for Existence*. Baltimore, MD: Williams and Wilkins.

Gilichinsky, D. A. (2007). Microbial populations in Antarctic permafrost: biodiversity, state age, and inplication for astrobiology. *Astrobiology*, **7**, 275–311.

Glatz, R. E., Lepp, P. W., Ward, B. B., and Francis, C. A. (2006). Planktonic microbial community composition across steep physical/chemical gradients in permanently ice-covered Lake Bonney, Antarctica. *Geobiology*, **4**, 53–67.
Glausiusz, J. (2007). Extreme culture. *Nature*, **447**, 905–906.
Goldman, C. R. (1964). Primary productivity studies in Antarctic lakes. In *Biologie Antarctique*. Proceedings of 1st SCAR symposium, ed. R. Carrick, M. Holgate, and J. Prevost. Paris: Hermann, pp. 291–299.
Goldman, C. R., Mason, D. T., and Hobbie, J. E. (1967). Two Antarctic desert lakes. *Limnology and Oceanography*, **12**, 295–310.
Gordon, D. A., Priscu, J., and Giovannoni, S. (2000). Origin and phylogeny of microbes living in permanent Antarctic lake ice. *Microbial Ecology*, **39**, 197–202.
Hall, J., Mitchell, K., Jackson-Weaver, O., et al. (2008). Molecular characterization of the diversity and distribution of a thermal spring microbial community using rRNA and functional genes. *Applied and Environmental Microbiology*, **74**, 4910–4922.
Hardin, G. (1960). The competitive exclusion principle. *Science*, **131**, 1292–1297.
Hogg, I. D., Craig Cary, S., Convey, P., et al. (2006). Biotic interactions in Antarctic terrestrial ecosystems: are they a factor? *Soil Biology and Biochemistry*, **38**, 3035–3040.
Hopkins, D. W., Sparrow, A. D., Novis, P. M., et al. (2006). Controls on the distribution of productivity and organic resources in Antarctic Dry Valley soils. *Proceedings of the Royal Society of London, Biological Sciences*, **273**, 2687–2695.
Horner-Devine, M. C., Carney, K. M., and Bohannan, B. J. M. (2003). An ecological perspective on bacterial biodiversity. *Proceedings of the Royal Society of London, Series B*, **271**, 113–122.
Horner-Devine, M. C., Silver, J. M., Leibold, M. A., et al. (2007). A comparison of taxon co-occurrence for macro- and microorganisms. *Ecology*, **88**, 1345–1353.
Horowitz, N. H., Cameron, R. E., and Hubbard, J. S. (1972). Microbiology of the Dry Valleys of Antarctica. *Science*, **176**, 242–245.
Howard-Williams, C., Schwarz, A. M., Hawes, I., and Priscu, J. C. (1998). Optical properties of the McMurdo Dry Valley lakes, Antarctica. *Antarctic Research Series*, **72**, 189–203.
Hughes, J. B., Hellmann, J. J., Ricketts, T. H., and Bohannan, B. J. (2001). Counting the uncountable: statistical approaches to estimating microbial diversity. *Applied and Environmental Microbiology*, **67**, 4399–4406.
Huston, M. (1979). A general hypothesis of species diversity. *American Naturalist*, **113**, 81.
Hutchinson, G. E. (1961). The paradox of the plankton. *American Naturalist*, **95**, 137–145.
Jones, J. and Simon, B. (1985). Interaction of acetogens and methanogens in anaerobic freshwater sediments. *Applied and Environmental Microbiology*, **49**, 944–948.
Jung, D. O., Achenbach, L. A., Karr, E. A., Takaichi, S., and Madigan, M. T. (2004). A gas vesiculate planktonic strain of the purple non-sulfur bacterium *Rhodoferax antarcticus* isolated from Lake Fryxell, Dry Valleys, Antarctica. *Archives of Microbiology*, **182**, 236–243.
Karr, E. A., Sattley, W. M., Jung, D. O., Madigan, M. T., and Achenbach, L. A. (2003). Remarkable diversity of phototrophic purple bacteria in a permanently frozen Antarctic lake. *Applied and Environmental Microbiology*, **69**, 4910–4914.
Karr, E. A., Sattley, W. M., Rice, M. R., et al. (2005). Diversity and distribution of sulfate-reducing bacteria in permanently frozen Lake Fryxell, McMurdo Dry Valleys, Antarctica. *Applied and Environmental Microbiology*, **71**, 6353–6359.

Karr, E. A., Ng, J. M., Belchik, S. M., et al. (2006). Biodiversity of methanogenic and other archaea in the permanently frozen Lake Fryxell, Antarctica. *Applied and Environmental Microbiology*, **72**, 1663–1666.

Kepner, Jr., R. L., Wharton, Jr., R. A., and Suttle, C. A. (1998). Viruses in Antarctic lakes. *Limnology and Oceanography*, **43**, 1754–1761.

Kepner, R. L., Wharton, Jr., R. A., and Coats, D. W. (1999). Ciliated protozoa of two Antarctic lakes: analysis by quantitative protargol staining and examination of artificial substrates. *Polar Biology*, **21**, 285–294.

Koob, D. D. and Leister, G. L. (1972). Primary productivity and associated physical chemical and biological characteristics of Lake Bonney: a perennially ice-covered lake in Antarctica. *Antarctic Research Series*, **20**, 51–68.

Krembs, C., Juhl, A. R., Long, R. A., and Azam, F. (1998). Nanoscale patchiness of bacteria in lake water studied with the spatial information preservation method. *Limnology and Oceanography*, **43**, 307–314.

Lancaster, N. (2002). Flux of eolian sediment in the McMurdo Dry Valleys, Antarctica: a preliminary assessment. *Arctic Antarctic and Alpine Research*, **34**, 318–323.

Laybourn-Parry, J. (1997). The microbial loop in Antarctic lakes. In *Ecosystems Processes in Antarctic Ice-Free Landscapes*, ed. W. B. Lyons, C. Howard-Williams, and I. Hawes. Rotterdam, Netherlands: A. A. Balkema, pp. 231–240.

Lee, P. A., Mikucki, J. A., Foreman, C. M., et al. (2004). Thermodynamic constraints on microbially mediated processes in lakes of the McMurdo Dry Valleys, Antarctica. *Geomicrobiology Journal*, **21**, 221–237.

Levins, R. (1968). *Evolution in Changing Environments*. Princeton, NJ: Princeton University Press.

Lisle, J. T. and Priscu, J. C. (2004). The occurrence of lysogenic bacteria and microbial aggregates in the lakes of the McMurdo Dry Valleys, Antarctica. *Microbial Ecology*, **47**, 427–439.

Lizotte, M. P. and Priscu, J. C. (1992). Spectral irradiance and bio-optical properties in perennially ice-covered lakes of the Dry Valleys (McMurdo Sound, Antarctica). *Antarctic Research Series*, **57**, 1–14.

Lizotte, M. P., Sharp, T. R., and Priscu, J. C. (1996). Phytoplankton dynamics in the stratified water column of Lake Bonney, Antarctica. I. Biomass and productivity during the winter-spring transition. *Polar Biology*, **16**, 155–162.

Lozupone, C. and Knight, R. (2005). UniFrac: a new phylogenetic method for comparing microbial communities. *Applied and Environmental Microbiology*, **71**, 8228–8235.

Lozupone, C. A. and Knight, R. (2007). Global patterns in bacterial diversity. *Proceedings of the National Academy of Sciences of the United States of America*, **104**, 11436.

Lyons, W. B., Welch, K. A., Neumann, K., et al. (1998). Geochemical linkages among glaciers, streams and lakes within the Taylor Valley, Antarctica. *Antarctic Research Series*, **72**, 77–92.

Lyons, W. B., Tyler, S. W., Wharton, R. A., McKnight, D. M., and Vaughn, B. H. (2004). A Late Holocene desiccation of Lake Hoare and Lake Fryxell, McMurdo Dry Valleys, Antarctica. *Antarctic Science*, **10**, 247–256.

Madigan, M. T., Jung, D. O., Woese, C. R., and Achenbach, L. A. (2000). *Rhodoferax antarcticus* sp. nov., a moderately psychrophilic purple nonsulfur bacterium isolated from an Antarctic microbial mat. *Archives of Microbiology*, **173**, 269–277.

Magurran, A. E. (2004). *Measuring Biological Diversity*. Oxford, UK: Blackwell.

Matsubaya, O., Sakai, H., Torii, T., Burton, H., and Kerry, K. (1979). Antarctic saline lakes: stable isotopic ratios, chemical compositions and evolution. *Geochimica et Cosmochimica Acta*, **43**, 7–25.

May, R. M. (1974). On the theory of niche overlap. *Theoretical Population Biology*, **5**, 297–332.

May, R. M. (1975). Patterns of species abundance and diversity. *Ecology and Evolution of Communities*, **1**, 81–120.

McKay, C. P., Andersen, D. T., Wayne, H., et al. (2005). Polar lakes, streams, and springs as analogs for the hydrological cycle on Mars. In *Water on Mars and Life*, ed. T. Tokano. Heidelberg, Germany: Springer, pp. 219–233.

McKnight, D. M., Alger, A., Tate, C. M., Shupe, G., and Spaulding, S. (1998). Longitudinal patterns in algal abundance and species distribution in meltwater streams in Taylor Valley, Southern Victoria Land, Antarctica. In *Ecosystem Dynamics in a Polar Desert: The McMurdo Dry Valleys, Antarctica*, ed. J. C. Priscu. Antarctic Research Series 72. Washington, D.C.: American Geophysical Union, pp. 109–127.

McKnight, D. M., Niyogi, D. K., Alger, A. S., et al. (1999). Dry valley streams in Antarctica: ecosystems waiting for water. *BioScience*, **49**, 985–995.

McKnight, D. M., Tate, C. M., Andrews, E. D., et al. (2007). Reactivation of a cryptobiotic stream ecosystem in the McMurdo Dry Valleys, Antarctica: a long-term geomorphological experiment. *Geomorphology*, **89**, 186–204.

Mikell, Jr., A. T., Parker, B. C., and Gregory, E. M. (1986). Factors affecting high-oxygen survival of heterotrophic microorganisms from an Antarctic lake. *Applied and Environmental Microbiology*, **52**, 1236–1241.

Mikucki, J. A. (2005). Microbial ecology of an Antarctic subglacial environment. Ph.D. thesis, Montana State University, Bozeman, MT.

Mikucki, J. A. and Priscu, J. C. (2007). Bacterial diversity associated with Blood Falls, a subglacial outflow from the Taylor Glacier, Antarctica. *Applied and Environmental Microbiology*, **73**, 4029–4039.

Mosier, A. C., Murray, A. E., and Fritsen, C. H. (2007). Microbiota within the perennial ice cover of Lake Vida, Antarctica. *FEMS Microbiology Ecology*, **59**, 274–288.

Nemergut, D. R., Costello, E. K., Meyer, A. F., et al. (2005). Structure and function of alpine and arctic soil microbial communities. *Research in Microbiology*, **156**, 775–784.

Niederberger, T. D., McDonald, I. R., Hacker, A. L., et al. (2008). Microbial community composition in soils of northern Victoria Land, Antarctica. *Environmental Microbiology*, **10**, 1713–1724.

Nkem, J. N., Wall, D. H., Virginia, R. A., et al. (2006). Wind dispersal of soil invertebrates in the McMurdo Dry Valleys, Antarctica. *Polar Biology*, **29**, 346–352.

Paerl, H. W. and Priscu, J. C. (1998). Microbial phototrophic, heterotrophic, and diazotrophic activities associated with aggregates in the permanent ice cover of Lake Bonney, Antarctica. *Microbial Ecology*, **36**, 221–230.

Paine, R. T. (1966). Food web complexity and species diversity. *American Naturalist*, **100**, 65–75.

Parker, B. C., Hoehn, R. C., Paterson, R. A., et al. (1977). Changes in dissolved organic matter, photosynthetic production and microbial community composition in Lake Bonney, southern Victoria Land, Antarctica. In *Adaptations within Antarctic Ecosystems*, ed. G. Llano. Washington, D.C.: Smithsonian Institution, pp. 859–872.

Parker, B. C., Simmons, G. M., Wharton, R. A., Seaburg, K. G., and Love, F. G. (1982). Removal of organic and inorganic matter from Antarctic lakes by aerial escape of bluegreen algal mats. *Journal of Phycology*, **18**, 72–78.

Poff, N. L. and Ward, J. V. (1989). Implications of streamflow variability and predictability for lotic community structure: a regional analysis of streamflow patterns. *Canadian Journal of Fisheries and Aquatic Sciences*, **46**, 1805–1817.

Porazinska, D. L., Fountain, A. G., Nylen, T. H., et al. (2004). The biodiversity and biogeochemistry of cryoconite holes from McMurdo Dry Valley glaciers, Antarctica. *Arctic Antarctic and Alpine Research*, **36**, 84–91.

Priscu, J. C. (1995). Phytoplankton nutrient deficiency in lakes of the McMurdo Dry Valleys, Antarctica. *Freshwater Biology*, **34**, 215–227.

Priscu, J. C., Downes, M. T., and McKay, C. P. (1996). Extreme supersaturation of nitrous oxide in a poorly ventilated Antarctic lake. *Limnology and Oceanography*, **41**, 1544–1551.

Priscu, J. C., Fritsen, C. H., Adams, E. E., et al. (1998). Perennial Antarctic lake ice: an oasis for life in a polar desert. *Science*, **280**, 2095–2098.

Priscu, J. C., Wolf, C. F., Takacs, C. D., et al. (1999). Carbon transformations in a perennially ice-covered Antarctic lake. *BioScience*, **49**, 997–1008.

Priscu, J. C., Vincent, W. F., and Howard-Williams, C. (2002). Inorganic nitrogen uptake and regeneration in perennially ice-covered Lakes Fryxell and Vanda, Antarctica. *Journal of Plankton Research*, **11**, 335–351.

Priscu, J. C., Christner, B. C., Dore, J. E., et al. (2008). Supersaturated N_2O in a perennially ice-covered lake: molecular and stable isotopic evidence for a biogeochemical relict. *Limnology and Oceanography*, **53**(6), 2439–2450.

Roberts, E. C. and Laybourn-Parry, J. (1999). Mixotrophic cryptophytes and their predators in the Dry Valley lakes of Antarctica. *Freshwater Biology*, **41**, 737–746.

Roberts, E. C., Laybourn-Parry, J., McKnight, D. M., and Novarino, G. (2000). Stratification and dynamics of microbial loop communities in Lake Fryxell, Antarctica. *Freshwater Biology*, **44**, 649–661.

Roberts, E. C., Priscu, J. C., Wolf, C., Lyons, W. B., and Laybourn-Parry, J. (2004). The distribution of microplankton in the McMurdo Dry Valley lakes, Antarctica: response to ecosystem legacy or present-day climatic controls? *Polar Biology*, **27**, 238–249.

Roth, R. R. (1976). Spatial heterogeneity and bird species diversity. *Ecology*, **57**, 773–782.

Runkel, R. L., McKnight, D. M., and Andrews, E. D. (1998). Analysis of transient storage subject to unsteady flow: diel flow variation in an Antarctic stream. *Journal of the North American Benthological Society*, **17**, 143–154.

Sattley, W. M. and Madigan, M. T. (2006). Isolation, characterization, and ecology of cold-active, chemolithotrophic, sulfur-oxidizing bacteria from perennially ice-covered Lake Fryxell, Antarctica. *Applied and Environmental Microbiology*, **72**, 5562–5568.

Sattley, W. M. and Madigan, M. T. (2007). Cold-active acetogenic bacteria from surficial sediments of perennially ice-covered Lake Fryxell, Antarctica. *FEMS Microbiology Letters*, **272**, 48–54.

Säwström, C., Lisle, J., Anesio, A. M., Priscu, J. C., and Laybourn-Parry, J. (2008). Bacteriophage in polar inland waters. *Extremophiles*, **12**, 167–175.

Schoener, T. W. (1974). Competition and the form of habitat shift. *Theoretical Population Biology*, **6**, 265–307.

Schwarz, A. M. J., Green, T. G. A., and Seppelt, R. D. (1992). Terrestrial vegetation at Canada Glacier, southern Victoria Land, Antarctica. *Polar Biology*, **12**, 397–404.

Scott, R. F. (1905). *The Voyage of the "Discovery"*. New York: C. Scribner's Sons.

Shivaji, S., Reddy, G. S. N., Suresh, K., et al. (2005). *Psychrobacter vallis* sp. nov. and *Psychrobacter aquaticus* sp. nov., from Antarctica. *International Journal of Systematic and Evolutionary Microbiology*, **55**, 757–762.

Skotnicki, M. L., Ninham, J. A., and Selkirk, P. M. (2000). Genetic diversity, mutagenesis and dispersal of Antarctic mosses: a review of progress with molecular studies. *Antarctic Science*, **12**, 363–373.

Smith, J. J., Tow, L. A., Stafford, W., Cary, C., and Cowan, D. A. (2006). Bacterial diversity in three different Antarctic cold desert mineral soils. *Microbial Ecology*, **51**, 413–421.

Sommer, U. (1985). Comparison between steady state and non-steady state competition: experiments with natural phytoplankton. *Limnology and Oceanography*, **30**, 335–346.

Spigel, R. H. and Priscu, J. C. (1998). Physical limnology of the McMurdo Dry Valley lakes. In *Ecosystem Dynamics in a Polar Desert: The McMurdo Dry Valleys, Antarctica*, ed. J. C. Priscu. Antarctic Research Series 72. Washington, D.C.: American Geophysical Union, pp. 153–187.

Stackebrandt, E., Brambilla, E., Cousin, S., Dirks, W., and Pukall, R. (2004). Culture-independent analysis of bacterial species from an anaerobic mat from Lake Fryxell, Antarctica: prokaryotic diversity revisited. *Cellular and Molecular Biology*, **50**, 517–524.

Stingl, U., Cho, J. C., Foo, W., et al. (2008). Dilution-to-extinction culturing of psychrotolerant planktonic bacteria from permanently ice-covered lakes in the McMurdo Dry Valleys, Antarctica. *Microbial Ecology*, **55**, 395–405.

Takacs, C. D. and Priscu, J. C. (1998). Bacterioplankton dynamics in the McMurdo Dry Valley lakes, Antarctica: production and biomass loss over four seasons. *Microbial Ecology*, **36**, 239–250.

Takacs, C. D., Priscu, J. C., and McKnight, D. M. (2001). Bacterial dissolved organic carbon demand in McMurdo Dry Valley lakes, Antarctica. *Limnology and Oceanography*, **46**, 1189–1194.

Takacs-Vesbach, C., Mitchell, K., Jackson-Weaver, O., and Reysenbach, A. L. (2008). Volcanic calderas delineate biogeographic provinces among Yellowstone thermophiles. *Environmental Microbiology*, **10**, 1681–1689.

Taton, A., Grubisic, S., Brambilla, E., De Wit, R., and Wilmotte, A. (2003). Cyanobacterial diversity in natural and artificial microbial mats of Lake Fryxell (McMurdo Dry Valleys, Antarctica): a morphological and molecular approach. *Applied and Environmental Microbiology*, **69**, 5157–5169.

Thompson, J. R., Pacocha, S., Pharino, C., et al. (2005). Genotypic diversity within a natural coastal bacterioplankton population. *Science*, **307**, 1311–1313.

Tilman, D., Kilham, S. S., and Kilham, P. (1982). Phytoplankton community ecology: the role of limiting nutrients. *Annual Reviews in Ecology and Systematics*, **13**, 349–372.

Tindall, B. J., Brambilla, E., Steffen, M., et al. (2000). Cultivatable microbial biodiversity: gnawing at the Gordian knot. *Environmental Microbiology*, **2**, 310–318.

Torsvik, V., Ovreas, L., and Thingstad, T. F. (2002). Prokaryotic diversity: magnitude, dynamics, and controlling factors. *Science*, **296**, 1064–1066.

Townsend, C. R., Scarsbrook, M. R., and Doledec, S. (1997). The intermediate disturbance hypothesis, refugia, and biodiversity in streams. *Limnology and Oceanography*, **42**, 938–949.

Treonis, A. M., Wall, D. H., and Virginia, R. A. (1999). Invertebrate biodiversity in Antarctic dry valley soils and sediments. *Ecosystems*, **2**, 482–492.

Urbach, E., Vergin, K. L., Young, L., et al. (2001). Unusual bacterioplankton community structure in ultra-oligotrophic Crater Lake. *Limnology and Oceanography*, **46**, 557–572.

Urbach, E., Vergin, K. L., Larson, G. L., and Giovannoni, S. J. (2007). Bacterioplankton communities of Crater Lake, Oregon: dynamic changes with euphotic zone food web structure and stable deep water populations. *Hydrobiologia*, **574**, 161–177.

Van Trappen, S., Mergaert, J., Van Eygen, S., et al. (2002). Diversity of 746 heterotrophic bacteria isolated from microbial mats from ten Antarctic lakes. *Systematic and Applied Microbiology*, **25**, 603–610.

Van Trappen, S., Vandecandelaere, I., Mergaert, J., and Swings, J. (2004). *Algoriphagus antarcticus* sp. nov., a novel psychrophile from microbial mats in Antarctic lakes. *International Journal of Systematic and Evolutionary Microbiology*, **54**, 1969–1973.

Vincent, W. F. (1981). Production strategies in Antarctic inland waters: phytoplankton eco-physiology in a permanently ice-covered lake. *Ecology*, **62**, 1215–1224.

Vincent, W. F. (1988). *Microbial Ecosystems of Antarctica*. New York: Cambridge University Press.

Vincent, W. F. and Howard-Williams, C. (1986). Antarctic stream ecosystems: physiological ecology of a blue-green algal epilithon. *Freshwater Biology*, **16**, 219–233.

Vincent, W. F. and Howard-Williams, C. (1989). Microbial communities in Southern Victoria Land streams (Antarctica). II. The effects of low temperature. *Hydrobiologia*, **172**, 39–49.

Vincent, W. F. and James, M. R. (1996). Biodiversity in extreme aquatic environments: lakes, ponds and streams of the Ross Sea Sector, Antarctica. *Biodiversity and Conservation*, **5**, 1451–1471.

Vincent, W. F., Rae, R., Laurion, I., Howard-Williams, C., and Priscu, J. C. (1998). Transparency of Antarctic ice-covered lakes to solar UV radiation. *Limnology and Oceanography*, **43**, 618–624.

Virginia, R. A. and Wall, D. H. (1999). How soils structure communities in the Antarctic Dry Valleys. *BioScience*, **49**, 973–983.

Vishniac, H. S. (1993). The microbiology of Antarctic soils. In *Antarctic Microbiology*, ed. E. I. Friedmann. New York: Wiley-Liss, pp. 297–341.

Voytek, M. A. and Ward, B. B. (1995). Detection of ammonium-oxidizing bacteria of the beta-subclass of the class Proteobacteria in aquatic samples with the PCR. *Applied and Environmental Microbiology*, **61**, 1444–1450.

Voytek, M. A., Priscu, J. C., and Ward, B. B. (1999). The distribution and relative abundance of ammonia-oxidizing bacteria in lakes of the McMurdo Dry Valley, Antarctica. *Hydrobiologia*, **401**, 113–130.

Ward, B. B. and Priscu, J. C. (1997). Detection and characterization of denitrifying bacteria from a permanently ice-covered Antarctic lake. *Hydrobiologia*, **347**, 57–68.

Ward, B. B., Granger, J., Maldonado, M. T., et al. (2005). Denitrification in the hypolimnion of permanently ice-covered Lake Bonney, Antarctica. *Aquatic Microbial Ecology*, **38**, 295–307.

Ward, D. M. (1998). A natural species concept for prokaryotes. *Current Opinion in Microbiology*, **1**, 271–277.

Ward, D. M., Cohan, F. M., Bhaya, D., et al. (2008). Genomics, environmental genomics and the issue of microbial species. *Heredity*, **100**, 207–219.

Wharton, Jr., R. A., McKay, C. P., Simmons, Jr., G. M., and Parker, B. C. (1985). Cryoconite holes on glaciers. *BioScience*, **35**, 499–503.

Wynn-Williams, D. D. and Edwards, H. G. M. (2000). Antarctic ecosystems as models for extraterrestrial surface habitats. *Planetary and Space Science*, **48**, 1065–1075.

Young, I. M. and Ritz, K. (2000). Tillage, habitat space and function of soil microbes. *Soil and Tillage Research*, **53**, 201–213.

Zeglin, L. (2008). Microbial diversity and function at aquatic-terrestrial interfaces in desert ecosystems. Ph.D. thesis, University of New Mexico, Albuquerque, NM.

Zeglin, L., Sinsabaugh, R., Barrett, J. E., Gooseff, M., and Takacs-Vesbach, C. (2009). Landscape distribution of microbial activity in the McMurdo Dry Valleys: linked biotic processes, hydrology and geochemistry in a cold desert ecosystem. *Ecosystems*, **12**(4), 562–573.

Zhou, J., Xia, B., Treves, D. S., et al. (2002). Spatial and resource factors influencing high microbial diversity in soil. *Applied and Environmental Microbiology*, **68**, 326–334.

9

Other analogs to Mars: high-altitude, subsurface, desert, and polar environments

NATHALIE A. CABROL, DALE T. ANDERSEN, CAROL R. STOKER, PASCAL LEE, CHRISTOPHER P. McKAY, AND DAVID S. WETTERGREEN

Abstract

The McMurdo Dry Valleys detailed in previous chapters represent one environment for life thought to have existed on Mars among many. This chapter illustrates other potential habitats and their significance: (1) high-altitude lakes subjected to rapid climate change in the Andes provide analogy to the Noachian/Hesperian transition on Mars; (2) Río Tinto, Spain, where conditions are reminiscent of Meridiani Planum, unravels an underground anaerobic chemoautotroph biosphere that could resemble a modern refuge for life on Mars; (3) the High Arctic hosts gullies analogous to those observed on Mars, whose fresh deposits could provide access to traces of past and/or present underground oases; it is also in this polar environment that the Haughton-Mars Project helps answer long-standing questions, revisiting classical assumptions, and sometimes reshaping our thinking on many issues in planetary science and astrobiology, in particular in relation to Mars; (4) the search for microbial life in the arid soils of the Atacama desert and its robotic detection characterize what role aridity plays in the distribution of life and how to search for evidence of rare and scattered biosignatures.

Introduction

Because of its geology and climate evolution, Mars is likely to have developed a diversity of potential habitats for life over time. The main ingredients for habitability (i.e., water, energy, and nutrients) were present early, as demonstrated by the Spirit and Opportunity rovers at Gusev crater and Meridiani

Life in Antarctic Deserts and Other Cold Dry Environments: Astrobiological Analogs, ed. Peter T. Doran, W. Berry Lyons and Diane M. McKnight. Published by Cambridge University Press.

Planum (Knoll et al., 2005; Des Marais et al., 2005, 2008). Data from orbiting missions suggest that (1) water could have flowed on the surface of Mars in the past seven years (Malin et al., 2006); (2) volcanic activity has occurred only 1–2 million years ago (Hartmann et al., 1999; Baker, 2005) when martian volcanoes where previously thought to have been extinct for hundreds of millions of years; (3) energy and water were colocated recently in the north polar regions where volcanic flows cover ice deposits; (4) evidence points to the possibility of a large frozen sea in the northern hemisphere as recently as 5 million years ago (Murray et al., 2005); and (5) traces of methane have been measured in the martian atmosphere (Krasnopolsky et al., 1997; Summers et al., 2002; Formisano et al., 2004; Lyons et al., 2005; Mumma et al., 2005). Whether its origin is cosmogonic (Krasnopolsky et al., 1997), geologic (Lyons et al., 2005; Oze and Sharma, 2005), and/or biologic (Pellenberg et al., 2003; Krasnopolsky et al., 2004) is still debated but its presence has undeniable astrobiological implications (Max and Clifford, 2000; Atreya et al., 2007).

It appears then that Mars could have been habitable throughout its history, although the type, scale, and location of niches would have changed over time in a magnitude that still has to be understood. It is generally accepted, however, that surface life was possible in the first 500 million years but unlikely in later geological times. If life ever appeared on Mars and endured both the rapid climate change of the Noachian/Hesperian transition (NHT) about 3.7–3.2 Ga ago (Golombek et al., 2006) and the climate instability inherent to the planet's orbital characteristics, it is likely that it was sheltered underground. The range of potential habitats to explore in the search for past and/or present life on Mars is, therefore, wide and multifaceted and they all have some level of analogy here on Earth.

Alongside robotic missions, the exploration of terrestrial analogs provides the only method so far for the scientific community to go back in time on Mars and learn about what life, its location, habitats, survival (or lack thereof), and evolution could have been at various geological epochs. It informs us on how life modifies its environment and how future missions could recognize its geo- and biosignatures and explore them.

Declining lake habitats

By analogy to Earth, lakes on early Mars would have been favorable sites for the inception and development of life (Des Marais and Farmer, 1995; Farmer and Des Marais, 1999; Cabrol et al., 2001b; Cabrol and Grin, 2004). Evidence abounds planetwide. Dry basins at the termini of fluvial valley networks were the first indicators of their existence. Other evidence includes delta-like landforms (De Hon, 1991, 1992, 2001; Cabrol and Grin, 2001; Malin and Edgett, 2003; Wood, 2005). When compared with terrestrial analogs, their morphology is a testimony of surface water stability over tens to hundreds of thousands of years, evidence supported by the presence of hydrated minerals showing abundant surface water at the same period (Bibring et al., 2005). This favorable period for surface water activity was relatively short lived, possibly 300–500 million years. From mineralogical observations (Christensen et al., 2003; Bibring et al., 2005) peak lacustrine activity might have taken place between 4.0–3.7 $\pm$ 0.2 Ga ago. The consequence of the loss of atmosphere is visible in the geological record at the tail end of this period which shows a significant decrease in surface flow erosion (Carr, 1996) and accumulation of sulfate deposits rather than clay, which is abundant earlier (Bibring et al., 2005). Atmosphere rarefaction had several critical consequences for lakes and other surface bodies of water, including: intense evaporation, shrinking water column, enhanced impact of ultraviolet radiation (UVR), and cooling temperatures leading to the formation of ice cover (McKay et al., 1985; Moore et al., 1995). The resulting pressure on putative aqueous ecosystems would continuously increase with time thereafter. There is no evidence in the surface record that at any time in later geological periods conditions returned close to what they were prior to the NHT. The lakes' dynamics at that particular time on Mars, the time elapsed before they disappeared, and aqueous life adaptation strategies to mounting environmental pressure all have fundamental astrobiological implications. They could point to whether or not putative life in martian lakes was given time to adapt and transition to new, more sheltered, habitats underground as environmental conditions were globally collapsing.

Ideally, a systemic understanding of martian lakes at the NHT would require the study of terrestrial environments where many of the parameters analogous to this climatic transition are present together: low air temperature (T), high daily and yearly ΔT, aridity, strong evaporation, thin atmosphere, high UVR, ice, reduced precipitation, and volcanic and hydrothermal activity. Such environment exists at high altitude on Earth. In the Andes, ancient lakes are evaporating rapidly due to climatic change initiated 18 000 years ago (Messerli et al., 1993; Wirrmann and Mourguiart, 1995; Abbott et al., 1997; Baucom and Rigsby, 1999; Sylvestre et al., 1999; Grosjean, 2001; Vuille et al., 2003) and evaporation currently accelerates because of global warming (Cabrol et al., 2005a; Hock et al., 2003, 2005b).

High-altitude lakes environment

The lakes investigated by the High-Lakes Project (HLP) are located in the Potosi region (southwest Bolivian Andes) and the neighboring Chilean altiplano at the contact with the Atacama desert (Fig. 9.1). This region provides many parallels to Mars at the NHT allowing the study of the interaction between multiple environmental factors in the evolution of declining lakes and their ecosystems. Several lakes were selected on the basis of plausible analogy with the physical and limnological diversity of martian lakes during the NHT as modeled from data. Laguna Blanca, Laguna Verde, and the summit lake of Licancabur are detailed here.

At the end of the Holocene, Laguna Blanca and Laguna Verde were originally a single 3 × 7 km, 50-m deep lake (Cabrol, 2006; Cabrol et al., 2007a). They are located in a tectonic basin at the foot of the Licancabur

Fig. 9.1. (a) The Licancabur volcano is shared by Chile and Bolivia. (b) Laguna Verde left and Laguna Blanca to the far right. (c) Close-up on the Licancabur northeastern slopes. The arrow shows the location of HLP summit camp at 5900 m. (d) The Licancabur summit lake viewed from the Bolivian rim. The lake is ~100 m × 90 m large and 5.5 m deep. (e) Colony of red copepods at the center of the Licancabur summit lake. (Photo credits: HLP, NAI/SETI/NASA Ames.)

volcano. As a result of climate change, evaporation, and topography, these two lakes are now separating and are only connected by a 20-m-long, 5-m-wide channel. They already display very different ecosystems (Acs et al., 2003; Kiss et al., 2004) and limnologic characteristics (Hock et al., 2003, 2005b; Cabrol et al., 2007a). The Licancabur lake is nested ~80 m below the summit crater rim of a 5970-m high dormant volcano, which last erupted ~10 000 years ago (Rudolf, 1955; Da Silva and Francis, 1991). The colocation of the three lakes allows the quantification of the impact of elevation on the atmospheric pressure (P_{Atm}), the air temperature (T), the daily and yearly air temperature magnitude (ΔT), relative humidity (RH), and ultraviolet radiation (UVR) flux on the lake dynamics and physical and biological evolution. Environment and parallels to Mars are summarized in Table 9.1.

The climate evolution provides another analogy to Mars at the NHT. These lakes were formed 18 000–10 000 years ago when the altiplano was experiencing 500 mm y^{-1} precipitation. At the Holocene transition, aridity set in (Messerli et al., 1993; Grosjean, 2001) with current precipitation $\leq$100 mm y^{-1} generating a strong negative water balance (Hock et al., 2003, 2005b). Sonar sounding and paleoterraces show that Laguna Verde lost 45 m of its water column in 6000 years and this trend continues today (Cabrol, 2006).

Investigation

The location of the study area implies that HLP must rely on high mountaineering, high-altitude diving, and other field techniques to collect data and samples. Point data acquisition, long-term logging, sampling, mapping, and *in situ* experiments are combined with geological, limnological, and biological laboratory analysis.

In 2003, meteorological stations were installed at Laguna Blanca and on the shore of the Licancabur lake. They have been logging (T, RH, wind speed) year round since. The characterization of UVR is performed at the surface of the water, near shore, and at depth. A lightweight hand-held probe is used to record conductive heat flux from the lake bottom and collect point measurements at the surface. For long-term monitoring and integrated experiments, two types of stations are deployed in the field. The first is an Eldonet, 3-channel (UVB: 280–315 nm, UVA: 315–400 nm, photosynthetically available radiation (PAR): 400–700 nm) submersible, logging ultraviolet dosimeter. Two stations were installed in the field in 2003, one at Laguna Blanca and one at Licancabur for continuous logging.

The second type of experiment is an array of gazer chambers positioned in shallow waters. The goal of this array is to understand the impact of increased

Table 9.1. *Lakes physical characteristics and analogy to ancient Mars environment*

Physical parameter	Laguna Blanca	Laguna Verde	Licancabur summit lake	Mars NHT
Latitude (°S)	22°47′.00″	22°47′.32″	22°88′.30″	
Longitude (°W)	67°47′.00″	67°49′.16″	67°88′.40″	
Location	Bolivia	Bolivia	Bolivia/Chile	
Elevation (m)	4340	4340	5916	
Surf. area (km^2)	3.5	7.5	0.01	Declining
Maximum depth (m)	0.5	5.5	5.5	Reducing
Atm. pressure (mb)	550–600	550–600	480	500–1000[(a)]
Air temp. (°C)	−28/+10	−28/+10	−40/+5	−50/+27
Av. water temp. (°C)	12[(b)]	13[(b)]	4.9	Relevant[(c)]
UVR (% sea level)	200	200	216	Archean Earth
UVA ($W\,m^{-2}$)	100/55	100/55	110/60	
UVB ($W\,m^{-2}$)	15/7[(d)]	15/7[(d)]	17/8[(d)]	[(d)]
RH (%)[(e)]	9–25	9–25	2–70	Low
Ice cover	Variable year round	Rare	April–September	Relevant
pH	7	8.2	6.9	[(e)]
TDS ($mg\,L^{-1}$)	23 050	199 000	1307	
Wind speed ($km\,h^{-1}$)	0–60	0–60	0–100	0–300
Geology	Basaltic andesite	Basaltic andesite	Basaltic andesite	Observed[(f)]
Hydrothermal input	12–36 °C	12–15 °C	TBD	Relevant
Snow ($mm\,y^{-1}$)	100–150	100–150	100–150	Relevant
Life	Abundant/ Diverse	Present	Abundant/ Diverse	TBD

Notes: NHT, Noachian/Hesperian transition; RH, relevant humidity; TDS, total dissolved solids; TBD, to be determined.

[a] From literature;

[b] Fed by hydrothermal springs;

[c] There is no direct observation so far on Mars but a combination of factors (e.g., association of colocalized evidence of past volcanic and aqueous activity makes hydrothermal activity highly probable) and/or modeling show that a range of temperatures, and pressure is relevant to early Mars;

[d] Values close to UVB in present-day Mars (Cockell and Raven, 2004);

[e] OMEGA on Mars Express (Bibring et al., 2005) suggests that conditions were alkaline on early Mars (phyllosilicate era) and became acidic later on (sulfate era). Acidic aqueous environments are suggested by both Spirit and Opportunity at Gusev and Meridiani (Knoll et al., 2005) on terrain ~3.7–3.5 Ga (Golombek et al., 2006);

[f] Both andesite and basalt have been identified from orbit.

UVR on life, especially on periphyton, which is attached to rocks or other substrates in the part of the water column where UVR impact is the most severe (<0.5 m). Each station is composed of two boxes made of one OP-3 UV filtering and one OP-4 UV transmitting sheets. The acrylite OP-3 absorbs 98% of the incident UV. The OP-4 transmits much of the radiation between 260 and 370 nm and light between 395 and 1000 nm. The first samples were retrieved in 2006 and are currently under analysis to study differences (if any) between microorganisms exposed to and those protected from UVR and for comparison with lower sites (Vinebrook and Leavitt, 1996).

Water samples are collected from each lake every year for fluid chemistry (Hock et al., 2005a) and limnology. Inorganic anion (e.g., sulfate, chloride, others) analysis is performed by Ion Chromatography (IC). Trace element concentrations are obtained from Inductively Coupled Plasma Mass Spectrometry (ICP-MS) and Atomic Emission Spectrometry (ICP-AES). ICP methods allow the analysis of the main rock-derived elements in volcanic lake fluids (e.g., Na, Ca, K, Mg, Fe, and Al from andesitic volcanoes). This is paired with IC analysis of the principal anions derived from magmatic gas – hydrothermal fluid interaction (e.g., Cl, SO_4, and F) to characterize the presence, type, and extent of any hydrothermal activity in the lakes.

The evolution of the lakes' dimensions is surveyed yearly through several approaches – photography, bathymetric profiles with hand-held sonar, and bathymetric mapping performed in 2006 using a Global Positioning System (GPS) located sonar depth finder and surface temperature profiler. This system was mounted on a 52 cm radio-controlled boat. The GPS unit logged the position, the depth, and surface temperature simultaneously of profiles covering the entire lake, producing a 3-D topographic map of the lake's bottom (Cabrol et al., 2007b, 2007c; Morris et al., 2007).

The High-Lakes Project also characterizes microbial life (past and present) and ecosystems, and the impact of environmental extremes and rapid climate change on habitat and life. Since 2002, biological sampling of living specimens has included plankton netting for taxonomy and molecular analysis; scuba diving; and organic geochemical and microbiological characterization (Acs et al., 2003; Fike et al., 2003; Kiss et al., 2004; Cabrol et al., 2007a). DNA is extracted from samples with sequencing of genes encoding from the 16S rRNA of organisms.

Evolution of habitat and life

Abundant cyanobacteria, diatoms, and ostracods are among the fossils identified in the ~100-km^2 microbialite field surrounding the lagunas that was sampled in 2004 and 2005 (Cabrol et al., 2007a). Their identification allowed the reconstruction of the paleolake environment from ~13 240 years before present (BP) and pointed to major climate changes and their impact on habitat and microbial life (Cabrol et al., 2005b, 2006).

Modeling for potential evaporation (Hamon, 1961) based on measurements of air temperature, relative humidity, and water temperature data predicts a strong negative water balance, with evaporation/precipitation ~1000 for Licancabur, and ~1400 for the lagunas (Hock et al., 2005a, 2008). This hydrologic disequilibrium is confirmed by water chemistry measurements from the 2002 and 2004 field campaigns of sodium and chloride concentrations in the lakes, which show an increase in Na (78–110‰) and Cl (130–170‰). The pH of all three lakes is also changing. Yearly measurements indicate a pH of 8.4 in 2002 and 6.9 in 2005 for Licancabur; 9.0 in 2002 and 8.19 in 2004 for Laguna Verde. The trend is reversed for Laguna Blanca with a pH increase from 7.2 to 8.42 between 2002 and 2004. Changes were also notable in amount of total dissolved solids (TDS) between 2002 and 2004 (in mg L^{-1}), when the TDS values went from 22 400 to 23 050; 117 500 to 199 000; and 1050 to 1368 for Laguna Blanca, Verde, and Licancabur, respectively. Laguna Blanca is becoming more basic and the other lakes more acidic. The chemical environment is dynamic and appears linked to annual climate trends and influenced by geothermal (spring) activity at both Laguna Blanca and Laguna Verde. In some cases, changes in pH appear positively correlated to changes in TDS, while in others this correlation is negative or not well determined.

The loss of water column impacts habitability, especially in its shielding role against UVR. Measured UVR is 155% of sea level at the lagunas and 170% at Licancabur. Instantaneous flux for the lagunas at solar noon ranges from 34–84 W m^{-2} in UVA year round and 1–12 W m^{-2} in UVB. In 2003, maximum UVB values in austral summer solstice exceeded 13 W m^{-2} (Hock et al., 2005a). At Licancabur, instantaneous UVB flux at solar noon for the solstice was 17 W m^{-2}. Ozone has a small effect on UV flux compared with solar angle because column abundance is relatively low. However, UVB increases at a slower rate than PAR with higher ozone abundance. Erythemally weighted irradiance exceeds 23 W m^{-2}, nearly a factor of 100 greater than sea level values for the same latitude observed by the Total Ozone Mapping Spectrometer, TOMS, 180–360 μW m^{-2} (Herman et al., 1996).

Radiation flux is crucial for microbial life. While PAR is beneficially used by photosynthetic organisms, high levels of UVB and UVC damage DNA and affect the ability of microbial organisms to repair it. This condition is exacerbated by high daily ΔT fluctuations (Williamson et al., 2002) and sudden temperature fluctuations associated with environmental phenomena

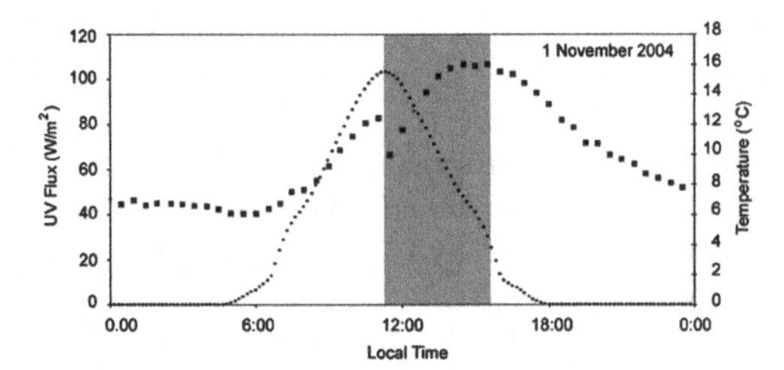

Fig. 9.2. Fluctuation of water temperature (grey squares) and UVR (black dots) at Laguna Blanca. The grey area shows daytime (lake unfrozen) when UVR and T diverge, generating challenging conditions for life (high UV:T ratio) in the shallow 50 cm deep lake.

such as wind, clouds, and hazes that are common in the area (Fig. 9.2). Instantaneous temperature drops of 10–15 °C are not uncommon. At the lagunas, daily $\Delta T = 25\,°C$ in summer and 30 °C in winter. At Licancabur, these values are 33 °C and 50 °C for summer and winter, respectively.

Radiation is attenuated in the water column in very different ways for the three lakes. Licancabur is a freshwater lake. Measurements showed that UVB is attenuated by ~50% within 53 cm (the optical depth), UVA within 1.91 m, and PAR within 10 m (beyond the lake's depth). These measurements were paired with laboratory-based modeling using samples of the summit lake water. Modeled attenuation fell within 1% of the field measurement (Hock et al., 2005a). Its waters are no more than 5% less transmissive than deionized water at all observed wavelengths. UVR is strongly attenuated (50%) in fluids from the hypersaline Laguna Blanca, i.e., UVA (28 cm), UVB (7 cm), and PAR (88 cm), beyond the lake's maximum depth of 50 cm. Cyanobacteria colonies anchored on rocks in shallow waters (≤20 cm) show efficient adaptation strategies to protect themselves against large daily UVR/T swings (Fleming and Prufert-Bebout, 2009).

The water temperature of the lagunas and their contents in Si is explained by cold-to-mild hydrothermal springs (Table 9.1) and large colonies of diatoms (Kiss et al., 2004; Cabrol et al., 2007a). The springs are sites of dense cyanobacterial and other algae concentration with a wide range of pigments. Laguna Blanca's ecosystem is the most diverse with worms, copepods, ostracods, diatoms, cyanobacteria, and heterotrophic bacteria. Laguna Verde is simpler with heterotrophic bacteria, cyanobacteria, and diatoms (Hustedt, 1927; Acs et al., 2003; Kiss et al., 2004; Cabrol et al., 2007a; Fleming and Prufert-Bebout, 2009).

In 2004 and 2006, the HLP diving team documented swarms of ~500–1000 μm red copepods in the Licancabur lake between 0.5–5.5 m depth. They were observed alongside with ostracods and other zooplankton species in what is so far the highest habitat reported for lacustrine copepods (Servant Vilardy et al., 2000). The samples collected in 2006 are being analyzed to document the specific species and their genetical blueprint. Copepods are abundant in Laguna Blanca but only a few dead specimens were found in Laguna Verde.

The documentation of Licancabur's ecosystem was limited up to 2005. Because of a 80-cm-thick ice cover that year, the lake could only be sampled near shore where the ice had thawed at depths of 2, 5, and 10 cm. Thus the samples did not represent the lake's complete depth profile (5.5 m), and were taken above the cut-off for UVR. Total aerobic heterotrophs were plated between 5 °C and 37 °C. Maximum abundance and diversity (colony types) were found at 17 °C and 25 °C (Cabrol et al., 2007b). Growth rate was similar at both temperatures. There was slightly less diversity and abundance at 5 °C, and a much slower growth rate (weeks versus days). While surprising on a first order because of the altitude and climate, these results were found to be in good agreement with the surface water thermal map completed in 2006, which shows warm spots near shore at the sample sites related both to shallow water and large rocks.

The DNA extracted from the summit lake sediment sample was used to prepare a polymerase chain reaction (PCR)-based 16S rRNA gene library using bacterial primers targeting nearly the entire gene. This preliminary survey revealed a microbial population dominated by Proteobacteria and Bacteriodetes. There were very few photosynthetic microorganisms (cyanobacteria), which may have resulted from the sample being taken at a depth above the cut-off of high-level UVR. Overall, 12% of the isolates are currently not classified at the phylum level and 70% are not classified at the genus level. More work is ongoing to establish the library of species living in Licancabur. In 2006, the bottom sediment was sampled over the entire profile of the lake to complete the microbial investigation (Morris et al., 2007).

The past four years of HLP investigation suggest that extreme factors do not always combine to generate a more hostile environment, as proven by abundant life. In specific cases, they mitigate each other's impact and prolong habitability, while in others feedback mechanisms subject life to new sets of extremes (Cabrol et al., 2005b, 2006). How long surface habitability was prolonged this way at the NHT on Mars is unknown. However, by the end of that period, Mars had become inhospitable at the surface. If life was present, it had to seek refuge underground. The MARTE project at Río Tinto explored such a possibility.

Subsurface aqueous habitats

The Río Tinto system

Río Tinto is recognized as a geochemical analog to iron and sulfur minerals found on Mars, particularly at Meridiani Planum where iron sulfates formed in aqueous acidic conditions were discovered. The river is sourced at the Peña Del Hierro in the core of the Iberian Pyritic Belt (IPB, southwestern Spain) and flows 100 km through the province of Huelva, Spain, until it arrives at the Atlantic Ocean (Fig. 9.3). The IPB is a geological formation of hydrothermal origin (Leistel et al., 1998) and one of the biggest deposits of metallic sulfides in the world. Massive bodies of iron and copper sulfides, as well as minor quantities of lead and zinc, constitute the main mineral ores. The IPB is an important mining district exploited for over 5000 years. It is amongst

Fig. 9.3. (Left) Location of Río Tinto in southwest Spain. The field site is at the headwaters on the north end of the river. (Right top) Aerial photograph of the field site location (black arrow). The drill site was located near the edge of a mine pit crater shown in cross section in the right lower panel. The drill perforated the red gossanized layer at the surface. Beneath and to the left of this site the grey deposit is massive sulfide sampled in wells 4 and 8. The light-toned unit to the right of the drill site is rich in sulfates and clays.

the "King Solomon's Mines" of the ancient world. Paleolithic inhabitants followed by Phoenicians and Romans mined intensively in the area for 3000 years and today mining for gold, zinc, and copper continues.

The river has a red wine color, a result of the high concentration of ferric iron maintained in solution by an acidic pH (mean 2.3). In addition to iron, the river has high concentrations of heavy metals (Cu, Zn, As, Mn, Cr) and sulfates (Amils et al., 2007). High summertime temperatures lead to significant evaporation rates and the deposited evaporates, dominated by sulfates such as jarosite and schwertzmannite (Fernández-Remolar et al., 2005), are unstable at higher pH. Iron mineral precipitation occurs on biofilms forming in the river, preserving a fossilized record of life (Amils et al., 2007). Although the acidic waters of the Río Tinto were once thought to result from environmental damage due to mining, research in the area (Fernández-Remolar et al., 2005, 2008) shows that a similar environment has existed for at least 100 000 years and thus predates human activity.

The extreme acidity and the high concentration of heavy metals found in the Río Tinto ecosystem are the direct consequence of biological processes of iron- and sulfur-oxidizing microbes (bacteria and archea) growing in the rich, complex, metallic sulfide deposits of the IPB (López-Archilla et al., 2001; González-Toril et al., 2003). The microbial ecosystem consists primarily of acidophilic iron- and sulfur-oxidizing microorganisms, some of which are facultative anaerobes (González-Toril et al., 2003).

In contrast with typical acid mine drainage systems, where acid is generated in the interaction of oxygenated meteoric water with mine tailings at the surface (Maier, 2000), Río Tinto is sourced with springs already laden with iron and heavy metals, and at low pH, indicating an underground aquifer and suggesting that a chemoautotrophic system based on the oxidation of iron and sulfur minerals might be found in the subsurface aquifer (Stoker et al., 2003, 2004; Fernández-Remolar et al., 2008). Such a system is an interesting model for chemoautotrophic life on Mars, a planet rich in iron and sulfur minerals but lacking significant oxygen.

Mars analogy

The results at Meridiani have been interpreted as the result of interdune inundation by shallow acidic waters that left behind sulfate evaporates (Klingelhofer et al., 2004; Squyres et al., 2004; Fernández-Remolar et al., 2005; Amils et al., 2007). Evaporation of acidic brines from Río Tinto during the dry season produces similar sulfate species to those observed at Meridiani (Fernández-Remolar et al., 2005). In fact, the similarity of many terrestrial mine deposits to the minerals found on Mars is notable (Burt et al., 2006). The presence of sulfur in the regolith has been recognized since the Viking mission conducted the first chemical analysis of Mars soils (Clark et al., 1982). A significant portion of this sulfur may be derived from oxidative weathering of iron sulfide minerals (Burns and Fisher, 1993). The jarosite deposits discovered in Meridiani Planum have been interpreted as a result of the aqueous alteration of pyrite derived from volcanic (Burns and Fisher, 1990) or hydrothermal activity (Hynek, 2004).

Río Tinto is sourced by artesian springs that emanate along steep slopes; at their source, many of these seeps are of very low pH (0–1) and contain high mineral concentrations indicating saline groundwater. These springs represent an interesting analog to martian gullies (Malin and Edgett, 2000). Sampling the groundwater system that feeds these springs would be an attractive target for a future Mars drilling mission in search of life. In that respect, the search for subsurface life at Río Tinto as an exploration of an unknown system also provides valuable lessons for understanding how to go about searching for evidence of past or present life on Mars. The study of life at Río Tinto provides parallels to help better understand life and its biosignatures in other iron-rich acidic environments such as may have been true on the Archean Earth or on Noachian Mars (Amils et al., 2007) and the existence of a modern living analog system may be crucial to the recognition of a fossil biota on Mars.

The MARTE project

Searching for present life on Mars is likely to involve drilling into the subsurface. The present-day Mars surface lacks liquid water. High UVR and strong oxidants in surface soils decompose organic compounds (Klein, 1979). However, geologic evidence shows that ground ice is prevalent in mid and high latitudes (e.g., Squyres and Carr, 1986; Boynton et al., 2002). The interaction of ground ice and volcanism, as evidenced by the large shield volcanoes, suggests that hydrothermal activity may have (or may still) occur in the subsurface. Theoretical and observational evidence (Clifford and Parker, 2001; Clifford, 2003) points to a global-scale subsurface aquifer that would be a prime location to search for present life, although the average depth of liquid water, estimated at a few to several kilometers, is not reachable by drilling with current robotic technology. However, gullies prevalent at mid and high latitudes hint to shallower aquifers possibly at depths of a few hundred meters in some locations (Heldmann and Mellon, 2004; Malin et al., 2006).

Drilling may be equally important for finding evidence of ancient life dating back when liquid water was stable at the surface. The oxidizing conditions on the martian surface may destroy organics in the soil to depths of up to several meters (Bullock et al., 1994). The rock surfaces are weathered by exposure to these oxidants and grinding them to remove the weathered coating has been important for understanding the unaltered rock composition (Arvidson et al., 2004). Drilling is the ideal method for obtaining clues to the geologic record through undisturbed stratigraphy.

The Mars Astrobiology Research and Technology Experiment (MARTE) (Stoker et al., 2003, 2004, 2005) is a collaborative project between NASA and Spain's Centro de Astrobiología. The project studies the subsurface at Río Tinto as an analog system for subsurface life on Mars. One motivation is to determine whether sulfide minerals can provide the key substrate for a subsurface chemolithotrophic biosphere living independently of oxygen, as this would be an important model system for life on present Mars (and likely for most of the past 3.5 billion years after the NHT period). Another important part of the project's motivation is the development of new drilling technology, instrumentation for life detection, and sampling strategies to search for life on Mars.

The project investigates the hypothesis that iron and sulfur minerals provide energy sources for chemoautotrophic metabolism in an anaerobic environment in the subsurface at Río Tinto. To test the hypothesis, rock and groundwater fluids were sampled from locations where they interact in a sulfide deposit and at downstream locations where the reactions have gone to completion. Using commercial coring drills, a series of boreholes were drilled and springs were sampled to characterize subsurface life, energy sources, and by-products (Stoker et al., 2004, 2005). Three boreholes were drilled and sampled using commercial coring. Two penetrated down to 165 m and intersected sulfide minerals and overlying volcanoclastic deposits; the third borehole reached a depth of 60 m, penetrating organic-rich shale downstream of the sulfide lenses.

Subsurface aquifer habitats were characterized using a combination of techniques including core mineralogy and petrography, leached ions from core samples and groundwater fluids, dissolved gases in groundwater fluids, and microbial assays using both culture and culture independent methods. The characterization of the subsurface microbiology is still in progress and suggests the presence of iron- and sulfur-oxidizing biota metabolizing in the absence of or at very low concentrations of oxygen measured in the water table (Stoker et al., 2005; Parro et al., 2008). The biological or geochemical alterations of sulfide minerals is responsible for the production of gases

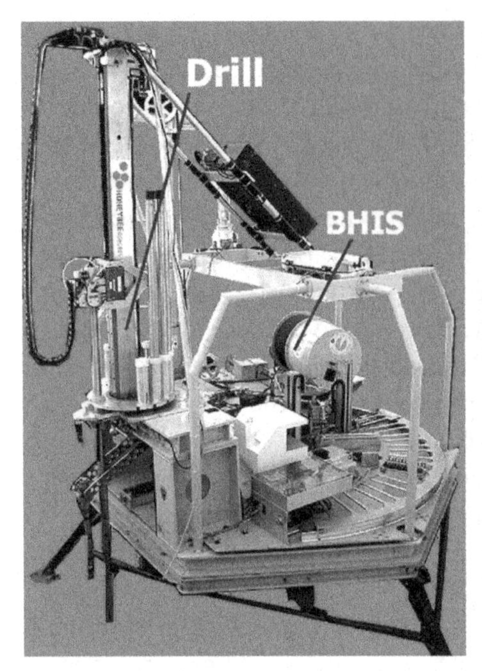

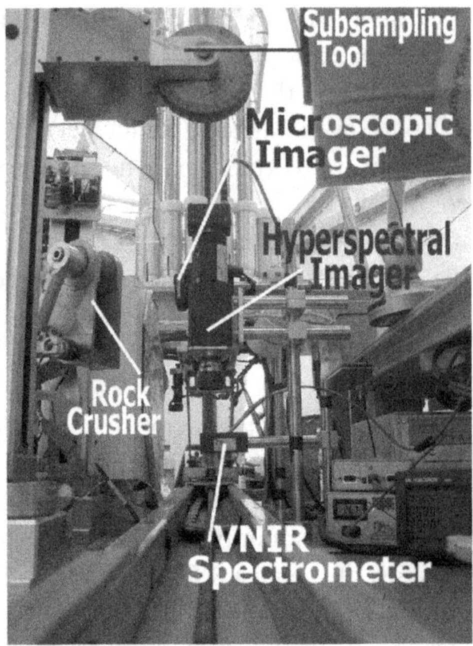

Fig. 9.4. Two views of the MARTE drill platform. (Left) Wide angle view of the complete lander platform. (Right) Lander deck showing the linear rail for the core clamp, the remote sensing instruments, and the sampling saw and rock crusher.

including H_2 and CH_4 and provide resources for microbes that depend on them while secondary minerals, iron oxyhydroxides and sulfates, result from metabolic products (Parro et al., 2008). Environmental Electron Scanning Microscopy and Energy-Dispersive X-ray (ESEM-EDAX) analyses of core samples provide evidence of microbes attacking the pyrite and leaving behind distinctive signatures that may be preserved in the geologic record (Parro et al., 2008).

The approach and investigation methods learned from this exploration guided the development of robotic drilling and sample handling systems, as well as an instrumentation suite used to detect subsurface life as part of a simulated Mars drilling mission (Parro et al., 2008), which used a fluid-less, 10-axis, fully autonomous coring drill mounted on a simulated lander (Fig. 9.4). The sample handling and science instrument systems were also fully automated. The MARTE drill penetrated into gossanized material, the mineral end-product of oxidation of sulfide, found at the surface of the area overlying the sulfide lenses. Cores were brought to the surface, cut open, and examined with color wide angle context imaging, color microscopic imaging, visible/near infrared point spectra, and (lower resolution) visible/near infrared hyperspectral images. Cores were then stored for further processing or ejected. When the drill was not in use a borehole inspection system was inserted in the hole and collected panoramic imaging and Raman spectra of the borehole walls. Furthermore, life detection was performed on full cores using an adenosine triphosphate luciferin-luciferase (ATP) bioluminescence assay, and on crushed core sections using a Sign-of-Life Detector (SOLID2), an antibody array based instrument (Parro et al., 2008). A blind remote science team analyzed the data and chose subsample locations for life detection assays.

In the 30 days of robotic operation, the drill penetrated to 6 m collecting 21 cores (Bonaccorsi and Stoker, 2008); the science team selected 15 samples for life assays and analyses looking for life signatures with SOLID2 – biosignatures were detected in 12 of them (Parro et al., 2008). The science teams correctly interpreted the nature of the deposits drilled as compared with the ground truth results. This mission demonstrated that robotic drilling on Mars is feasible in the next generation of missions and that searching for life signatures on such missions could be scientifically rewarding.

After the robotic mission, cores were returned to the lab for subsequent analysis including assays for organic compounds. Although the borehole was shallow and drilled in a wooded area, total organic carbon varied by up to four orders of magnitude between the surface layers and a depth of 6 m. In this depth range, small pockets of clay were found to be ten times higher in plant-derived organics compared with the hematite-rich material (Bonaccorsi and Stoker, 2008). This suggests that the highly oxidizing nature of iron mineralogy similar to that found on Mars may provide a poor substrate for preservation of organic compounds even on Earth in an environment where organics are being produced biologically.

Arctic environments

While drilling will provide access to deep aqueous habitats, biogenic underground material could have been deposited recently at the surface of Mars by natural processes such as springs. Terrestrial analogs to these springs in the polar regions have provided quantitative models that show how liquid water could have persisted on Mars even if mean annual temperatures have been below freezing for most of its history.

Springs on Earth and Mars

Small-scale features resembling terrestrial water-carved gullies are observed on Mars (Malin and Edgett, 2000). The superposition of gullies on geologically young surfaces such as dunes and polygons as well as the extreme scarcity of superposed impact craters indicate their relative youth. They suggest a formation within the past few million years (Malin and Edgett, 2003, 2000) and possibly in present days (Malin and Edgett, 2006).The gullies are found exclusively poleward of 30° latitude in both the northern and southern hemispheres (Edgett and Williams, 2003; Heldmann et al., 2005b). These regions correspond with the areas of current ground ice stability on Mars (Mellon and Jakosky, 1995) hinting that gully formation may be intrinsically tied to the presence of subsurface ice in the martian polar desert environment.

Hydrothermal convection driven by magmatic activity or impact melt may have provided a mechanism for replenishing water in a martian aquifer. Springs could have formed early in Mars' history as a result of volcanism and meteoritic impacts. Dissolved salts are likely to be present and may enhance the persistence of liquid water environments by depressing the freezing point. On Earth, highly mineralized brines are found in sub-Arctic Canadian Shield wells and in high latitude springs of the Canadian High Arctic (Pollard et al., 1999; Andersen et al., 2002; Andersen, 2004) and Svalbard (Amundsen et al., 2004). Brines flowing onto the surface would form large icings and eventually would deposit the salts via freeze fractionation and/or evaporation.

Few examples of springs in high latitudes or regions of thick continuous permafrost exist. Several groups have reported the occurrence of springs in the High Arctic. Spitsbergen, the largest island in the Svalbard archipelago, resides between 77° and 80° N where permafrost depths are estimated at 100–400 m. Roughly 60% of the island is covered in glaciers and a number of springs occur from Bockfjord in the northernmost part of the island to Sørkapp in the south (Lauritzen and Bottrell, 1994; Banks et al., 1999). The Trollosen spring, a thermoglacial karst spring in south Spitsbergen, has a reported discharge rate of about 18 $m^3 s^{-1}$ of a turbid, 4 °C water during the summer. Its flow is thought to originate from moulins on a glacier about 5 km distant. Nearby springs discharge 12–15 °C water and seem to have a separate hydrothermal source with a subsurface temperature in the deep aquifer of at least 30 °C (Lauritzen and Bottrell, 1994). The thermal springs at Bockfjord are associated with local volcanic sources and geochemical evidence suggests that temperatures at depth reach between 130 and 180 °C. The use of these springs as an analog site for testing life detection technologies has been undertaken by several international research groups (Amundsen et al., 2004; Steele et al., 2004).

Supraglacial sulfur springs are located at 81°019′ N, 81°359′ W in Borup Fiord Pass on northern Ellesmere Island in the Canadian High Arctic

(Grasby et al., 2002). Ten spring outlets were observed discharging from the surface of a 200-m-thick glacier with active discharges estimated at $17\,m^3\,s^{-1}$ at some locations and diffuse seeps elsewhere. The spring outflows are located ~500 m from the terminus of the glacier depositing native sulfur, gypsum, and calcite onto precipitate mounds on top of the ice surface. Outflow temperatures of 1–2 °C were measured (Grasby et al., 2002) and report of H_2S smell was made in and around the outlets.

The water chemistry of the springs is different from local meltwater with higher pH and conductivity values than local melt. Light and electron microscopy and culture-independent molecular methods provide evidence of a microbial community associated with the springs. Grasby (2003) reports the formation of a metastable carbonate known as vaterite, a rare hexagonal $CaCO_3$ polymorph, that precipitates from the spring water as spheres 0.5–10 cm in diameter. The low solute concentration of the spring water implies that these are seasonal springs. Other glaciers in the region exhibit this same type of flow during the summer melt season and are not that uncommon. However, what sets these supraglacial outflows apart from the others is the large amount of mineral precipitates that are carried up from beneath the glacier, indicating the water is flowing well below the surface and is interacting with the local geology underlying the glacier.

The only other known perennial springs at these high latitudes are those located on Axel Heiberg Island in the Canadian High Arctic (Fig. 9.5). The springs occur in a region with a mean annual air temperature of −15.5 °C. Spring flow rates and discharge temperatures are constant throughout the year. Filamentous bacteria, biofilms, and mineral precipitates occur in association with the emergent, anoxic brine flowing from the springs (Andersen, 2004; Omelon et al., 2006; Perreault et al., 2007).

The first report of saline springs at the base of Gypsum Hill on the north side of Expedition Fiord was made in 1964 (Beschel, 1963). Their location, size, distribution, discharge rates, and temperatures were described as well as the basic chemistry assessed.

Although several authors mentioned the springs, they were not the focus of their studies (e.g., Heldmann et al., 2005a). The first detailed study (Pollard et al., 1999) documented the hydrologic and geomorphic characteristics of the springs and provided additional information on a second set of springs 11 km to the west at Colour Peak (Fig. 9.6). Initial measurements of spring discharge, water chemistry, and descriptions of the mineral precipitates associated with the springs were among the data presented which showed that they are perennial in nature and that the outflow temperatures and flow rates remain constant throughout the year. The formation of ikaite ($CaCO_3{\cdot}6H_2O$)

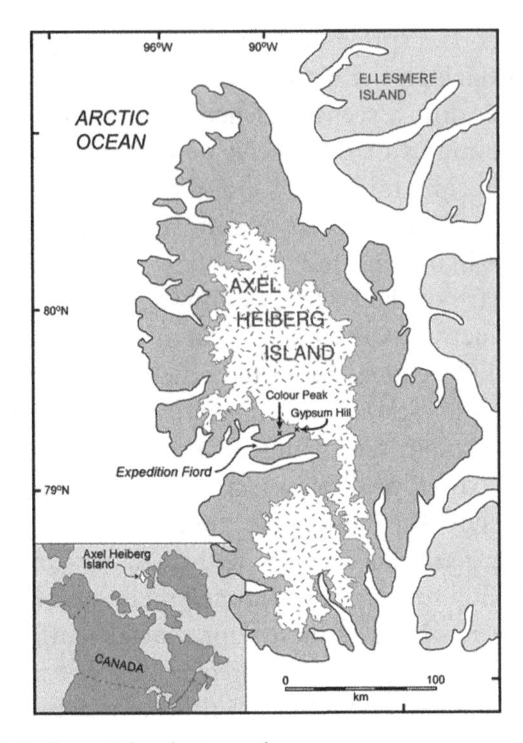

Fig. 9.5. Axel Heiberg Island research area.

Fig. 9.6. Colour Peak Springs, Axel Heiberg Island.

was reported (Omelon et al., 2001), a metastable carbonate mineral that develops at low temperature. Since the springs on Axel Heiberg Island flow all year with little variation in temperature, a thermal model was developed to explain how the springs are able to maintain perennial flow (Andersen et al., 2002). The ultimate source of this brine solution has been enigmatic for many years. It is suggested that the source of the water is a combination from basal melting of the Muller ice sheet and water from glacially dammed Phantom and Astro Lakes located ~350 m above the outlet of the springs (Andersen et al., 2002). Phantom and Astro lakes are sufficiently large to sustain through taliks (unfrozen ground above, below, or within discontinuous frozen soil). Faulting beneath the lakes may also provide conduits into the subsurface. Water from the two lakes could flow beneath the permafrost via the underlying evaporite strata and return to the surface as a spring discharge flowing through the evaporite piercement structures (Andersen et al., 2002). The resulting brine is warmed to the geothermal temperature within the underlying evaporite layer and loses heat to the surrounding permafrost as it flows upward to the Gypsum Hill and Colour Peak spring outlets. Large icing results from the flow of the springs after freeze-up in the fall (Beschel, 1963). The evolution of the icing occurs on an annual basis and can be divided into four distinct phases. The icing formation, icing melt, and icing washout processes are dependent upon ambient air temperatures. The icing deposit begins to form in late September once the spring water cools from its outlet temperature to the freezing point of −7 °C (Heldmann et al., 2005a). The ice grows in lateral extent via channelized brine flow under the insulating ice cover. Mass balance calculations show that the total mass of the icing is consistent with the amount of total discharge from the spring outlets.

In summary, water is abundant at the Earth's polar regions but in general it is not available as an exploitable resource by life because it is normally found as ice. The north and northwest Arctic archipelago and much of the Antarctic are characterized by polar desert conditions with mean annual temperatures at or below −15 to −20 °C. Periods of summer melt are short and the winters are long and cold. Most surface water is derived from the melting of local ice caps and glaciers during the short summers. The current glacial and periglacial environments have existed for tens of thousands to millions of years developing thick, continuous permafrost. Nevertheless, life has exploited these environments, both in the Antarctic and the High Arctic.

Springs, residual icings, and massive ground-ice deposits located in the Arctic and Antarctic thus provide opportunities to study microbial ecosystems in extreme polar environments. The ranges in temperature, pH, redox, nutrient availability, and the large seasonal variations in light have undoubtedly shaped the structure and function of the ecosystem as well as impacting the biological record left in the sediments. Mars may have once hosted microbial ecosystems in a physical setting not too dissimilar to these regions.

Haughton Crater and Devon Island

Studies of habitats and their microbial ecosystems in regions of thick, continuous permafrost undoubtedly aid in preparing the future search for evidence of life in the martian permafrost. Here, the Haughton Mars Project (HMP) exemplifies their rich diversity and evolution in a geologic and climatic environment very similar to what Mars has been throughout its history.

Haughton Crater (Fig. 9.7) on Devon Island, High Arctic (75.2° N, 89.5° W) is ~20 km in diameter and ~39 Ma old (Late Eocene). It is the second highest-latitude terrestrial impact crater on land. Devon Island is the largest uninhabited island on Earth. Haughton has been the subject of a number of scientific investigations since its identification as an impact structure in the early 1970s (e.g., Grieve, 1988; Lee and Osinski, 2005).

Impact craters are one of the most common geologic features on planetary bodies with solid surfaces, and the result of a fundamental and universal process in planetary development and evolution – collisions. The study of terrestrial impact craters helps understand the physics of impact cratering and the key role this process has played in shaping planetary environments and life through time.

Although Haughton is merely one of more than 180 impact craters known on Earth, it is the only crater set in a polar desert and the best-preserved impact structure in its size class. Most of the well-preserved impact craters on Earth are relatively young and small. They involved low formation energies and lack many of the geologic complexities found only at large impact sites

Fig. 9.7. ASTER satellite image of Haughton Crater, Devon Island. (Image: HMP.)

(e.g., central uplifts, collapsed/terraced rims, hydrothermal systems). Large craters form more rarely than small ones, and those found on Earth tend to be older and less well preserved. But due to its polar location, Haughton is an exception to this rule. It has experienced a predominantly cold and dry climate throughout its history and remains well preserved in spite of its age. Haughton's present desert setting also allows it to be well exposed. From these considerations alone, Haughton is an outstanding planetary analog.

With regard to Mars, the analog value of the Haughton site resides in two distinct domains: Haughton Crater itself; and Devon Island beyond Haughton Crater. The second domain includes valley networks, canyons, gullies, rock glaciers, ice-cored mounds, patterned ground, and many other features of aqueous, fluvial, glacial, periglacial, lacustrine, depositional, and erosional origin found across Devon Island.

Since Haughton's formation in the Late Eocene, the climate of the High Arctic has to first order experienced a jagged trend toward cooling. Much of Devon Island, including Haughton Crater, was glaciated during Last Glacial Maximum which ended ~10 000 years ago. At present, Devon Island is a polar desert with an annual mean temperature of −16 °C and annual precipitation of <13 mm. While not as extreme as Mars or Antarctica, the present climate does support continuous permafrost under mostly rocky and unvegetated terrain. Liquid H_2O, while present, occurs only seasonally, transiently, and/or locally.

The Haughton-Mars Project

The unique combination on Earth of an impact structure and a polar desert is what makes Haughton attractive *a priori* as a potential analog for Mars (Lee, 1997). In 1997, a NASA-supported pilot study was initiated and a field party of four deployed to Devon Island (Lee et al., 1998). Both Haughton Crater and areas of Devon Island outside the crater were found to offer a wide range of planetary analogs. Research opportunities were identified not only in science, but also in exploration. Given the far-reaching potential of the site, the HMP was established as a long-term framework for analog studies on Devon Island (Fig. 9.8).

The HMP was designed from the outset as an international multidisciplinary planetary analog field research project with both science and exploration components. The Science program studies the site itself, in particular topics relevant to planetary sciences and astrobiology. The Exploration program makes use of the site as a planetary test bed (Moon or Mars) and takes advantage of actual ongoing field science activities (not simulations thereof) to develop technologies, strategies, and the operational experience needed for

Fig. 9.8. The Haughton-Mars Project Research Station (foreground) and Haughton Crater in the distance. (Image credit: HMP/P. Lee.)

planning future human and robotic planetary exploration missions. The project also engages government and private sector partners, international participants, and students. The HMP thus fulfills all four functions or usages of analogs (Lee et al., 2004): to help (1) learn; (2) test; (3) train; and (4) engage.

Haughton Crater and Mars

Analogies between Haughton Crater and craters on Mars are clearest when either H_2O, environmental evolution, and/or the question of life are considered. For instance, while Haughton's massive impact breccia deposits are of interest for understanding impact ejecta on any planet, it is their cooling history, interaction with groundwater and ice, weathering style, erosion rate, and ability to preserve an organic or biological record through time that present specific relevance to, and thus analog value for, Mars.

Selected results from the studies of geology and biology of Haughton Crater and its relevance to Mars (Lee and Osinsky, 2005) are summarized below.

The age of Haughton Crater was reevaluated using high precision ^{40}Ar–^{39}Ar laserprobe dating (Sherlock et al., 2005). The new study yields an age of ~39 Ma, greater than previously thought (~23 Ma). Haughton's age has far-reaching implications for investigating post-impact modification rates and climate evolution on Mars (Lee and McKay, 2003). Given what is known about climate evolution in the Arctic over the past 39 Ma and the degradation experienced by Haughton, it was inferred that integrated denudation rates on Mars ($\sim 10^{-3}$–10^{-2} μm y^{-1}) are two to three orders of magnitude slower than classically believed (Lee et al., 2005), casting doubt on the widely held view that Early Mars was wet and warm for substantial lengths of time.

Osinski et al. (2005b) provide a geological overview of Haughton, including a refined model for the crater's formation and evolution. The response of sedimentary rocks, in particular carbonates, to hypervelocity impact, was examined (Osinsky et al., 2005c) and it was shown that they underwent melting during the Haughton impact event, a fact not recognized prior to HMP studies. Considering the potential importance of carbonates as target rocks on Mars, the finding has relevance beyond Haughton itself, for instance as input into Mars impact modeling studies.

Beyond the excavation of a crater, the Haughton impact event was accompanied by several important post-impact processes, for example impact-induced hydrothermalism. The identification of impact-induced hydrothermal signatures at Haughton has been a priority of the HMP (Lee, 1997; Lee et al., 1998). Such systems arise only in association with relatively large impact events and are usually poorly preserved and expressed. Mapping and studying hydrothermal systems at Haughton helps understand their nature, distribution, and duration at this impact site, but also determine where analogous signatures might be found at impact sites on Mars (Lee et al., 1998; Osinsky et al., 2001). Initial findings and classifications of hydrothermal signatures at Haughton were reported (Osinsky et al., 2001). A moderate- to low-temperature hydrothermal system was generated by the interaction of groundwaters with the hot impact melt breccias that filled the interior of the crater. Temperature constraints for the hydrothermal systems were provided (Osinsky et al., 2005a) and they established the first map showing the complete distribution of different hydrothermal deposits around a mid-size complex impact structure. Continuing with post-impact processes, Haughton's hydrocarbon system was investigated (Parnell et al., 2005) and shows how five different types of geochemical signatures (thermal effect of impact on target rocks, hydrocarbons in impact-induced hydrothermal system, organic signatures in post-impact sedimentary crater-fill, modern sediment in and around the crater, and contamination by later life) may remain recorded at the site of a relatively large impact. Organics were traced from pre-impact country target rocks to debris resulting from the erosion and weathering of impact-processed materials. The results have important implications for the preservation and recoverability of potential ancient biosignatures in heavily impact-processed terrains on Mars.

Post-impact paleolacustrine deposits (Haughton Formation) are also present inside Haughton Crater (Hickey et al., 1988). A series of new intracrater sedimentary units at Haughton were reported (Osinski and Lee, 2005). They show that the Haughton Formation was deposited up to several million years after the formation of Haughton Crater rather than in the immediate aftermath of the impact event. This study also identifies a glacial record at Haughton. The investigation of intracrater sedimentary deposits helps understand not only processes operating inside Haughton through time, but the origin, nature, preservation, and accessibility of potentially analogous deposits inside craters on Mars (Lee et al., 1998).

Finally, post-impact biological succession and recovery at Haughton was investigated (Cockell et al., 2005). A synthesis view of the effects of impacts on habitats for microbial lithophytic organisms was developed. An interesting astrobiological insight gained is the possibility that impacts, rather than being strictly life destroying, may also represent habitat-generating opportunities for microbial life (Cockell and Lee, 2002). Implications include the possibility that the development of microbial life on Early Earth (and Early Mars?) was facilitated by impacts.

Devon Island and Mars

A wide array of potential Mars analog features on Devon Island are found outside Haughton Crater. Some of them have potentially profound implications for Mars.

Lee et al. (1999) draw an analogy between the small valley networks (SVNs) of Devon Island and the SVNs of Mars. Those on Devon exhibit all the distinctive morphologic traits characterizing SVNs on Mars. Classical interpretations for SVNs on Mars are that they formed under relatively slow H_2O discharge regimes, that is, under a wet and warm climate, the latter being actually difficult to achieve (because of the Early Faint Sun paradox, and low global atmospheric pressures later in Mars' history). However, many SVNs on Devon Island are glacial meltwater channel networks formed subglacially under confined flow (Lee et al., 2001). An analogous origin, if applied to SVNs on Mars, would explain both their peculiar morphologies and context, and more importantly, remove any requirement for a warm climate. Thus, HMP studies suggest that Mars may have been climatically frigid since ~4 Ga ago, consistent with HMP estimates of low integrated denudation rates on Mars.

A climatically cold Mars is also the consistent picture emerging from analog studies of canyons on Devon Island. Large canyons on Devon offer high fidelity and unique morphologic analogs for some canyons on Mars, down to peculiar attributes (Lee et al., 2001). A classical interpretation for those martian canyons is that they formed by sapping under a relatively warm climate (Sharp and Malin, 1975). The canyons on Devon, however, are glacial trough valleys incised by local dynamic ice–stream flow in the context of cold-based glaciations (Lee et al., 2002). The resulting landscape of selective linear erosion is actually diagnostic of cold-based glaciation, at least on Earth. When applied to Mars, such an interpretation provides an alternative explanation for the formation of canyons on Mars with no warm climate requirement.

Finally, a wide range of gully systems are found on Devon Island that offer both morphologic and contextual analogs for several types of gully systems on Mars (Lee et al., 2002, 2006). The gullies on Devon grow mainly by seasonal melting of transient deposits of surface snow or ice (depending on gully type). Subsurface contributions (seepage from an aquifer, thawing of ground ice) are minor. On the basis of how gullies form on Devon, a possible explanation for the formation of most gullies on Mars is considered to be via melting of transient surface snow and ice deposits during episodes of atmospheric pressure increase, probably from obliquity variations of Mars on timescales of $\sim 10^5$–10^7 years (Lee et al., 2001, 2002, 2004, 2006).

After a decade of analog studies on Devon Island, it is anticipated that the wealth of analog aspects still left to be explored will continue to attract a wide range of investigations for many years to come.

Aridity and desert analogs

In contrast with protected habitats, the surface environment on Mars is inhospitable due to extreme cold and aridity, low atmospheric pressure, and strong ultraviolet light. As shown in the previous sections, different analog environments on Earth provide examples of one or more of these environmental stresses. Extreme arid deserts provide the best environments to study life under extreme dry stress, which parallels conditions that Mars has experienced for the most part of the past 3.5 billion years.

The three candidates for the driest desert on Earth are the Atacama desert in Chile (McKay et al., 2003; Navarro-Gonzalez, 2003), the Lud desert in Iran, and the region of the Darb el Arba'in desert (Haynes, 2001) in the eastern Sahara desert near the border of Egypt, Lybia, and Sudan. In all three the level of rain in the hyperarid core is virtually indistinguishable from zero.

For practical and logistics reasons the Atacama has been studied the most. Studies in its hyperarid core have provided insights into the dry limit of life, the relationship between dehydration and radiation resistance, and the detectability of organic remains of life at extreme low levels.

The Atacama is a temperate desert, and although it is never extremely hot, it is one of the driest deserts in the world. Both the Lut and the eastern Sahara deserts have hot summers. Over a four-year period of meteorological recording the highest temperature recorded in the hyperarid core of the Atacama was 37.9 °C, and the minimum was −5.7 °C (Haynes, 2001).

The Atacama thus provides a site where the ecological effects of dehydration can be studied separately from temperature extremes – either hot or cold.

The Atacama extends for over 1000 km from central Peru into Chile along the Pacific coast from 17° S to 27° S between 69° W and 71° W. Several factors contribute to the lack of rain. The cold ocean water off the coast resulting from the cold north-flowing Humboldt ocean current leads to low-lying coastal fogs but no coastal rain. The generally stable position of the strong Pacific anticyclone generates southwesterlies that bring cool dry air inland. The Andes mountains block most moisture from the east but snowmelt from the Andes does provide moisture in the eastern parts of the Atacama, forming, for example, the Salar de Atacama.

The factors that create the arid conditions in the Atacama (the cold Humbolt current, the Pacific anticyclone, and the Andes mountains) have been present for many millions of years, making the Atacama one of the oldest deserts on Earth. Recent studies suggest that conditions there have been arid for 90 Ma (e.g., Hartley and Chong, 2002; Hartley et al., 2005) and regions within the desert have been hyperarid for 10–15 Ma (Eriksen, 1983; Berger and Cooke, 1997; Houston and Hartley, 2003). Thus, unlike the hyperarid core of the Sahara, there are no significant legacy effects in the soil organic material or water table from earlier more clement times. The system has had time to come to equilibrium with the arid conditions.

The age and aridity of the Atacama are probably directly responsible for its large nitrate accumulations. The nitrates are likely to be of atmospheric origin (Böhlke et al., 1997; Michalski et al., 2002, 2004) and are not biologically decomposed or carried away by water flow due to the extreme aridity. They have accumulated into significant concentrations over the long age of the desert.

In the present climate, there is practically no rain anywhere northward of 27° S; however, there are significant variations in the density of vegetation in the desert due to variations in the level of moisture provided by fogs (e.g., Caceres et al., 2007). Along the coast, fog creates a zone of vegetation from about 100 m to 700 m above sea level (Rundel et al., 1991). In locations where the coastal mountains are less than ~700 m, the thick marine fog penetrates inland and provides for lichens, algae, and even cacti.

The driest parts of the Atacama are located where the marine fog cannot penetrate. The main area is between approximately 22° S to 26° S in the broad valley formed by the Coastal Range and the Cordillera de Domeyko medial range (McKay et al., 2003) (Fig. 9.9). In the early 1900s nitrate mining operations were conducted in this area but most are now depleted. Here the coastal mountains average about 2.5 km in elevation and block the inland movement of fog.

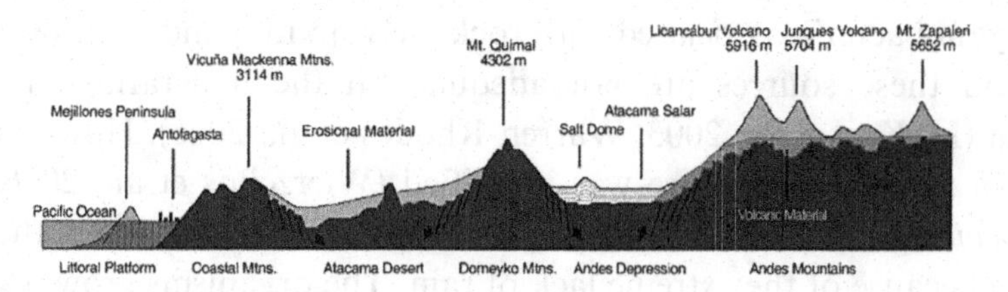

Fig. 9.9. Cross section of the Atacama near 24° S, in the vicinity of Antofagasta, Chile. (From McKay et al., 2003.)

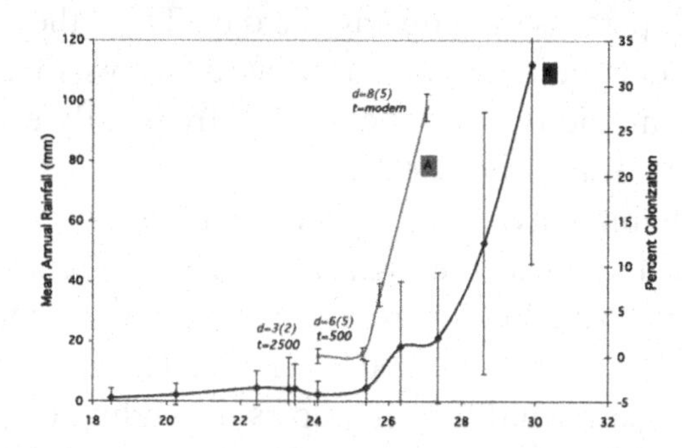

Fig. 9.10. Historical average rainfall (R) and the abundance (as percent colonization, A), diversity (*d*), and steady-state residence time (*t*) of hypolithic communities along a north–south transect. The diversity is shown as the total number of distinct sequences recovered and as the number of distinct cyanobacterial sequences recovered (in parentheses). (From Warren-Rhodes et al., 2006.)

For any ecosystem to be possible there must be a source of primary productivity. Thus, a key question in any environment, as well as for Mars, is the possibility of primary production, in specific photosynthesis. In extreme deserts no plants or bushes are present and the only photosynthetic organisms are found under diaphanous rocks or within porous sandstones. By being under, or within, rocks, these organisms live in an environment that holds moisture longer relative to the ambient conditions.

The distribution of quartz rocks colonized by cyanobacteria (Warren-Rhodes et al., 2006) shows that the fraction of rocks colonized drops sharply northward of 25° S and closely follows the drop in rain (Fig. 9.10). As mean rainfall declines from 21 to $<2\,\text{mm y}^{-1}$, the fraction of rocks colonized by cyanobacteria decreases from 28% to <0.1%, molecular diversity declines threefold, and organic carbon residence times increase by three orders of magnitude, reaching 2600 years in the hyperarid core. Communities contained a single *Chroococcidiopsis* morphospecies of cyanobacteria with heterotrophic associates, leading to the conclusion that for these organisms the dry limit for life had been crossed (Warren-Rhodes et al., 2006).

The cyanobacteria associated with rocks rely on rain and fog as sources of water and these sources are not adequate in the hyperarid core of the Atacama (McKay et al., 2003; Warren-Rhodes et al., 2006). However, novel habitat in the hyperarid core was identified (Wierzchos et al., 2006) where *Chroococcidiopsis* is found within halite "rock." Halite is stable in the hyper-arid core because of the extreme lack of rain. The organisms grow just below the surface of the halite. Halite can absorb atmospheric water when the relative humidity rises above ~75%, creating a saturated salt solution within the halite rock that persists even during the day. Thus, the organisms benefit from the relative deliquescence of halite as a means of extracting water. Nighttime relative humidity rises above 75% frequently even when no rain or fog is present (McKay et al., 2003).

The extent of habitat available under quartz rocks or within halite is limited. The main habitat in the area is the soil. In general the soils have low organic material and low microbial concentration (Drees et al., 2006; Caceres et al., 2007; Lester et al., 2007). The microbial concentration found varies considerably between different studies depending on the sampling site and the method used for enumerating the bacterial population. In some areas there are virtually no culturable soil bacteria (Lester et al., 2007) and only a small number of fungi have been cultured (Bagaley, 2006). These fungi are closely related to resistant-spore-forming fungal groups that are distributed by wind and associated with plant degradation and material decay (Conley et al., 2006). For all organisms, DNA extraction yields low levels compared with the soils in the southern, wetter, parts of the Atacama or along the coast where marine fog provides moisture for hypolithic cyanobacteria, lichen, and even cacti. There are very low levels of organic material and the organics that are present are refractory. An oxidizing agent present in the soil equally oxidizes L and D amino acids and L and D sugars (Navarro-Gonzalez et al., 2003).

Indeed, if the Viking Landers had sampled in the driest regions of the Atacama they would have returned results similar to those they returned from Mars (Klein, 1978, 1979), making the Atacama a good place to test instrumentation destined for Mars (Quinn et al., 2005; Skelley et al., 2005; Navarro-Gonzalez et al., 2006).

Robotic exploration of habitats and life

The aridity of the Atacama and its rare and scattered life are ideal for developing and validating robotic capabilities for planetary exploration. The desert consists of features important for analog experiments including craters from meteoritic impacts, volcanic and alluvial flows, rocky pavements, sandy dunes, and salars (salt flats). Its heavily eroded topography and total lack of vegetation combine to create a landscape analogous to that found on Mars.

Experiments with prototype rovers have been conducted in the Atacama for over a decade. In 1997, the Atacama Desert Trek (Wettergren et al., 1999) deployed the Nomad rover for long-distance traverse. Studies were conducted of operating modes from direct teleoperation to complete autonomy in simulation of martian, lunar, and Antarctic investigations. Nomad went on to successfully hunt meteorites in Antarctica (Apostolopoulos et al., 2001). In 2003, the Hyperion solar-powered robot performed experiments in endurance navigation and resource-constrained planning after having successfully demonstrated sun-synchronous navigation on Devon Island in Canada. In 2005, the Zoë astrobiology rover enabled autonomous large-area survey in an investigation of the distribution of microbial life in the desert.

These rover experiments have demonstrated capabilities necessary for robotic planetary exploration. Important technologies for rough-terrain mobility, autonomous navigation and localization, safeguarded teleoperation and immersive visualization, high-bandwidth communication, and science autonomy were validated in Mars analog field settings. In addition, instruments including neutron detectors, thermal and visible wavelength spectrometers, and fluorescence imagers have been tested with robots. Finally, field experiments in the Atacama have been vital in understanding the human aspect of remote experience and developing new methodologies for exploration.

Terrain morphology and appearance

The topographic profile of the Atacama (Fig. 9.9) holds much of the terrain morphology observed on Mars. The nearly static erosion processes have left the desert at very shallow slope angles. Flat and level dry lakebeds, salt flats, and salars are found throughout the desert. Alluvial fans and debris aprons offer some changing elevation and are occasionally broken by faulting and ancient streambeds that present considerable obstacles. In the core desert a rover encounters both loosely consolidated sands with a thin gypsum crust and packed desert pavement. Craters and sand-filled basins are also found, for example in Monturaqui (Bunch and Cassidy, 1972) and Valle de la Luna.

In many areas aeolian activity has scoured the surface leaving a pavement with scattered 0.5–3.0 m obstacles. This presents the planetary rover with dense, discrete obstacles commonly seen in impact debris fields. Contrary to the initial observation of unimpeded travel, experience has shown that rarely can travel over 100 m be achieved without encountering an obstacle that requires some amount of path deviation. Rover-scale obstacles vary in distribution but present the variety of challenges found on Mars (Fig. 9.11).

The high altitude (2000–3000 m) results in abundant solar energy. Insolation reaching 1000 W m^{-2} has been measured on the ground (Calderón et al., 2007). This amount of solar flux is more than encountered on Mars by a factor of approximately 3 when martian atmosphere is considered (insolation on the Earth's upper atmosphere is 1368 W m^{-2} compared with 589 W m^{-2} at Mars). Gravity being greater on Earth than Mars (9.81 m s^{-2} versus 3.69 m s^{-2}), a similar scaling factor emerges making the insolation condition in the Atacama for a terrestrial rover a good analog to the insolation condition for a martian rover.

Only in the most remote and inhabited regions does one encounter such a vast emptiness. It is important to consider the appearance of terrain for two reasons: the perception of scientists and the perception of robotic sensors. The former allows an easier immersion in mission simulation and the development of operational methods; the latter is necessary for testing, sensing, modeling, and planning technologies that can be translated to operation on Mars. The purely geologic appearance – rocks, soils, and near total lack of vegetation or human impact – is important to suspending disbelief and aiding scientists in envisioning their rover on Mars. The lack of visual cues allows better simulation, or conversely whenever vegetation or artificial structures are observed, they are immediately apparent and distracting. Equally important for the testing of robotic systems is that any observable features that are not analogous to the martian setting require tuning, filtering, or compensation that is not relevant to an actual flight system. The robot's perceptual system can be made to detect and avoid the fine, branching structure of a dry bush, but this may require different sensing modes, higher resolution, or complex interpretation. None of this will ultimately apply to a Mars rover so it is far more effective to use an analog setting in which the terrain closely approximates Mars in its appearance and specific details.

Planetary rover prototypes

In the past decade there have been three robotic systems tested in Mars analog experiments in the Atacama: the Nomad rover near Salar de Atacama, Hyperion in Salar Grande, and Zoë in five locations along the Coastal and Domeyko ranges (Fig. 9.11; Table 9.2). These planetary rover prototypes were created by the Robotics Institute at Carnegie Mellon University. The technical objectives of the field experiments varied but broadly focused on enabling long-distance mobility in planetary terrain with sufficient autonomy for navigation and science investigations.

Fig. 9.11. Rovers in the Atacama. (a) Nomad, 1997, demonstrated long-distance teleoperated traverse with high-bandwidth communication of immersive visual imagery. (b) Hyperion, 2003, was a solar-powered platform for autonomous navigation in obstacle-strewn terrain. (c) In 2005, Zöe with its instrument payload conducted multi-kilometer autonomous traverses to map the distribution of life. (d–g) Zöe traversing a variety of terrain types in the Atacama. On dry mud flats (d) sinkholes are the only obstacles. Much of the desert is loose unconsolidated soil with a thin crust, typically gypsum. (e) Faults bisecting the terrain. (f) Dense obstacle fields are common in slope breakdown areas. (g) Ancient drainages present insurmountable embankments that must be circumnavigated.

Table 9.2. *Rover specifications*

	Nomad	Hyperion	Zöe
Date, location	1997, Chile: Salar de Atacama 1999, Antarctica: Patriot Hills 2000, Antarctica, Elephant Moraine	2001, Canada: Devon Island 2003, Chile: Salar Grande	2004, Chile: Salar Grande, Buscuñan Hill 2005, Chile: Salar Grande, Salar de Navidad, Llano del Guanaco
Mass (kg)	725	181	198
Dimensions (m)	1.8 × 1.8 × 2.4 stowed 2.4 × 2.4 × 2.4 deployed	1.8 width (axles), 2.0 length (between axles), 1.8 height, 0.42 ground clearance	1.63 width (axles), 2.2 length (between axles), 1.8 height, 0.35 ground clearance
Wheel diam. (m)	0.76	0.75	0.75
Turning radius (m)	0.0	4.5	2.5
Speed ($m\,s^{-1}$)			
Nominal	0.3	0.25	0.90
Maximum	0.5	0.30	1.10
Power (V bus)	24	24	72
	150 W steady state + 100–600 W locomotion (straight to skidded turn)	120 W steady state + 90–200 W locomotion	120 W steady state + 90–260 W locomotion
Generator	3.5 kW internal combustion	Silicon, 12.8% conversion efficiency, 3.45 m^2	Triple junction, GaAs, 23% efficiency (average), 2.4 m^2
Storage	N/A	Lead-acid, 300 Whr capacity (×2)	Lithium-Polymer, 1500 Whr capacity (×2)
Computing (MHz)	200, Dual Pentium Pro 133, Pentium (×2) 50, MC 68040 40, MC 68030	500, Pentium 3	2.4 GHz, Pentium 4, 1 GB RAM (×2) 800, Celeron, 256 MB RAM
Sensors	Panospheric camera, high resolution stereo, magnetometer, temperature, humidity, wind speed	High resolution stereo camera, VNIR spectrometer (unmounted), temperature, humidity	High resolution trinocular stereo camera, VNIR spectrometer (350–2500 nm), TIR spectrometer, (unmounted), neutron detector (unmounted), fluorescence imager, temperature, humidity, condensation, wind speed, insolation
Operation	Direct teleoperation, safeguarded teleoperation, autonomous path following	Direct teleoperation, supervised autonomy	Direct teleoperation, navigational autonomy, science autonomy

Desert Trek

The deployment of Nomad in the Atacama (June 15–July 31, 1997) coincided with the landing of Pathfinder on Mars (July 4, 1997). Pathfinder was considered a technology demonstration. The rover had a reduced exploration range, few science instruments, and before the mission experience on remote science operations using rovers was limited to a handful of field simulations. But Pathfinder literally opened the path to what the science community knew would be future long-range rover missions with specialized science payload, and the Desert Trek was designed to test this concept.

The objectives of the Desert Trek included multiple technology evaluations and remote science investigations. The goal was for Nomad to travel 200 km during the 45-day field season. Ultimately a total distance of 223.5 km was achieved.

Five days were dedicated to remote science. The science objectives were to provide a realistic experience for remote operators through immersive imagery and virtual environment interfaces; to evaluate near-term planetary missions (to the Moon, Mars, and Antarctica) including operations environment and processes, exploration strategies, and interaction in the science teams; to evaluate various imaging techniques including panospheric imaging, foveal-resolution stereo imaging, image mosaicing, and textured terrain models; and to understand the reasons for correct and incorrect scientific interpretation by collecting ground-truth and reviewing scientists' methods and conclusions.

Two Mars mission simulations provided training for site characterization and sample caching. The site characterization exercise, in which scientists tried to characterize the climate, geology, and evidence of past life, was conducted without long-range or aerial imagery, in simulation of Mars missions. Scientists collaborated to analyze images from the science cameras, resulting in a slow but thorough process. The sample caching exercise utilized all available imagery and resulted in nearly four times the area covered with a number of distinct rock types selected as samples.

In the lunar mission simulation, planetary scientists from NASA and the U.S. Geological Survey performed "geology-on-the-fly," in which they assessed trafficability and surveyed gross geology while keeping the robot in motion 75% of the time. This strategy is appropriate for long-distance exploration or for traverse between sites of scientific interest. In a record for remote exploration at the time, Nomad traversed 1.3 km and examined ten science sites in a single day. During the lunar mission simulation Nomad's high-resolution color stereo cameras revealed a rock outcrop dating from the

Jurassic era. Geologists identified the outcrop as a target of interest in aerial images and navigated to it using rover panoramic imagery. Close inspection of exposed rocks indicated fossilized material. Later ground-truth analysis did confirm fossils in sampled material and the rock outcrop was correctly characterized as a fossil bed (Cabrol et al., 2001a). This was the first time geologists using a robot surrogate had made such a discovery, something that had been widely thought to be too difficult to do remotely. This has important implications to future planetary exploration. While noting that the demands of remote geology require practice, geologists concurred that even without a full suite of tools for geologic exploration (e.g., rock hammer, hand lens), useful and accurate analysis can be performed remotely.

Biogeologic mapping

Life in the Atacama is sparse overall and distributed in localized habitats on a scale of 10–100 m to kilometers. While some have been studied in detail (e.g., Navarro-Gonzalez et al., 2003; Wierzchos et al., 2006), little is quantified about the extent and distribution of life and habitats across the desert. This was the focus of the Life in the Atacama project (LITA), to investigate the regional distribution of life. The approach was to create biogeologic maps using survey traverses across the desert with biologic and geologic instruments. These surveys were accomplished in a method technologically relevant to Mars exploration using an autonomous astrobiology rover.

In the first LITA field investigation in 2003, testing focused on component technologies. Hyperion, a solar-powered rover designed to exploit the advantages of sun-synchrony in polar regions (Wettergren et al., 2005c) was reconfigured for an equatorial environment. An outcome of testing this prototype in the field was to develop the requirements for a terrestrial life-seeking rover. This led to the creation of Zoë for the second and third field seasons. The LITA investigation also saw prototyping and evaluation of three rover instruments: a trinocular stereo imager with high resolution and wide baseline, a visible-to-near-infrared spectrometer pointed from the rover mast, and an imager capable of detecting mineral and biologic fluorescence in daylight.

The concept of survey-traverse was formulated and the autonomous navigation algorithms were designed to achieve over a kilometer of travel per command cycle. Zöe eventually traversed over 257 km autonomously and of 602 traverses 75 exceeded 1 km in a command cycle (Fig. 9.12). New methods and procedures were developed so that remote scientists could utilize this over-the-horizon navigation capability. The change in approach from detailed observations with minimal mobility to quick survey with maximal

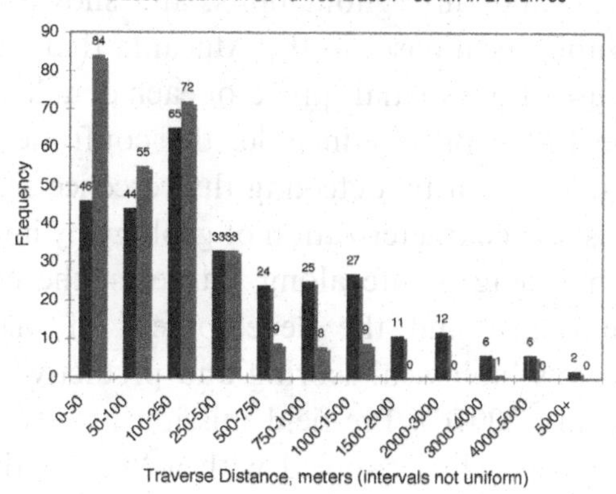

Fig. 9.12. Summary of every autonomous traverse conducted by Zöe (2004 and 2005). In 2004, shorter traverses were more common. The median distance in 2004 was 97 m but that improved in 2005 to 672 m. The total number of traverses over 1 km is 74.

distance traveled supported the challenges of life-seeking in the desert but required new thinking about how to plan and prioritize science. Scientists had to balance mobility and observation in an effort to survey the largest possible area without missing critical observation about life and habitats (Cabrol et al., 2007d). An exploration strategy and robotic technology that support this is "science-on-the-fly' (SOTF).

Employing the SOTF strategy a rover, for the first time, was able to make decisions on its own about what targets were worth fully documenting and which could be dismissed (Smith et al., 2005; Thompson et al., 2005; Wettergren et al., 2005a, 2005b). Zöe decided when the observation was significant enough to warrant additional measurements, in this case using cholorophyll abundance as the evaluation. The rover stopped at waypoints planned by remote scientists and quickly acquired an image of Fluorescence Imager (FI) workspace. After application of water, a FI image was acquired to check for chlorophyll response. A probabilistic evaluation of the abundance and structure of the fluorescence image triggered the entire suite of FI operations (carbohydrate, protein, lipid, and DNA dyes) when chlorophyll signatures were present.

As a result of SOTF more environmental units were characterized daily than otherwise possible. Both positive and negative detections were mapped and Zoë collected sufficient data to enable a characterization of those "non-habitat" (no detectable life, past or present) during its surveys. Time and in-depth analysis were focused on the main priority, finding life and characterizing habitats through the climatic gradient. SOTF brought into focus the relatively subtle changes within a unit and throughout units, and showed the characteristic patchiness of life within local oases in the Atacama that was not captured in previous years because of the spatial spread of each detailed sampling.

Results from the LITA project include: the confirmed identification of microbial habitats in daylight by detecting fluorescence signals from chlorophyll and dye probes; the characterization of geology by imaging and spectral measurement; the mapping of life along transects; the characterization of environmental conditions; and the development of mapping techniques including homogeneous biological scoring and predictive models of habitat location (Cabrol et al., 2007d). Pre-field mission assessments of geological orbital data, particularly when coupled with a "follow-the-water" strategy accurately identified many promising macroscale habitats within regions and locales, and predicted geological diversity at the microscale ($10\,cm^{-2}$) by broad type and composition, as obtained from microscopic imaging. Although scattered, habitats often had common mineralogical, morphological, geological, and topographical characteristics. For instance, at the microscale, heave-type substrates were shown to have significantly higher microbial abundance than pebble-type habitats (Warren-Rhodes and Weinstein, 2007).

Overall, desert pavement was the dominant habitat at all sites. Heaved gypsum crust was prevalent as well in both Coastal Range sites. At meso- to microscale (m to μm) heterogeneity was significantly greater and niche microbial populations, such as lichens, moss, and endolithic microbiota, were observed. Similar to the findings for chlorophyll-based microbial communities, rover FI data also showed strong variations in the percent positive ratings for nonphotosynthetic populations (i.e., the DNA-Protein-Lipid-Carbohydrate biosignature rating). Results indicate abundance ranging from 10 to 1000 colony-forming units/gram-soils (CFU) determined as Most Probable Number (MPN) enumerations on 1/10 strength PCA medium (Navarro-Gonzalez et al., 2003). One site exemplified the patchiness of life in the Atacama with the highest and lowest abundance of CFU within a few meters with 10 000 CFU/g-soil at locale 5 and 1 CFU/g-soil at locale 6 (Cabrol et al., 2007d; Warren-Rhodes and Weinstein, 2007).

While no site will ever be a perfect analog to Mars, the search for microbial life in the Atacama (or anywhere the conditions are extreme enough to be called a Mars analog) and the biologic exploration of Mars have an essential similarity: the necessity for an unambiguous detection of sparsely distributed biogenic material. Evidence of past or present life (if any) will not be uniformly distributed on Mars or easy to find, neither is it usually on terrestrial analogs with clues found at all scales. The key to future success on Mars is to learn how to integrate all this information and requires a sustained and diverse program of analog mission field studies.

Acknowledgments

HLP is funded by the NASA Astrobiology Institute (NAI) through the SETI lead team grant NNA04CC05A; MARTE is funded by the NASA Astrobiology Science and Technology for Exploring Planet (ASTEP) program as are the two projects of life characterization in the Atacama desert (Long Term Extreme Environment and Mars Analog Studies in the Atacama Desert, and LITA); the High-Arctic research is funded by grants from NASA's Astrobiology program (grants NAGW-12395; NNX07AD90A). The authors want to thank SERNAP (Bolivian National Park Services), the Universidad Catolicà del Norte, Antofagasta (Chile), the Canadian Polar Continental Shelf Project, and the Director of the McGill Arctic Research Station (M.A.R.S.), Dr. W.H. Pollard, for their continued support. The Haughton-Mars Project, its research programs, and the HMP Research Station are managed by the Mars Institute in collaboration with the SETI Institute, with research support from NASA and the Canadian Space Agency. Special thanks are owed to the United States Marine Corps, the Air National Guard, the Polar Continental Shelf Project of Natural Resources Canada, the Nunavut Research Institute, the Qikiqtani Inuit Association, Indian and Northern Affairs Canada, and the Arctic Communities of Resolute Bay and Grise Fiord for their participation and support.

References

Abbott, M., Brenner, M. W., and Kelts, K. R. (1997). A 3500 ^{14}C yr high-resolution record of water-level changes in Lake Titicaca, Bolivia/Peru. *Quaternary Research*, **47**, 169–180.

Acs, E., Cabrol, N. A., Grigorszky, I., et al. (2003). Similarities and dissimilarities in biodiversity of three high-altitude mountain lakes (Andes, Bolivia). In *6th Hungarian Ecological Congress*, ed. M. Dombos and G. Lakner. Godollo, Hungary: St. Stephan University, Publishers, 305 pp.

Amils, R., González-Toril, E., Fernández-Remolar, D., et al. (2007). Extreme environments as Mars terrestrial analogs: the Río Tinto case. *Planetary and Space Science*, **55**, 370–381.

Amundsen, H., Steele, A., Fogel, M., et al. (2004). Life in a Mars analogue: microbial activity associated with carbonate cemented lava breccia from NW Spitsbergen. *Geochimica et Cosmochimica Acta*, **68**(11, Suppl. 1), A804.

Andersen, D. (2004). Perennial springs in the Canadian High Arctic: analogues of hydrothermal systems on Mars. Ph.D. thesis. McGill University, Montreal, Canada.

Andersen, D. T., Pollard, W. H., McKay, C. P., and Heldmann, J. (2002). Cold springs in permafrost on Earth and Mars. *Journal of Geophysical Research*, **107**, 1–7.

Apostolopoulos, D., Pedersen, L., Shamah, B., et al. (2001). Robotic Antarctic meteorite search: outcome, In *Proceedings of IEEE International Conference on Robotics and Automation (ICRA)*, Seoul, Korea, pp. 4174–4179.

Arvidson, R. E., Anderson, R. C., Barlett, P., et al. (2004). Localization and physical properties experiments conducted by Spirit at Gusev crater. *Science*, **305**, 821–824.

Atreya, S. K., Mahaffy, P. R., and Wong, A. S. (2007). Methane and related trace species on Mars: origin, loss, implications for life and habitability. *Planetary and Space Science*, **55**, 358–369.

Bagaley, D. R. (2006). Uncovering bacterial diversity on and below the surface of a hyper-arid environment, the Atacama Desert, Chile. M. S. thesis, Louisiana State University, Baton Rouge, LA.

Baker, V. R. (2005). Picturing a recently active planet. *Nature*, **434**, 280–283.

Banks, D., Siewers, U., Sletten, R. S., et al. (1999). The thermal springs of Bockfjorden, Svalbard. II. Selected aspects of trace element hydrochemistry. *Geothermics*, **28**(6), 713–728.

Baucom, P. C. and Rigsby, C. A. (1999). Climate and lake-level history of the northern Altiplano, Bolivia, as recorded in Holocene sediments of the Rio Desaguadero. *Journal of Sedimentary Research*, **69**(3), 597–611.

Berger, I. A. and Cooke, R. U. (1997). The origin and distribution of salts on alluvial fans in the Atacama Desert, northern Chile. *Earth Surface Processes and Landforms*, **22**, 581–600.

Beschel, R. E. (1963). Sulfur springs at Gypsum Hill. In *Jacobsen-McGill Arctic Research Expedition, 1959–1962. Preliminary Report 1961–1962*, ed. F. Muller. Montreal, Canada: McGill University, pp. 183–187.

Bibring, J. P., Langevin, Y., Gendrin, A., et al. (2005). Mars surface diversity as revealed by the OMEGA/Mars Express observations. *Science*, **307**(5715), 1576–1581.

Böhlke, J. K., Ericksen, G. E., and Revesz, K. (1997). Stable isotope evidence for an atmospheric origin of desert nitrate deposits in northern Chile and southern California, USA. *Chemical Geology*, **136**, 135–152.

Bonaccorsi, R. and Stoker, C. (2008). Science results from a Mars drilling simulation (Río Tinto, Spain), and ground truth for remote science observations. *Astrobiology*, **8**(5), 967–985.

Boynton, W. V., Feldman, W. C., Squyres, S. W., et al. (2002). Distribution of hydrogen in the near surface of Mars: evidence for subsurface ice deposits. *Science*, **297**(5578), 81–85.

Bullock, M. A., Stoker, C. R., McKay, C. P., and Zent, A. P. (1994). A coupled soil atmosphere model of H_2O_2 on Mars. *Icarus*, **107**(1), 142–154.

Bunch, P. E. and Cassidy, W. (1972). Petrographic and electron microprobe study of the Monturaqui impactite. *Contributions to Mineralogy and Petrology*, **36**, 95–112.

Burns, R. G. and Fisher, D. S. (1990). Iron-sulfur mineralogy on Mars: magmatic evolution and chemical weathering products. *Journal of Geophysical Research*, **95**, 14 415–14 421.

Burns, R. G. and Fisher, D. S. (1993). Rates of oxidative weathering on the surface of Mars. *Journal of Geophysical Research*, **98**(E2), 3365–3372.

Burt, D. M., Wohletz, K. H., and Knauth, L. P. (2006). Mars and mine dumps. *Eos Transactions, AGU*, **87**(49), 549, doi: 10.1029/2006EO490003.

Cabrol, N. A. (2006). Habitability and life survival potential on early Mars: clues from the red and the blue planets. *Astrobiology Science Conference*, Abstract 16.

Cabrol, N. A. and Grin, E. A. (2001). The evolution of lacustrine environments on Mars: is Mars only hydrologically dormant? *Icarus*, **149**, 291–328.

Cabrol, N. A. and Grin, E. A. (2004). Ancient and recent lakes on Mars. In *Water and Life on Mars*, ed. T. Tokano. Berlin: Springer-Verlag, Chapter 10, pp. 181–205.

Cabrol, N. A., Bettis, III, E. A., Glenister, B., et al. (2001a). Nomad Rover field experiment, Atacama Desert (Chile). II. Identification of paleolife evidence using a robotic vehicle: lessons and recommendations for a Mars sample return mission. *Journal of Geophysical Research*, **106**(E4), 7639–7663.

Cabrol, N. A., Wynn-Williams, D. D., Crawford, D. A., and Grin, E. A. (2001b). Recent aqueous environments in impact crater lakes on Mars 2001: an astrobiological perspective. 2nd Mars Polar Conference Special Issue. *Icarus*, **154**, 98–112.

Cabrol, N. A., Hock, A. N., Grin, E. A., Kovacs, G. T., and Parazynski, S. (2005a). Can the combination of extremes protect life: clues from altiplanic lakes and implication for early Mars. *American Geophysical Union Fall Meeting*, Abstract 7353.

Cabrol, N. A., Hock, A. N., Grin, E. A., Kovacs, G. T., and Parazynski, S. (2005b). Combination of environmental extremes in altiplanic lakes and the past habitability of Mars. *Salt Lake City Annual Meeting*, **37**(7), Abstract 90 041.

Cabrol, N. A., Hock, A. N., Sunagua, M., and Grin, E. A. (2006). Evolution of aqueous habitat and life in high-altitude lakes during rapid climate change: astrobiological methods and geo and biosignatures. *Lunar and Planetary Science Conference*, **37**, Abstract 1016.

Cabrol, N. A., Grin, E. A., Kiss, K. T., et al. (2007a). Signatures of habitats and life in Earth's high-altitude lakes: clues to Noachian aqueous environments on Mars. In *Geology of Mars*, ed. M. Chapman. Cambridge, UK: Cambridge University Press, Chapter 14, pp. 349–370.

Cabrol, N. A., Minkley, Jr., E. G., Youngeob, Y., et al. (2007b). Unraveling life's diversity in Earth's highest volcanic lake. Paper presented at 2007 Bioastronomy Conference, Puerto Rico.

Cabrol, N. A., Minkley, Jr., E. G., Youngeob, Y., et al. (2007c). 2006 HLP diving expedition in the highest volcanic lake on earth and characterization of its ecosystem. Paper presented at SPIE Astrobiology Conference, San Diego, CA.

Cabrol, N. A., Wettergreen, D. S., Warren-Rhodes, K., et al. (2007d). Life in the Atacama: searching for life with rovers (science overview). *Journal of Geophysical Research, Biogeosciences*, **112**, G04S02.

Caceres, L., Gómez-Silva., B., Garró, X., et al. (2007). Relative humidity patterns and fog water precipitation in the Atacama Desert and biological implications. *Journal of Geophysical Research, Biogeosciences*, **112**, GO4S14.

Calderón, F., Lüders, A., Wettergreen, D., Teza, J., and Guesalaga, A. (2007). Analysis of high-efficiency solar cells in mobile robot applications. *Journal of Solar Energy Engineering*, **129**(3), 343–346.

Carr, M. H. (1996). *Water on Mars*. New York: Oxford University Press.

Christensen, P. R., Bandfield, J. L., Bell, III, J. F., et al. (2003). Morphology and composition of the surface of Mars: Mars Odyssey THEMIS results. *Science*, **300**, 2056–2061.

Clark, B. C., Baird, A. K., Weldon, R. J., et al. (1982). Chemical composition of Martian fines. *Journal of Geophysical Research*, **87**(B12), 10 059–10 067.

Clifford, S. M. (2003). Mars H_2O: limits of theoretical modeling and geomorphic interpretation in assessing the present distribution of subsurface H_2O on Mars. *Lunar and Planetary Science Conference*, **34**, Abstract 2118.

Clifford, S. M. and Parker, T. J. (2001). The evolution of the Martian hydrosphere: implications for the fate of a primordial ocean and the current state of the northern plains. *Icarus*, **154**, 40–79.

Cockell, C. S. and Lee, P. (2002). The biology of terrestrial impact craters: a review. *Biological Reviews*, **77**, 279–310.

Cockell, C. S. and Raven, J. A. (2004). Zones of photosynthetic potential on Mars and the early Earth. *Icarus*, **169**(2), 300–310.

Cockell, C. S., Lee, P., Broady, P., et al. (2005). Effects of asteroid and comet impacts on lithophytic habitats: a synthesis. *Meteoritics and Planetary Science*, **40**, 1901–1914.

Conley, C. A., Ishkhanova, G., McKay, C. P., and Cullings, K. (2006). A preliminary survey of non-lichenized fungi cultured from the hyperarid Atacama Desert of Chile. *Astrobiology*, **6**(4), 521–526.

Da Silva, F. and Francis, J. (1991). *Volcanoes of the Central Andes*. Berlin: Springer-Verlag.

De Hon, R. A. (1991). Classification of Martian lacustrine basins. *Lunar and Planetary Science Conference*, **22**, 293–294.

De Hon, R. A. (1992). Martian lake basins and lacustrine plains, *Earth Moon and Planets*, **56**(2), 95–122.

De Hon, R. A. (2001). Sedimentary provinces of Mars. *Lunar and Planetary Science Conference*, **32**, Abstract 1361.

Des Marais, D. J. and Farmer, J. D. (1995). The search for extinct life. In *An Exobiological Strategy for Mars Exploration*, ed. M. Meyer and J. Kerridge. NASA Special Publication 530. New York: NASA, pp. 21–25.

Des Marais, D. J., Clark, B. C. Crumpler L. S., et al. (2005). Astrobiology and the basaltic plains in Gusev crater. *Lunar and Planetary Science Conference*, **36**, Abstract 2353.

Des Marais, D. J., and the Athena Science team (2008). MER Spirit assessed potential ancient habitable environments in Gusev crater, Mars (Abstract). Astrobiology Science Conference, April 14–17, 2008, Santa Clara, CA. *Astrobiology*, **8**(2), 433.

Drees, K. P., Neilson, J. W., Betancourt, J. L., et al. (2006). Bacterial community structure in the hyperarid core of the Atacama Desert, Chile. *Applied and Environmental Microbiology*, **72**, 7902–7908.

Edgett, K. S, Malin, M. C., Williams, R. M. E., and Davis, S. D. (2003). Polar and middle-latitude martian gullies: a view from MGS MOC after two Mars years in the mapping orbit. *Lunar and Planetary Science Conference*, **34**, Abstract 1038.

Ericksen, G. E. (1983). The Chilean nitrate deposits. *American Science*, **71**, 366–375.

Farmer, J. D. and Des Marais, D. J. (1999). Exploring for a record of ancient martian life. *Journal of Geophysical Research*, **104**(E11), 26 977–26 995.

Fernández-Remolar, D. C., Prieto-Ballesteros, O., Rodríguez, N., et al. (2005). Río Tinto faulted volcanosedimentary deposits as analog habitats for extant subsurface biospheres on Mars: a synthesis of the MARTE drilling geobiology results. *Lunar and Planetary Science Conference*, **36**, Abstract 1360.

Fernández-Remolar, D. C., Prieto-Ballesteros, O., Rodríguez, N., et al. (2008). Underground habitats found in the Río Tinto Basin: a model for subsurface life habitats on Mars. MARTE Project special issue. *Astrobiology*, **8**(5), 1023–1047.

Fike, D., Cabrol, N. A., Grin, E. A., et al. (2003). Exploring the limits of life: microbiology and organic geochemistry of the world's highest lake atop the Licancabur volcano (6000 m) and adjacent high altitude lakes. EGS–AGU–EUG Joint Assembly, Nice, France, Abstract 13 201.

Fleming, E. D. and Prufert-Bebout, L. (2009). Characterization of cyanobacteria from a natural high ultraviolet radiation environment in Laguna Blanca, Bolivia. High Lakes Project Special Issue. *Journal of Geophysical Research, Biogeosciences*, in press.

Formisano, V., Encrenaz, T., Ignatiev, N., and Giuranna, M. (2004). Detection of methane in the atmosphere of Mars. *Science*, **306**, 1758–1761.

Golombek, M., Crumpler, L. S., Grant, J. A., et al. (2006). Geology of the Gusev cratered plains from the Spirit rover traverse. *Journal of Geophysical Research*, **111**, doi: 10.1029/2005JE002503.

González-Toril, E. F., Llobeet-Brossna, E., Casamayor, E. O., Amann, R., and Amils, R. (2003). Microbial ecology of an extreme acidic environment, the Tinto River. *Applied and Environmental Microbiology*, **69**, 4853–4865.

Grasby, S. E. (2003). Naturally precipitating vaterite (m-$CaCO_3$) spheres: unusual carbonates in an extreme environment. *Geochimica et Cosmochimica Acta*, **67**, 1659–1666.

Grasby, S. E, Allen, C. C., Longazo, T. G., et al. (2002). Supraglacial sulfur springs and associated biological activity in the Canadian High Arctic: signs of life beneath the ice. *Astrobiology*, **3**(3), 583–596.

Grieve, R. A. F. (1988). The Haughton impact structure: summary and synthesis of the results of the HISS project. *Meteoritics*, **23**, 249–254.

Grosjean, M. (2001). Mid-Holocene climate in the south-central Andes: humid or dry? *Science*, **292**, 2391–2392.

Hamon, W. R. (1961). Estimating potential evapotranspiration. In *Proceedings of the American Society of Civil Engineering, Journal of Hydology Division*, **87**(HY3), 107–120.

Hartley, A. J. and Chong, G. (2002). Late Pliocene age for the Atacama Desert: implications for the desertification of western South America. *Geology*, **30**, 43–46.

Hartley, A. J., Chong, G., Houston, J., and Mather, A. (2005). 150 million years of climatic stability: evidence from the Atacama Desert, northern Chile. *Journal of Geological Society*, **162**, 421–424.

Hartmann, W. K., Malin, M., McEwen, A., et al. (1999). Evidence for recent volcanism on Mars from crater counts. *Nature*, **397**, 586–589.

Haynes, Jr., C. V. (2001). Geochronology and climate change of the Pleistocene–Holocene transition in the Darb el Arba'in Desert, Eastern Sahara. *Geoarchaeology*, **16**, 119–141.

Heldmann, J. L. and Mellon, M. T. (2004). Observations of Martian gullies and constraints on potential formation mechanisms. *Icarus*, **168**, 285–304.

Heldmann, J. L., Pollard, W. H., McKay, C. P., Andersen, D. T., and Toon, O. B. (2005a). Annual development cycle of an icing deposit and associated perennial spring activity on Axel Heiberg Island, Canadian High Arctic. *Arctic Antarctic and Alpine Research*, **37**(1), 127–135.

Heldmann, J. L., Toon, O. B., Pollard, W. H., et al. (2005b). Formation of Martian gullies by the action of liquid water flowing under current Martian environmental conditions. *Journal of Geophysical Research*, **110**, E05004, doi: 10.1029/2004JE002261.

Herman, J. R., Bhartia, P. K., Ziemke, J., Ahmad, Z., and Larko, D. (1996). UV-B increases (1979–1992) from decreases in total ozone. *Geophysical Research, Letters*, **23**, 2117–2120.

Hickey, L. J., Johnson, K. R., and Dawson, M. R. (1988). The stratigraphy, sedimentology, and fossils of the Haughton formation: a post-impact crater-fill, Devon Island, Canada. *Meteoritics*, **23**, 221–231.

Hock, A. N., Cabrol, N. A., Grin, E. A., Fike, D. A., and Paige, D. A. (2003). 2002 Licancabur Expedition Team: hydrothermal circulation at the world's highest lake? An environmental study of the Licancabur Volcano crater lake as a terrestrial analog to martian paleolakes. *Geophysical Research, Abstracts*, **5** (13 586).

Hock, A. N., Cabrol, N. A., Grin, E. A., and Rothschild, L. (2005a). Ultraviolet radiation and life at high-altitude: Licancabur 2004. *NASA Astrobiology Institute 2005 Biennial Meeting*, University of Colorado, Boulder, Abstract 1043.

Hock, A. N., Cabrol, N. A., Grin, E. A., et al. (2005b). Mars-relevant conditions at the lakes of Licancabur volcano, Bolivia. *2005 American Geophysical Union Fall Meeting*, San Francisco, Abstract P41D-06.

Houston, J. and Hartley, A. J. (2003). The central Andean west-slope rainshadow and its potential contribution to the origin of hyper-aridity in the Atacama Desert. *International Journal of Climatology*, **23**, 1453–1464.

Hustedt, F. (1927). Die Diatomeen der interstadialen Seekreide. *International Review of Hydrobiology*, **18**, 317–320.

Hynek, B. M. (2004). Implications for hydrologic processes on Mars from extensive bedrock outcrops throughout Terra Meridiani. *Nature*, **431**, 156–159.

Kiss, K. T., Acs, E., Boris, G., et al. (2004). Habitats extrêmes pour les communautés de diatomées dans les lacs de haute altitude (Laguna Blanca et lac de cratère du volcan Licancabur, Bolivie). *23ème Colloque de l'Association des Diatomistes de Langue Française*, Orléans, France, p. 55.

Klein, H. P. (1978). The Viking biological experiments on Mars. *Icarus*, **34**, 666–674.

Klein, H. P. (1979). The Viking Mission and the search for life on Mars. *Review of Geophysics and Space Physics*, **17**, 1655–1662.

Klingelhofer, G., Morris, R. V., Berhhardt, B., et al. (2004). Jarosite and hematite at Meridiani Planum from Opportunity's Mossbauer spectrometer. *Science*, **306**, 1741–1745.

Knoll, A. H., Carr, M., Clark, B., et al. (2005). An astrobiological perspective on Meridiani Planum. *Earth and Planetary Science Letters*, **240**, 179–189.

Krasnopolsky, V. A., Bjoraker, G. L., Mumma, M. J., and Jennings, D. E. (1997). High resolution spectroscopy of Mars at 3.7 and 8 mm: a sensitive search for H_2O_2, H_2CO, HCl and CH_4 and detection of HDO. *Journal of Geophysical Research*, **102**(E3), 6525–6534.

Krasnopolsky, V. A., Maillard, J. P., and Owen, T. C. (2004). Detection of methane in the Martian atmosphere: evidence for life? *Icarus*, **172**, 537–547.

Lauritzen, S. E. and Bottrell, S. (1994). Microbiological activity in thermoglacial karst springs, south Spitsbergen. *Geomicrobiology Journal*, **12**, 161–173.

Lee, P. (1997). A unique Mars/Early Mars analog on Earth: the Haughton impact structure, Devon Island, Canadian Arctic. In *Conference on Early Mars: Geologic and Hydrologic Evolution, Physical and Chemical Environments, and the Implications for Life*. Lunar and Planetary Institute Contribution, **916**, p. 50.

Lee, P. and McKay, C. P. (2003). Mars: always cold, sometimes wet? *Lunar and Planetary Science Conference*, **34**, Abstract 2127.

Lee, P. and Osinski, G. R. (2005). Haughton-Mars Project: overview of science investigations at the Haughton impact structure, Devon Island, High Arctic. *Meteoritics and Planetary Science*, **40**, 1755–1758.

Lee, P., Bunch, T. E., Cabrol, N. A., et al. (1998). Haughton: Mars 97. I. Overview of observations at the Haughton impact crater, a unique Mars analog site in the Canadian High Arctic. *Lunar and Planetary Science Conference*, **38**, Abstract 1973.

Lee, P., Rice, Jr., J., Bunch, T. E., et al. (1999). Possible analogs for small valleys on Mars at the Haughton impact crater site, Devon Island, Canadian High Arctic. *Lunar and Planetary Science Conference*, **30**, Abstract 2033.

Lee, P., Cockell, C., Marinova, M., McKay, C., and Rice, Jr., J. W. (2001). Snow and ice melt slope flow features on Devon Island, Nunavut, Arctic Canada, as possible analogs for recent slope flow features on Mars. *Lunar and Planetary Science Conference*, **32**, Abstract 1809.

Lee, P., McKay, C., and Matthews, J. (2002). Gullies on Mars: clues to their formation timescale from possible analogs from Devon Island, Nunavut, Arctic Canada. *Lunar and Planetary Science Conference*, **33**, Abstract 2050.

Lee, P., Cockell, C., and McKay, C. (2004). Gullies on Mars: origin by snow and ice melting and potential for life based on possible analogs from Devon Island, High Arctic. *Lunar and Planetary Science Conference*, **35**, Abstract 2122.

Lee, P., Boucher, M., Desportes, C., et al. (2005). Mars, always cold, sometimes wet: new constraints on Mars denudation rates and climate evolution from analog studies at Haughton Crater, Devon Island, High Arctic. *Lunar and Planetary Science Conference*, **36**, Abstract 2270.

Lee, P., Gass, B. J., Osinski, G. O., et al. (2006). Gullies on Mars: fresh gullies in dirty snow, Devon Island, High-Arctic, as end-member analog. *Lunar and Planetary Science Conference*, **37**, Abstract 1818.

Leistel, J. M., Marcoux, E., and Duchamps, Y. (1998). The volcano-hosted massive sulphide deposits of the Iberian Pyrite Belt: review and preface to the thematic issue. *Mineralium Deposita*, **33**, 82–97.

Lester, E. D., Satomi, M., and Ponce, A. (2007). Microflora of extreme arid Atacama Desert soils. *Soil Biology and Biochemistry*, **39**, 704–708.

López-Archilla, A. I., Marín, I., Gonzáles, A., and Amils, R. (2001). Microbial community composition and ecology of an acidic aquatic environment. *Microbial Ecology* **41**, 20–35.

Lyons, J. R., Manning, C., and Nimmo, F. (2005). Formation of methane on Mars by fluid-rock interaction in the crust. *Geophysical Research, Letters*, **32**, L13201.

Maier, R. M. (2000). Biogeochemical cycling. In *Environmental Microbiology*, ed. R. M. Maier, I. L. Pepper, and C. P. Gerba. San Diego, CA: Academic Press, Chapter 14.

Malin, M. C. and Edgett, K. S. (2000). Evidence for recent groundwater seepage and surface runoff on Mars. *Science*, **288**(5475), 2330–2335.

Malin, M. C. and Edgett, K. S. (2003). Evidence for persistent flow and aqueous sedimentation on early Mars. *Science*, **302**(5652), 1931–1934.

Malin, M. C., Edgett, K. S., Posiolova, L. V., McColley, S. M., and Noe Dobrea, E. Z. (2006). Present-day impact cratering impact and contemporary gully activity on Mars. *Science*, **314**(5805), 1573–1577.

Max, M. D. and Clifford, S. M. (2000). The state of potential distribution and biological implications of methane in the martian crust. *Journal of Geophysical Research*, **105**, 4165–4171.

McKay, C. P., Clow, S. S., Wharton, Jr., R. A., and Squyres, S. W. (1985). Thickness of ice on perenially frozen lakes. *Nature*, **313**, 561–562.

McKay, C. P., Friedman, E. I., Gómez-Silva, B., et al. (2003). Temperature and moisture conditions for life in the extreme arid region of the Atacama Desert: four years of observation including the El Niño of 1997–1998. *Astrobiology*, **3**(2), 393–406.

Mellon, M. T. and Jakosky, B. M. (1995). The distribution and behavior of martian ground ice during past and present epochs. *Journal of Geophysical Research*, **100**, 11 781–11 799.

Messerli, B., Grosjean, M., Bonani, G., et al. (1993). Climate change and dynamics of natural resources in the Altiplano of northern Chile during late glacial and Holocene time: first synthesis. *Mountain Research and Development*, **13**(2), 117–127.

Michalski, G., Savarino, J., Böhlke, J. K., and Thiemens, M. (2002). Determination of the total oxygen isotopic composition of nitrate and the calibration of delta ^{17}O nitrate reference material. *Analytical Chemistry*, **74**, 4989–4993.

Michalski, G., Böhlke, J. K., and Thiemens, M. (2004). Long term atmospheric deposition as the source of nitrate and other salts in the Atacama Desert, Chile: new evidence from mass independent oxygen isotopic compositions. *Geochimica et Cosmochimica Acta*, **68**, 4023–4028.

Moore, J. M., Janke, D. R., Clow, G. D., et al. (1995). The circum-Chryse region as a possible example of a hydrologic cycle on Mars: geologic evidence and theoretical evaluation. *Journal of Geophysical Research*, **100**, 5433–5448.

Morris, R. L., Berthold, R., and Cabrol, N. (2007). Diving at extreme altitude: dive planning and execution during the 2006 High-Lakes science expedition. In *Diving for Science: Proceedings of 26th AAUS Scientific Symposium*, ed. N. W. Pollock and J. M. Godfrey. Dauphin Island, AL: American Academy of Underwater Sciences.

Mumma, M. J., Novak, R. E., Hewagama, T., et al. (2005). Absolute abundances of methane and water on Mars: spatial maps. *Bulletin of the American Astronomical Society*, **37**, 669–670.

Murray, J. B., Muller, J.-P., Neukum, G., et al. (2005). Evidence from the Mars Express high resolution stereo camera for a frozen sea close to Mars' equator. *Nature*, **434**, 352–356.

Navarro-Gonzalez, R., Rainey, F. A., Molina, P., et al. (2003). Mars-like soils in the Atacama Desert, Chile and the dry limit of microbial life. *Science*, **302**, 1018–1021.

Navarro-Gonzalez, R., Navarro, K. F., de la Rosa, J., et al. (2006). The limitations on organic detection in Mars-like soils by thermal volatilization-gas chromatography-MS and their implications for the Viking results. *Proceedings of the National Academy of Sciences of the USA*, **103**, 16 089–16 094.

Omelon, C. R., Pollard, W. H., and Marion, G. M. (2001). Seasonal formation of ikaite ($CaCO_3{\cdot}6H_2O$) in saline spring discharge at Expedition Fiord, Canadian High Arctic: assessing conditional constraints for natural crystal growth. *Geochimica et Cosmochimica Acta*, **65**, 1429–1437.

Omelon, C. R., Pollard, W. H., and Andersen, D. T. (2006). A geochemical evaluation of perennial spring activity and associated mineral precipitates at Expedition Fjord, Axel Heiberg Island, Canadian High Arctic. *Applied Geochemistry*, **21**, 1–15.

Osinski, G. R. and Lee, P. (2005). Intra-crater sedimentary deposits at the Haughton impact structure, Devon Island, Canadian High Arctic. *Meteoritics and Planetary Science*, **40**, 1887–1899.

Osinski, G. R., Spray, J. G., and Lee, P. (2001). Impact-induced hydrothermal activity within the Haughton impact structure, Arctic Canada; generation of a transient, warm, wet oasis. *Meteoritics and Planetary Science*, **36**, 731–745.

Osinski, G. R., Lee, P., Parnell, J., Spray, J. G., and Baron, M. (2005a). A case study of impact-induced hydrothermal activity: the Haughton impact structure, Devon Island, Canadian High Arctic. *Meteoritics and Planetary Science*, **40**, 1859–1877.

Osinski, G. R., Lee, P., Spray, J. G., et al. (2005b). Geological overview and cratering model of the Haughton impact structure, Devon Island, Canadian High Arctic. *Meteoritics and Planetary Science*, **40**, 1759–1776.

Osinski, G. R., Spray, J. G., and Lee, P. (2005c). Impactites of the Haughton impact structure, Devon Island, Canadian High Arctic. *Meteoritics and Planetary Science*, **40**, 1789–1812.

Oze, C. and Sharma, M. (2005). Have olivine, will gas: serpentinization and the abiogenic production of methane on Mars. *Geophysical Research Letters*, **32**, L10203.

Parnell, J., Bowden, S., Lee, P., Osinski, G. R., and Cockell, C. S. (2005). Application of organic geochemistry to detect signatures of organic matter in the Haughton impact structure. *Meteoritics and Planetary Science*, **40**, 1879–1885.

Parro, V., Fernández-Calvo, P., Rodríguez Manfredi, J. A., et al. (2008). SOLID2: an antibody array-based life detector instrument in a Mars drilling simulation Experiment (MARTE). *Astrobiology*, **8**(5), 987–999.

Pellenbarg, R. E., Max, M. D., and Clifford, S. M. (2003). Methane and carbon dioxide hydrates on Mars: potential origins, distribution, detection, and implications for future in situ resource utilization. *Journal of Geophysical Research, Planets*, **108**(E4), 231–235.

Perreault, N., Andersen, D. T., Pollard, W. H., Greer, C. W., and Whyte, L. G. (2007). Culture independent analyses of microbial biodiversity in cold saline perennial springs in the Canadian High Arctic. *Applied and Environmental Microbiology*, **73**(5), 1532–1543.

Pollard, W., Omelon, C., Andersen, D., and McKay, C. (1999). Perennial spring occurrence in the Expedition Fiord area of western Axel Heiberg Island, Canadian High Arctic. *Canadian Journal of Earth Science*, **36**, 105–120.

Quinn, R. C., Zent, A. P., Grunthaner, F. J., et al. (2005). Detection and characterization of oxidizing acids in the Atacama Desert using the Mars Oxidation Instrument. *Planetary and Space Science*, **53**, 1376–1388.

Rudolf, W. E. (1955). Licancabur: mountain of the Atacamenos. *Geographical Review*, **45**, 151–171.

Rundel, P. W., Dillon, M. O., Palma, B., et al. (1991). The phytogeography and ecology of the coastal Atacama and Peruvian deserts. *Aliso*, **13**(1), 1–49.

Servant Vilardy, S., Risacher, F., Roux, M., Landre, J., and Cornee, A. (2000). Les diatomées des milieux salés (Ouest Lipez, SW de l'Altiplano bolivien). www.mnhn.fr/mnhn/geo/diatoms/.

Sharp, R. P. and Malin, M. C. (1975). Channels on Mars. *Geological Society of America Bulletin*, **86**(5), 593–609.

Sherlock, S., Kelley, S., Parnell, J., et al. (2005). Re-evaluating the age of the Haughton impact event. *Meteoritics and Planetary Science*, **40**, 1777–1787.

Skelley, A. M., Scherer, J. R., Aubrey, A. D., et al. (2005). Development and evaluation of a microdevice for amino acid biomarker detection and analysis on

Mars. *Proceedings of the National Academy of Sciences of the USA*, **102**, 1041–1046.

Smith, T., Niekum, S., Thompson, D., and Wettergreen, D. (2005). Concepts for science autonomy during robotic traverse and survey. Paper presented at IEEE Aerospace Conference, Big Sky, MT. Washington, D.C.: IEEE.

Squyres, S. W. and Carr, M. H. (1986). Geomorphic evidence for the distribution of ground ice on Mars. *Science*, **231**, 249–252.

Squyres, S. W., Arvidson, R. E., Bell, III, J. F., et al. (2004). The Opportunity rover's Athena science investigation at Meridiani Planum, Mars. *Science*, **306**, 1698–1703.

Steele, A., Schweizer, M., Amundsen, H. E. F., and Wainwright, N. (2004). In-field testing of life detection instruments and protocols in a Mars analogue arctic environment. *International Journal of Astrobiology, Supplement*, **1**, 24.

Stoker, C., Mandell, L., McKay, C., et al. (2003). Mars Analog Research and Technology Experiment (MARTE): a simulated Mars drilling mission to search for subsurface life at the Rio Tinto, Spain. *Lunar and Planetary Science Conference*, **34**, Abstract 1076.

Stoker, C. R., Dunagan, S., Stevens, T., et al. (2004). Mars Analog Río Tinto Experiment (MARTE): 2003 drilling campaign to search for a subsurface biosphere at Río Tinto, Spain. *Lunar and Planetary Science Conference*, **35**, Abstract 2025.

Stoker, C. R., Stevens, T., Amils, R., et al. (2005). Characterization of a subsurface biosphere in a massive sulfide deposit at Río Tinto, Spain: implications for extant life on Mars. *Lunar and Planetary Science Conference*, **36**, Abstract 1534.

Summers, M. E., Lieb, B. J., Chapman, E., and Yung, Y. L. (2002). Atmospheric biomarkers of subsurface life on Mars. *Geophysical Research, Letters*, **29**(24), 2171, doi: 10.1029/2002GL015377.

Sylvestre, F., Servant, M., Servant-Vildary, S., Causse, C., and Fournier, C. (1999). Lake-level chronology on the southern Bolivian Altiplano (18 degrees – 23 degrees S) during late-glacial time and the early Holocene. *Quaternary Research*, **51**, 54–66.

Thompson, D. R., Smith, T., and Wettergreen, D. (2005). Autonomous detection of novel biologic and geologic features in Atacama Desert rover imagery. *Lunar and Planetary Science Conference*, **37**, Abstract 2085.

Vinebrook, R. R. and Leavitt, P. R. (1996). Effects of ultraviolet radiation on periphyton in an alpine lake. *Limnology and Oceanography*, **41**(5), 1035–1040.

Vuille, M., Bradley, R. S., Werner, M., and Keimig, F. (2003). 20th century climate change in the tropical Andes: observations and model results. *Palaeogeography Palaeoclimatology Palaeoecology*, **194**, 123–138.

Warren-Rhodes, K. A., Rhodes, K. L., Pointing, S. B., et al. (2006). Hypolithic cyanobacteria, dry limit of photosynthesis and microbial ecology in the hyper-arid Atacama Desert. *Microbial Ecology*, **52**(3), 389–398.

Warren-Rhodes, K. A., Weinstein, S., Dohm, J., et al. (2007). Robotic ecological mapping: habitats and the search for life on Mars in the Atacama desert. *Journal of Geophysical Research, Biogeosciences*, **112**, G04S02.

Wettergreen, D., Bapna, D., Maimone, M., and Thomas, G. (1999). Developing Nomad for robotic exploration of the Atacama Desert. *Robotics and Autonomous Systems Journal*, **26**(2–3), 127–148.

Wettergreen, D. S., Cabrol, N. A., Baskaran, V., et al. (2005a). Second experiment in the robotic investigation of life in the Atacama Desert of Chile. Paper

presented at International Symposium on Artificial Intelligence, Robotics and Automation in Space (i-SAIRAS), Munich, Germany.
Wettergreen, D., Cabrol, N., Teza, J., et al. (2005b). First experiments in the robotic investigation of life in the Atacama Desert of Chile. *Proceedings of the 2005 IEEE International Conference on Robotics and Automation*. Washington, D.C.: IEEE, pp. 873–878.
Wettergreen, D., Tompkins, P., Urmson, C., Wagner, M. D., and Whittaker, W. L. (2005c). Sun-synchronous robotic exploration: technical description and field experimentation. *Journal of Robotics Research*, **24**(1), 3–30.
Wierzchos, J., Ascaso, C., and McKay, C. P. (2006). Endolithic cyanobacteria in halite rocks from the hyper-arid core of the Atacama Desert. *Astrobiology*, **6**(3), 1.
Williamson, G., Grad, H., De Lange, J., and Gilroy, S. (2002). Temperature-dependent ultraviolet responses in zooplankton: implications of climate change. *Limnology and Oceanography*, **47**(6), 1844–1848.
Wirrmann, D. and Mourguiart, P. (1995). Late Quaternary spatio-temporal limnological variations in the Altiplano of Bolivia. *Quaternary Research*, **43**, 344–354.
Wood, L. J. (2005). Geomorphology of the Mars northeast Holden delta. *Search and Discovery Article*, 110 027.

Index

Printed in the United States
by Baker & Taylor Publisher Services

Printed in the United States
by Baker & Taylor Publisher Services